2nd International Conference on Mining, Materials and Metallurgical Engineering (ICMMME 2017)

IOP Conference Series: Materials Science and Engineering Volume 191

Bangkok, Thailand
17 – 18 March 2017

ISBN: 978-1-5108-3978-6
ISSN: 1757-8981

Printed from e-media with permission by:

Curran Associates, Inc.
57 Morehouse Lane
Red Hook, NY 12571

Some format issues inherent in the e-media version may also appear in this print version.

Printed by Curran Associates, Inc. (2017)

For permission requests, please contact the Institute of Physics
at the address below.

Institute of Physics
Dirac House, Temple Back
Bristol BS1 6BE UK

Phone: 44 1 17 929 7481
Fax: 44 1 17 920 0979

techtracking@iop.org

Additional copies of this publication are available from:

Curran Associates, Inc.
57 Morehouse Lane
Red Hook, NY 12571 USA
Phone: 845-758-0400
Fax: 845-758-2633
Email: curran@proceedings.com
Web: www.proceedings.com

TABLE OF CONTENTS

POLYMERS AND COMPOSITES

NANO-MATERIALS AND BIOMEDICAL MATERIALS

MATERIAL PHYSICS AND CONDENSED MATTER PHYSICS

MATERIAL PROCESSING

MECHANICAL ENGINEERING AND MANUFACTURING

METALLURGY AND ALLOY PROPERTIES

BUILDING MATERIALS AND STRUCTURES

ENERGY AND COLLIERY ENGINEERING

IOP Conf. Series: Materials Science and Engineering **191** (2017) 011001 doi:10.1088/1757-899X/191/1/011001

PREFACE

It was a great pleasure to welcome all the participants in the 2017 2nd International Conference on Mining, Materials and Metallurgical Engineering (ICMMME 2017) held in Bangkok, Thailand, 17 - 19 March 2017. ICMMME is dedicated to new research developments and advances in the fields of Mining, Materials and Metallurgical Engineering.

The key goal of the conference was to bring the academic scientists, engineers, and industry researchers together to exchange and share their expertise, experience and research results, and discuss the challenges and future directions in their specialized areas of research in these fields. A professor from USA and two professors from South Korea delivered the invited keynote lectures about state of the art research in their areas of expertise. The conference provided a golden opportunity for the students, researchers and engineers to interact with the experts and specialists to discuss with them on technical matters and future directions of research.

The proceeding of the conference consists of papers from the universities, research institutes and industry. The papers were selected after a rigorous peer-review by conference committee members and international reviewers. The papers should provide the readers an overview of many recent advances in the fields related to materials especially the composite materials and metallurgical engineering including the modeling, analysis, computation and optimization as well as the experimental methods.

I would like to thank all the authors who have contributed to this volume and also to the members of the organizing committee, reviewers, speakers, chairpersons, sponsors and the conference participants for their support to ICMMME 2017.

Prof. Ramesh K. Agarwal
Washington University in St. Louis, USA
22 March 2017

IOP Conf. Series: Materials Science and Engineering **191** (2017) 011002 doi:10.1088/1757-899X/191/1/011002

Peer review statement

All papers published in this volume of *IOP Conference Series: Materials Science and Engineering* have been peer reviewed through processes administered by the proceedings Editors. Reviews were conducted by expert referees to the professional and scientific standards expected of a proceedings journal published by IOP Publishing.

IOP Conf. Series: Materials Science and Engineering **191** (2017) 012001 doi:10.1088/1757-899X/191/1/012001

Reducing friction and miscibility studies of FEP dispersion/ PDMS fluid blends

S Buapool*, N Thavarungkul and N Srisukhumboworncha i

Division of Materials Technology, School of Energy, Environment and Materials King Mongkut's University of Technology Thonburi, Bangkok, Thailand

* Email: sorasutee@gmail.com

Abstract. To develop new polymer blends having reduced friction force of fluorinated ethylene propylene (FEP) dispersion and improved adhesion of polydimethylsiloxane (PDMS) fluid, FEP dispersion was blended with PDMS fluids at different viscosities of 20 cSt and 100 cSt by using solution mixing method. The FEP/PDMS blends were coated on short hollow tubes and examined by penetrating the tubes into the rubber stoppers. It was found that the tubes coated with the blends showed reduced penetration and friction forces and improved adhesion. The tubes coated with the 100 cSt-PDMS blend in the ratio of 5:1.5 demonstrated the penetration and average friction forces as low as 3828 mN and 1524 mN, respectively. The formation of physical blends was characterized and confirmed by FTIR and DSC analyses.

1. Introduction

Fluorinated ethylene propylene (FEP) dispersion are commonly used as non-stick coating in many applications that require friction-reducing property, electrical insulation, biosafety, chemical resistance, and wear resistance [1]. An appropriate choice of the second polymer with specific properties can be blended for a variety of physical and chemical properties. Properties of the blend will depend largely on the degree of compatibility, miscibility of the polymers at the molecular level, and the characteristics of components and compositions [2]. Polydimethylsiloxane (PDMS) fluid is a hydrophobic material used for self-cleaning and lubricant coating in many applications [3]. PDMS fluid readily dissolves in solvent-born paints and, as a result, reduces the surface tension of the coating materials. PDMS fluid is, however, non-curable and able to be removed from the substrate [4]. Commonly, PDMS fluid is blended with organic coating materials such as alkyd, epoxy, and phenolic resins for reduced coefficient of friction of coating materials and improved adhesion and hardness, drying properties and durability of some exterior maintenance coating [5].

To combine the advantages of FEP dispersion and PDMS fluid, the aim of this study was to develop a polymer blend that could reduce friction force of FEP dispersion and improve adhesion of PDMS fluid. Blending of FEP dispersion and two different viscosities of PDMS fluids using a solution mixing technique was investigated. Miscibility of the FEP/PDMS blends was characterized and analyzed using Fourier transform infrared spectroscopy and differential scanning calorimetry.

2. Experimental

2.1. Materials

As-received FEP dispersion was mainly colloidal FEP particles, polyamine imide (PAI), chromium (III) oxide, and solvents of mainly methyl isobutyl ketone (MIBK), N-methyl-2-pyrrolidone (NMP), ethanol, and aromatic hydrocarbon. FEP dispersion was purchased from DuPont®. Chemical structures of FEP, PAI, and PDMS were shown in figure 1. Viscosities of PDMS fluids were 20 cSt and 100 cSt. PDMS fluids were purchased from Dow Corning®. Prior to the blending process, the as-received FEP dispersion in the container was rolled at 30 rpm for 30 minutes using a roller mixer. Polymer agglomerates and other contaminants were strained using 100-mesh wire sieves.

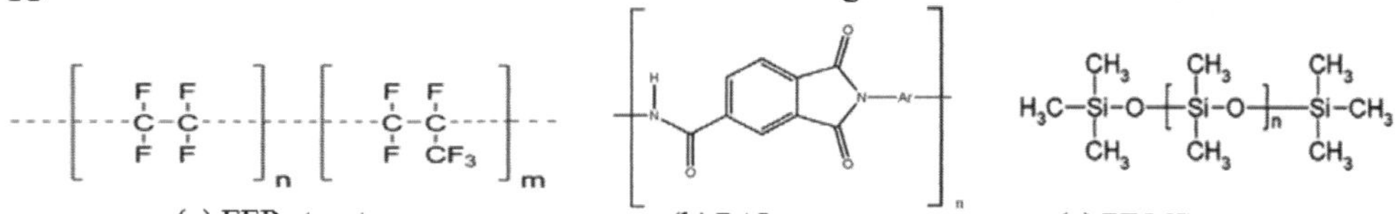

(a) FEP structure (b) PAI structure (c) PDMS structure

Figure 1. Chemical structure of polymers.

2.2. Coating preparation

Using a solution blending method, the as-receive PDMS fluid was added into the FEP dispersion. The same ratios were applied to the blending of FEP dispersion to PDMS fluids at viscosities of 20 cSt and 100 cSt were 5:1.0 and 5:1.5. The container of mixtures were rolled at 30 rpm for 5 minutes using a roller mixer at room temperature to ensure complete mixing. Pure FEP dispersion, pure PDMS fluid, and the blend solutions were coated on substrates, which were short and small hollow tubes made of 316 stainless steel. The tubes were 4 mm long and had 450 μm diameter. The distal end of the tubes had a bevel of 15 degrees. The substrates were cleaned to remove surface contaminations.

An in-house, dip-spin coating equipment was specifically designed for simultaneously spinning the substrate after withdrawing the substrate from a coating solution. The process parameters for this investigations were as follows. The dipping and withdrawal velocities of the substrate were 25 mm/s. The spinning speed of the substrate was 600 rpm. The immersion time in polymer solution was 10 s. Compressed air pressure for polymer solution inner the tube removal was set 400 kPa. The coating environment was at 25°C and 80% relative humidity. As PDMS fluid is non-curable, this study then set curing temperature and time for pure FEP dispersion and FEP/PDMS blend solutions at 205±10 °C and for 15 minutes, respectively, following FEP dispersion factsheet. As recommended in PDMS fluid factsheet, the drying temperature was at ambient temperature.

2.3. Method of analysis

2.3.1. Lubricity property testing. The penetration and friction forces of coatings were measured and controlled by using a texture analyzer (Shimadzu UH & AG-IS). Signals of both forces from the 50 N load cell penetration of texture analyzer were recorded using Trapezium software. The tube penetration depth was about 20 mm into the rubber stopper. The tube penetrations were made in a constant speed of 25 mm/s. The tests were conducted on a total of eight types of coatings on the tube samples including uncoated tube, pure FEP dispersion, pure 20 cSt-PDMS fluid, pure 100 cSt-PDMS fluid, FEP/20 cSt-PDMS blends with the ratio of 5:1.0 and 5:1.5, and FEP/100 cSt-PDMS blends with the ratio of 5:1.0 and 5:1.5.

2.3.2. Fourier transforms infrared spectroscopy (FTIR) analysis. In order to investigate the interaction between two polymers, FTIR analysis was carried out. FTIR spectra of functional groups in the polymers could be used to determine main chemical compositions of the samples. For solution characterization, FTIR spectra of the pure PDMS fluids were taken using Perkin Elmer Spectrum One FTIR Spectrometer. The light source was the middle range infrared in the range of 4000-600 cm^{-1}. The wavelength resolution was 4 cm^{-1}. For coating surface characterization, FTIR spectra of pure FEP dispersion and blends of FEP/PDMS with the ratio of 5:1.5 of the PDMS fluid having viscosities of 20 cSt and 100 cSt were taken using Perkin Elmer Spectrum Spotlight FTIR Imaging System. The samples were examined using Micro-ATR technique. The light source was the same range as the above and the wavelength resolution also the same.

2.3.3. Differential scanning calorimetry (DSC) analysis. DSC analysis was carried out in a Mettler Toledo DSC 1. In this study, the samples were the coating of pure FEP dispersion and FEP/PDMS blends with the ratio of 5:1.5 of the PDMS fluid having viscosities of 20 cSt and 100 cSt. The samples of 5.000 mg were scratched from the substrates, sealed in an aluminum pan, and heated from 25°C to 340°C at a heating rate of 10°C/min to investigate the degree of crystallinity at first heating scan. The samples were cooled down from 340°C to 25°C at a cooling rate of 10°C/min to investigate the cold crystallization temperature (T_{cc}). The samples were then heated up the second time from 25°C to 340°C at a heating rate of 10°C/min. The degree of crystallinity is the percentage of crystallinity content in a semicrystalline substrance. The percentage crystallinity of FEP crystalline in the blends were estimated according to the following equation $\%Crystallinity = (\Delta H_f / \Delta H_f^{\circ}) x100\%$, where ΔH_f and ΔH_f° are the enthalpy (J/g) of fusion of the blends and the enthalpy of 100% of FEP crystalline with a value of 24.3 J/g [6], respectively.

3. Results and discussion

3.1. Lubricity property testing

For lubricity property, a distal point of the tube was brought to touch the surface of the rubber stopper. The distal point stretched the rubber stopper, then cut into the surface, and began to penetrate into the rubber stopper. Uninterruptedly, the shaft of the tube passed through the original cut into the rubber stopper. The forces that the distal point cut the rubber stopper were termed the penetration force (measured at the maximum force) while the forces required to continue penetrating the shaft of the tube were termed friction force (measured at a stable force after penetration) [7]. The friction forces were measured at penetration depth after 8 mm to a total distance of 20 mm. The results of the penetration and friction forces of all types of coatings on the tube samples are shown in figure 2 and summarized in table 1. The average, minimum, and maximum friction forces were included. From the results, the blends reduced the penetration and the friction forces of pure FEP dispersion and could reduce both forces further with higher viscosity and increasing proportion of PDMS fluids. The FEP/100 cSt-PDMS fluid blends had the lowest penetration and friction forces. As expected, both forces measured from the uncoated tube sample were the highest.

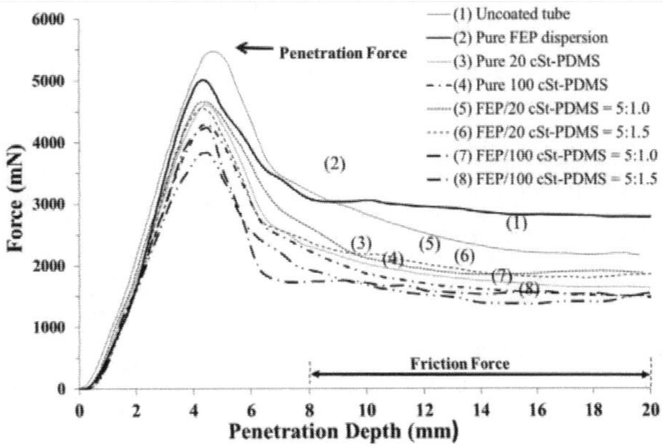

Figure 2. Forces of coating on the tube samples.

Coatings of the pure PDMS fluids on the tube samples were partially detached from the tube shafts during penetrating the rubber stoppers. Possible cause of poor adhesion of coating on the tube shafts was due to the fact that the PDMS fluid was non-curable and did not contain any functional groups that would allow the fluid to attach solidly on the substrate [3]. As a consequence, PDMS could be easily removed from the substrate [5]. Due to some partial detachment, both forces obtained from the

IOP Conf. Series: Materials Science and Engineering **191** (2017) 012001 doi:10.1088/1757-899X/191/1/012001

pure PDMS fluids might not represent correct values of the PDMS coatings. On the other hand, the coatings of pure FEP dispersion and the blends remained attached with the tube shafts during penetrating the rubber stoppers. Good adhesion of coatings might be contributed to the dissolution of PDMS fluid in the nonpolar FEP dispersion solvent [8]. PDMS fluid could then be blended well in FEP dispersion. At curing temperature, the solvents in the FEP/PDMS blends evaporate, colloid FEP particles deform, and PAI solution forms crosslink in the film [9]. It might be inferred that PDMS remained dispersed between FEP particles and PAI film. As a result, FEP/PDMS blended coatings possessed reduced penetration and friction forces and improved adhesion. Characterization using FTIR and DSC for understanding the interaction of two polymers were pursued.

Table 1. Penetration force and friction force.

Sample	Penetration force (mN)	Friction Force (mN)	
		Average	Min - Max
1)	5472	2196	1785-3178
2)	5009	2899	2786-3075
3)	4606	1829	1639-2329
4)	4290	1677	1479-2213
5)	4652	2002	1849-2684
6)	4550	1968	1801-2383
7)	4234	1599	1492-1758
8)	3828	1524	1377-1929

3.2. FTIR analysis

Figure 3 showed FTIR spectra indicating the peaks of pure PDMS fluid, pure FEP dispersion, and FEP/PDMS blends. FTIR spectra of pure PDMS fluid showed at 710, 810, 1025, 1096, 1264, and 2964 cm^{-1}. FTIR spectra of pure FEP dispersion showed at 646, 1164, 1222, 1384, 1517, 1725, 2964, and 3339 cm^{-1}. When PDMS was blended into FEP dispersion, FTIR spectra showed at 646, 710, 810, 1025, 1097, 1164, 1222, 1264, 1384, 1517, 1725, 2964, and 3339 cm^{-1}. The FTIR spectra of FEP/PDMS blends did not indicate the existence of other interaction between the component of polymers. Peaks of FEP in FEP/PDMS blend spectra remained at the original positions and had no sign of intensity change. Whereas those of PDMS in the blends showed decreases in transmittance, however, no shift of any peak. The decreases in transmittance of PDMS were in accordance with the ratio of FEP:PDMS, which was on the less side of 5:1.5. The results indicated the formation of physical blends.

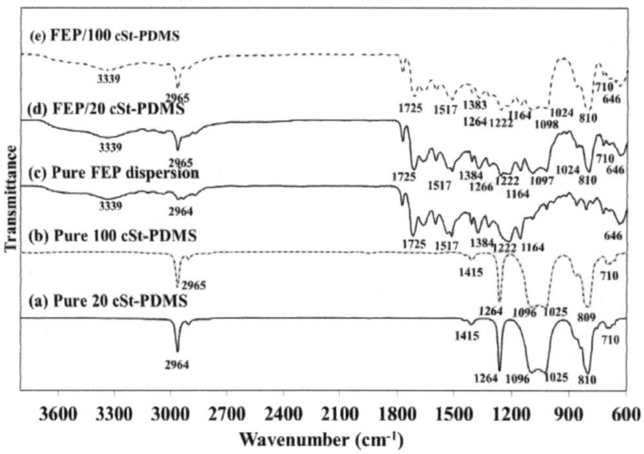

Figure 3. FTIR spectra of pure PDMS fluid, pure FEP dispersion, and FEP/PDMS (5:1.5) blends.

IOP Conf. Series: Materials Science and Engineering **191** (2017) 012001 doi:10.1088/1757-899X/191/1/012001

3.3. DSC analysis

Enthalpy of fusion and %crystallinity of FEP crystal of all samples were shown in figure 4. The crystallization temperatures of FEP and FEP/PDMS blends were about 237°C. The %crystallinities were estimated by using the equation on the peaks obtained from the first scans. The %crystallinities were decreased when the PDMS fluids were blended into the FEP dispersion. PDMS molecular chains might entangle with FEP. From the PDMS factsheet, the polymer chains of 100 cSt-PDMS fluid were longer than those of 20 cSt-PDMS fluid [4]. It was implied that the shorter chains of 20 cSt-PDMS fluid could disturb FEP in molecular level and result in a lesser amount of crystallization. In addition, the much smaller peaks observed on the second heating scans of the blends might be affected by insufficient crystallization during cooling from the first scans. Equivalent to annealing the samples, the curing at 205±10 °C for 15 minutes, as mentioned in section 2.2, that was 30 degrees lower than the crystallization temperatures also allowed the FEP crystals to grow and could result in higher amount of crystallinity. Higher %crystallinity could enhance hardness and adhesion of the coatings [10]. With DSC results, it could be indicated that the PDMS fluid had no interaction with the pure FEP dispersion.

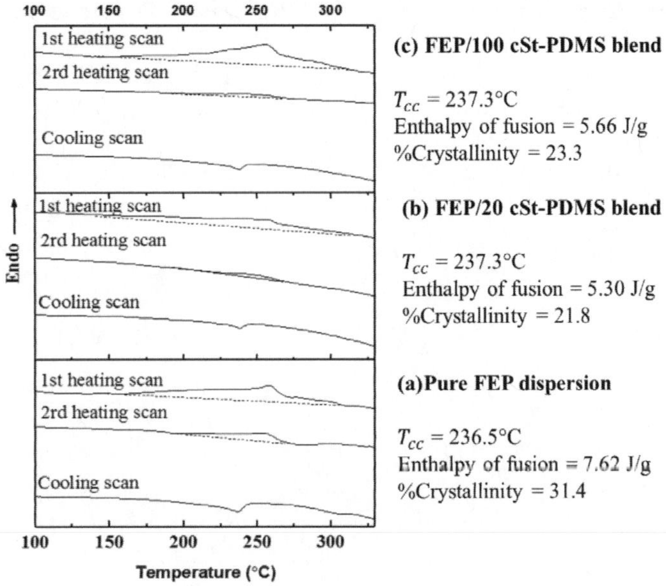

Figure 4. DSC results of pure FEP dispersion and FEP/PDMS (5:1.5) blends.

4. Conclusion

This study was to develop new FEP/PDMS blends. Several types of FEP/PDMS blends were coated on short hollow tubes and examined by penetrating the tubes into the rubber stoppers. Their lubricity properties, as a result, were compared with the tubes coated with pure FEP dispersion and pure PDMS fluid. The blends reduced penetration and friction forces and demonstrated improved adhesion. The samples coated with the blend at PDMS viscosity of 100 cSt in the ratio of 5:1.5 possessed the lowest penetration and friction forces, which were less than those of pure FEP dispersion. Miscibility of FEP/PDMS blend was observed with the formation of physical blends characterized and confirmed by FTIR and DSC analyses. This study was limited to lubricity properties testing. Other properties testing are of interested for coating application. Future work may include different the viscosity of PDMS and different ratio of blends.

Acknowledgements

The authors gratefully acknowledge the financial support of this work provided by National Research Council of Thailand (NRCT) and Thailand's Health Systems Research Institute (HSRI).

IOP Conf. Series: Materials Science and Engineering **191** (2017) 012001 doi:10.1088/1757-899X/191/1/012001

References

[1] Rowan, H.I., 1988. Thermoplastic fluoropolymers of engineering plastics, Engineered Materials Handbook vol. 2, ASM International, Metals Park, OH, 115-119.
[2] Singh, R.P., Das C.K., Mustafi, S.K., 2002. Polymer blends and alloys, Asian books private limited, New Delhi, 1-18.
[3] Andriot, M, DeGroot, J.V., Jr. and Meeks R., 2013. Silicones in industrial applications, Dow Corning Corporation.
[4] Dow Corning Corporation, Dow Corning® 360 medical fluid, Frequently asked questions.
[5] Wicks, Zeno W., Jr., Frank N. Jones, S. Peter Pappas, Dauglas A. Wicks., 2007. Organic coating: science and technology, 3rd ed., John Wiley&Sons, Inc., Hoboken, New Jersey, 326-333.
[6] Martienssen, W., Warlimont, H., 2005. Springer handbook of condensed matter and materials data, springer science and business media, 496-498.
[7] Lee, M., Lin, C., and Huang, C., 2002. The study on lubricity properties of arterial venous fistula needles, Biomedical engineering applications, basis & communications, 14(3), 127-130.
[8] Dow Corning Corporation, Dow Corning® 360 medical fluid, Product Information.
[9] McKeen, L.W., 2006. Fluorinated coatings and finishes handbook, William Andrew Publishing, New York, 49-50.
[10] Crompton, T.R., 2006. Polymer reference book, Rapra Technology Limite, UK, Publishing, 433-444.

IOP Conf. Series: Materials Science and Engineering **191** (2017) 012002 doi:10.1088/1757-899X/191/1/012002

Effect of EPDM-g-MAH on properties of HDPE/OBC blends

M Li[1,2], L Y Yu[3], P F Li[1,2], Y H Bin[1,2], and H J Zhang[4]

[1]Guangdong Zhuhai Supervision Testing Institute of Quality and Metrology, China
[2]National Quality Supervising Test Center for Materials of Ship and Marine Engineering Equipment, China
[3]Gree Electric Appliances Inc. of Zhuhai, China
[4]School of materials science and engineering, Fujian Normal University, China

E-mail: 18928088172@163.com

Abstract. In this paper, we take the HDPE as the matrix material, OBC as the toughening material, and EDPM-g-MAH as the compatibility agent, HDPE/OBC/EPDM-g-MAH blends were prepared by high speed mixing, melt extrusion, injection molding and so on. The effects of OBC and EPDM-g-MAH on mechanical properties, crystalline properties, fracture surface structure and rheological properties of HDPE were analyzed by universal tensile tester, melt mass flow rate test machine, DSC and SEM. Experimental results show that: with the addition of EPDM-g-MAH, the notched impact strength of the blends increased first and then decreased; HDPE/OBC blend containing 4% EPDM-g-MAH, OBC dispersion in the matrix is more uniform, particle size is significantly refined, melt flow has some improvement, Compared with HDPE/OBC blend materials, notched impact strength and elongation at break increased by 41.07% and 107.28% respectively, the toughness of the blend was greatly improved.

1. Introduction

HDPE is a kind of non polar and high crystallinity of the thermoplastic resin, inexpensive, lightweight, excellent moisture resistance and chemical stability and other advantages, is one of the most widely used polymer materials. However, due to the high crystallinity of HDPE and its own structural characteristics, it has many disadvantages, such as poor impact strength, etc. In order to widen the application field of HDPE, it is necessary to modify HDPE.

OBC has a unique “soft” and “hard” alternating block structure model of ethylene-octene block copolymer [1], compared with the other regulation of polyolefin copolymer, OBC has high temperature, excellent elasticity and compression deformation properties [2], is a new and efficient general plastic toughening agent.

In this paper, HDPE/OBC/EPDM-g-MAH was prepared with high speed mixing, melt extrusion and injection molding. The effect of EDPM-g-MAH content on the mechanical properties, melt flow rate, section structure, crystallization and melting behavior of HDPE/OBC blends was investigated.

2. Experimental part

2.1. Raw materials and reagents

HDPE, grade: DMDA 8920; OBC, grade: 9000; EDPM-g-MAH, grade: E-B12; Antioxidant1010.

2.2. *Instruments and equipment*

Extruder, XMTD-3001, Nanjing Zhicheng rubber machinery company; Injection molding, JN55-E, Chen Hung Machinery Factory Co., Ltd.; Electronic tension testing machine, CMT4204, Shenzhen SUNS Metering Technology Co. Ltd; Impact testing machine, ZBC1400-2, Shenzhen SUNS Metering Technology Co. Ltd; SEM, JSM-7500F, Japan JEOL company; DSC, DSC822e, Swiss Mettler-Toledo company; Melt mass flow rate tester, ZRZ1452, Shenzhen SUNS Metering Technology Co. Ltd.

2.3. *Preparation of HDPE/OBC blend*

2.3.1. Sample formulation. 0#: 99.5%HDPE(Mass fraction, the same below)+0.5% antioxidant1010; 1#: 81.5%HDPE+18%OBC+0.5% antioxidant1010; 2#: 99%1#+1%EPDM-g-MAH; 3#: 98%1#+2%EPDM-g-MAH; 4#: 97%1#+3%EPDM-g-MAH; 5#: 96%1#+4%EPDM-g-MAH.

2.3.2. Sample preparation. The HDPE and OBC were placed in the drying box at 75 °C for drying 4h, Then HDPE, OBC, EPDM-g-MAH, antioxidant 1010 according to a certain ratio at 65 °C evenly mixed in a high-speed mixer; then the initial mixture is placed in a double screw extruder for melt mixing, the extruder temperature was 132 °C, 137 °C, 142 °C, 150 °C, 155 °C and 152 °C, and then after cooling water cooling, granulator by extrusion pellets; blend in injection molding machine injection molding granulation after injection into standard samples, the temperature of 180 °C ~200 °C.

2.4. *Performance testing*

2.4.1. Mechanical property test. Tensile strength and elongation at break were tested by ISO 527-2: 2012; flexural strength were tested by ISO 178: 1993; notched impact strength were tested by ISO 179-2: 1997.

2.4.2. Melt mass flow rate measurement. Melt mass flow rate was measured by ISO 1133-1: 2011. Test temperature is 190 °C, load: 2.16 kg.

2.4.3. SEM test. The samples were frozen in liquid nitrogen for ten minutes, then etching at 80 °C of n-heptane for 5h, then using ethanol and distilled water cleaning section, by spraying treatment, observe the morphology of the sample section in scanning electron microscope.

2.4.4. DSC test. In nitrogen atmosphere, the gas flow rate is 50 ml/min, take the 3~6 mg in DSC thermal stage, rose from 170 °C to 30 °C at 10 °C /min, and then stay at 170 °C for 5min to eliminate the thermal history, and then at a rate of 10 °C /min from 170 °C to 30 °C, then heating at rate of 10 °C /min from 170 °C to 30 °C, record the crystallization and melting curve. $X(\%)=(\Delta Hc/\Delta H_m^0)\times 100\%$, Among ΔH_m^0 is the Melting enthalpy at 100% degree of crystallinity, ΔH_m^0 =287.3 J/g [3].

3. Results and discussion

3.1. *Effects of OBC and EPDM-g-MAH on mechanical properties of HDPE*

Effect of EPDM-g-MAH content on Mechanical property of HDPE/OBC blends are shown in Table 1 and figure 1. We can see from Table 1, adding 18%OBC (mass fraction, the same below), the notched impact strength of blends was 21.50kJ/m^2, while the elongation increased from 165.85% to 558.00%, the toughening effect is obvious, this is because the special structure of OBC ethylene octene copolymer, which makes the HDPE and OBC have good compatibility.

IOP Conf. Series: Materials Science and Engineering **191** (2017) 012002 doi:10.1088/1757-899X/191/1/012002

Table 1. Effect of OBC and EPDM-g-MAH content on mechanical properties of HDPE

Sample	Tensile strength/MPa	Elongation at break/%	Flexural strength/MPa	Notched impact strength/kJ.m^{-2}
0#	22.08	165.85	12.15	4.85
1#	15.35	558.00	7.82	21.50
2#	15.24	574.00	7.80	21.90
3#	15.12	879.00	7.75	22.70
4#	14.90	936.60	7.71	24.68
5#	14.60	1157.60	7.54	30.33

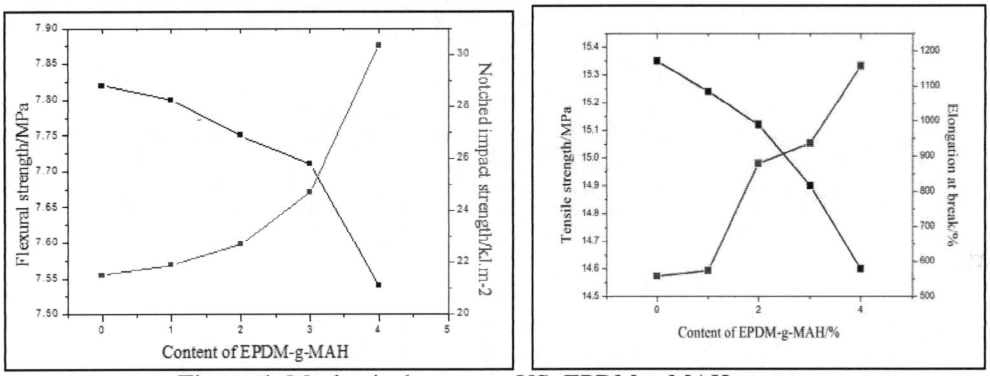

Figure 1. Mechanical property VS. EPDM-g-MAH content

From Figure 1 we can see that the notched impact strength of blends increased with the increase of EPDM-g-MAH content, when adding 4% EPDM-g-MAH, the notched impact strength of blends reached 30.33 kJ/m^2, higher than 1# 41.07%, the elongation is 1156.60% , higher than 1# 107.28%, the tensile strength and flexural strength were declined slightly. The reason, mainly the following two aspects: first, the non-polar part of EPDM-g-MAH is ethylene, propylene and non conjugated dienes copolymer, according to the similarity principle of compatibility, non polar part with HDPE and OBC has good compatibility [4]; Second, EPDM-g-MAH is an excellent synergistic toughening agent, so the impact strength and elongation of the blends increased, while the tensile strength and the bending strength does not change much.

3.2. Effects of HDPE and EPDM-g-MAH content on the melt flow rate of HDPE

The melt mass flow rate of OBC is 0.5 g/10min (Provided by DOW), the melt mass flow rate of pure HDPE is 19.7 g/10min, and the melt mass flow rate of 1# is 5.33g/10min, down by 72.94%, this is due to the low flow rate of OBC, melt the addition of OBC increases the blends in friction resistance, lead to the relative motion between polymer chains becomes difficult. From Figure 2 we can also see that the increase of EPDM-g-MAH content, melt flow rate of blends increases, this is because EPDM-g-MAH its good liquidity, driven by other components of the flow induced by [5].

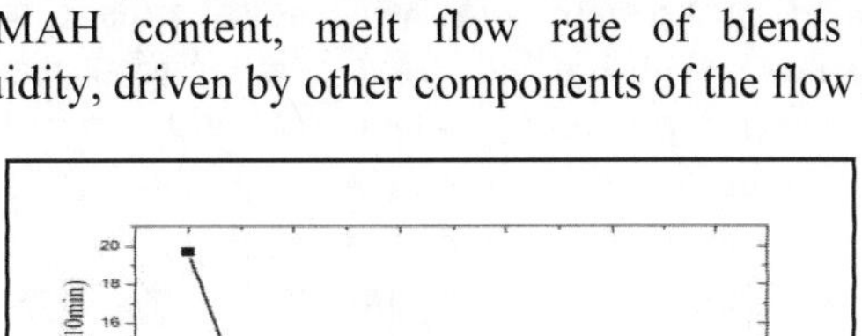

Figure 2. Melt mass-flow rate of HDPE and its blends

IOP Conf. Series: Materials Science and Engineering **191** (2017) 012002 doi:10.1088/1757-899X/191/1/012002

3.3. SEM analysis of HDPE and its blends

Figure 3 is the SEM of HDPE and its blends. 0# is smooth, the cross section is smooth and the crack is not cross, which indicates that the toughness of pure HDPE is not good; 1# is HDPE/OBC blends after etching, etching holes is OBC particles left, 1# showed that OBC particles can be homogeneously dispersed in the matrix, indicating that HDPE and OBC dispersed phase compatibility between certain, but the pore size of different sizes, indicating that HDPE and OBC are partially compatible; 5# is refined and the pore distribution is more uniform, the addition of EPDM-g-MAH, the distribution of dispersed phase OBC is more uniform, improved HDPE and OBC interface compatibility, improve the toughness of the composites.

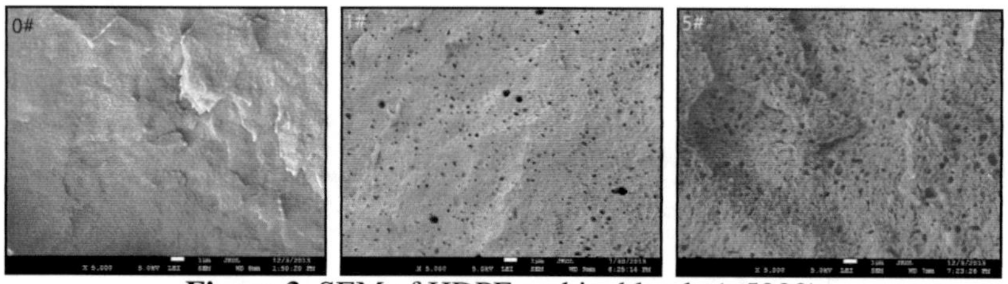

Figure 3. SEM of HDPE and its blends (×5000)

3.4. DSC analysis of HDPE/OBC/EPDM-g-MAH blends

Table 2 shows the pure HDPE (0#) crystallinity (X) 67.29%, HDPE/OBC (1#) crystallinity (X) 56.14%, this is due to the low crystallinity of block copolymer OBC; The crystallinity (X) of HDPE/OBC/EPDM-g-MAH (5#) was 51.93%, Crystallinity is still declining, This is because the addition of EPDM-g-MAH and EPDM segments in non conjugated dienes, the double bond under heat and force, resulting in a slight crosslinking, the flexibility of the molecular chain fell.

Table 2. DSC date of HDPE and HDPE/OBC/EPDM-g-MAH

Samples	T_c^f /°C	T_c^p /°C	T_c^o /°C	X/%	T_m^o /°C	T_m^p /°C	T_m^f /°C	ΔT
0#	112.17	117.88	119.52	67.29	126.33	131.15	135.49	13.27
1#	112.64	117.63	119.30	56.14	123.99	129.91	134.45	16.82
5#	112.73	117.33	119.12	51.93	124.40	130.96	134.25	13.63

Melting temperature $T_m = \Delta H/\Delta S$, among them, ΔH is the melting enthalpy change, ΔS is the melting entropy change [6]. From the above type: And the increase of ΔH or reduction of ΔS will increase the Tm. From Table 2 and Figure 4 shows that pure HDPE (0#) sample melting temperature is 131.15 °C, HDPE/OBC (mass fraction of OBC is 18%) the melting temperature of blends is 129.91 °C, decreased 1.24 °C, this is because the addition of OBC, reduce the flexibility of the system, thereby reducing the molecular chain in the crystallization process of ordered surface nucleation rate, then reduce the crystallinity of blends. HDPE/OBC/EPDM-g-MAH (5#) sample melting temperature is 130.96 °C, rise 1.05 °C than the 1# sample, this is because the addition of EPDM-g-MAH, increase the interaction between molecular blends, the melting temperature increases. Degree of cooling ΔT is used to characterize the nucleation ability of the polymer molecular chain. The smaller the value is, the higher the nucleation ability of the polymer is, the faster the crystallization rate [7]. HDPE /OBC/EPDM-g-MAH blends have the biggest ΔT , the results show that the addition of EPDM-g-MAH can improve the nucleation ability of HDPE/OBC blends, and the blends are easier to crystallize.

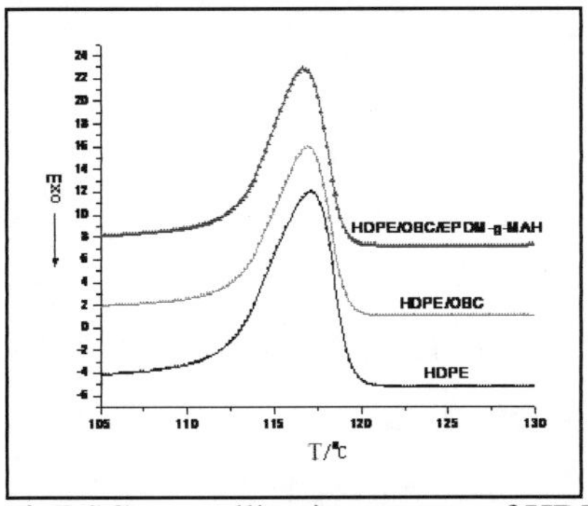

Figure 4. DSC crystallization curves of HDPE and its blends

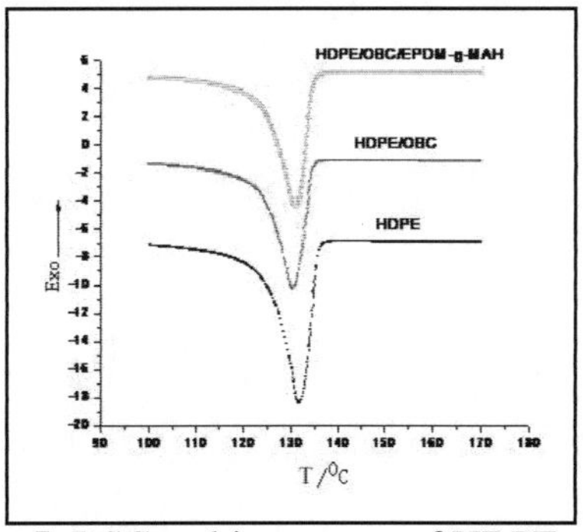

Figure 5. DSC melting curves of HDPE and its blends

4. Conclusion

1) New thermoplastic elastomer OBC and matrix HDPE has good compatibility, when adding 18% OBC, the notched impact strength of blends increased by 343.30%, the toughening effect is obvious.
2) When adding 4% EPDM-g-MAH, the notched impact strength of blends reached 30.33kJ/m^2, elongation was 1156.60%, compatibilization effect is obvious.
3) The addition of EPDM-g-MAH enhance the interface compatibility, making OBC more evenly distributed in the HDPE matrix, the size is refined and the crystallinity decreased.
4) With the addition of OBC and EPDM-g-MAH, the tensile strength and flexural strength of the blends decreased, the focus of future work is to find effective reinforcing material for HDPE/OBC.

Reference

[1] Daniel J. Arriola, Edmund M. Camahan, Phillip D. Hustad, at al. Catalytic Production of Olefin Block Copolymers via Chain Shuttling Polymerization [J]. Science, 2006, 312(5774): 714-718.
[2] Cheng Li, Hong Fan, Lie Lu, etc. Research on Toughening Modification of Copolymerized Polypropylene Using Polyolefin Elastomers [J]. Journal of Chemical Engineering of chinese Universities, 2009, 23(5): 813-815.
[3] Wenqiang Qiao, Xiaodong Fan, Jie Kong, etc. Effect of Polymeric Nucleating Agent on Crystllization of HDPE [J]. CHINA PLASTICS INDUSTRY, 2005, 33(2): 58-61.
[4] Weihua Fan. The Study of Toughened HDPE Engineering Plastics with Excellent Rigidity[D]. Zhengzhou: Zhengzhou University, 2006: 49-51.
[5] Xinming Liu, Yuan Zhou, Junxiao Wang, etc. Effect of Compatibilizer on properties of HDPE /PC mixed gold [J]. Modern plastics processing and applications, 2005, 17(1): 47-50.
[6] Fengqi Liu, Xinyi Tang. Polymer physics (Second Edition) [M]. Beijing. Higher Education Press, 2004: 134-137.
[7] Zheng Cui. Study on crystallization properties and formulation optimization of flame retardant reinforced PET [D]. Beijing: Beijing University of Chemical Technology, 2010: 38-44.

IOP Conf. Series: Materials Science and Engineering **191** (2017) 012003 doi:10.1088/1757-899X/191/1/012003

A study on flammability and moisture absorption behavior of sisal/coir fiber reinforced hybrid composites

Akash[1], Girisha K G[1], N S Venkatesha Gupta[2], K V Sreenivas Rao[3]

[1]Research Scholars, [2]Associate Professor, [3]Professor - Department of Mechanical Engineering, Siddaganga Institute of Technology, Tumakuru-572103, Karnataka, India

E-mail: akashbiradar@gmail.com

Abstract. Cellulosic fiber reinforced Polymer Matrix Composites (PMC's) are more frequently applied in construction industry and transportation, in which their flammability and water absorption behaviors are important. Fire resistance of cellulosic fiber reinforced composites is important parameter that often limits the application of composites in a given area. This work presents experimental results of a fire retardant behavior and moisture absorption behavior of different weight percentage (10, 20, 30, 40 and 50 wt. %) of sisal/coir fiber reinforced epoxy resin hybrid composites. Traditional cold pressing method was used to fabricate hybrid composites. Flammability behavior of the hybrid composite was studied by using vertical and horizontal burning rates as per standard UL-94. Addition of the cellulosic fiber increases the flammability since natural fiber supports fire. It proves as a bad flame retardant due to the generation of a surface layer during pyrolysis of the cellulosic fiber which exhibits poor fire retardant nature. This layer acts as supporter of fire, which spreads the heat from being transferred to the un-pyrolised material. The speed of flame is much faster in vertical burning position compared to horizontal burning position due to preheating of the specimen. Moisture absorption of sisal/coir fiber reinforced epoxy resin hybrid composites are studied according to ISO 62:1999 standard procedure. Absorption of moisture increases with increasing in the reinforcement weight percentage of cellulosic fiber in fabricated hybrid composite.

1. Introduction

Greater attention has been given to the cellulosic fiber reinforced Polymer Matrix Composite (PMC) materials in recent times due to persistent discussions on the climate changes and regulatory demands underlining the need for clean environment and utilization of renewable resources. This is mainly because the cellulosic fibers are de-composing in soil after use. Apart from decompose in nature, the cellulosic fibers involves number of advantages over petroleum based fibers such as high strength to weight ratio, abundantly available, low cost, easy process ability at relatively less energy requirement in tooling and assembly costs, higher strength and stiffness have made cellulosic fiber reinforced bio-composites widely acceptable in structural applications (1). Another distinct advantage is their ability to be engineered to obtain required properties in different directions by appropriate fiber placing in different layers of the laminated structure. Cellulosic fibers like sisal, hemp, coir, bamboo, banana, etc., have gained substantial important as reinforcements in PMC's. Among all cellulosic fibers sisal and coir fibers are most widely used cellulosic fibers and very easily cultivated. Sisal fiber is a hard fiber extracted from leaves of sisal plant. Coconuts are abundantly growing in coastal area of India. Coir fiber is one of the major wastes found in the southern coastal regions of India. Sisal and coir fibers are hydrophilic in nature and large amount of hydroxyl groups will give hydrophilic properties. This property will lead to a very poor interfacial bonding between fiber and hydrophobic matrix and very poor moisture absorption resistance. To enhance interfacial bonding between fiber and matrix fiber surface treatment must be carried out. Libo Yan et al. (2) studied the effect of sodium hydroxide treatment on microstructure and mechanical properties of natural fiber reinforced polymer composites. Noorunnisa Khanamet et al. (3) studied the effect of different types of surface modification of sisal fibers on the tensile and flexural properties of sisal fiber reinforced composites. Many authors have studied the effect of coir and sisal fiber based composites for physical and mechanical properties (4-5). These properties are based on the nature and morphology of the fiber, matrix properties and fiber-matrix interaction.

Environmental awareness motivates the use of cellulosic fibers, but their composites shows worse in flammability behavior compared to glass, carbon, etc., fiber reinforced composites. The application of advanced PMC's is constantly growing, especially in automotive, structural applications and industrial sectors. Flammability is a censorious matter in numerous industrial applications especially in the region of transformation where small spaces make fires a noticeable hazard. M.W.Chai (6) et al. compared the fire retardant behavior of glass fiber and flax fiber reinforced epoxy resin composites. Flax fiber reinforced composites burst into flames earlier, releases higher amount of heat and their structure deforms notably during combustion. K.Ramanaiah (7) et al. studied fire retardant behavior of polyester resin and sansevieria fiber reinforced polyester resin composite. In results, sansevieria fiber reinforced composite catches fire earlier, release high amount of carbon dioxide and total smoke during combustion compared to un-reinforced polyester resin. Although in practice the commercial applications of natural fiber based plastics are constantly growing, there is still poor interpretation of certain aspects of their behavior, such as their fire retardant behavior. With the above background, the goal of present research work is to fabricate sisal/coir fiber reinforced epoxy resin bio-composite and conduct the burning rates test and evaluate the response of the composite to fire and water absorption behavior.

2. Experimental details

2.1. Used materials

Sisal and coir fibers were used in this study. These cellulosic fibers were collected from plants grown in Karnataka (India). Epoxy resin araldite (AW106) and corresponding hardener (HV 953) were used to fabricate hybrid composite sheets, which is a low temperature curing epoxy system. Epoxy resin and hardener at a ratio of 10:1 was mixed to prepare matrix material.

2.2. Hybrid composite preparation

A simple cold pressing method was used to prepare the hybrid composite sheets in metallic (cast iron) mold at high pressure. Were hydraulic pressing machine was used to fabricate hybrid composite sheet. Both female and male dies are coated with a thin layer of Poly Vinyl Alcohol (PVA) to ensure that the sheet will not adhere to the mold. Sisal and coir fibers were mixed and added in to matrix material in a boul. 10, 20, 30, 40, and 50 wt.% of sisal/coir (1:1) fibers are used to fabricate the hybrid composite sheet. A thoroughly-mixed mixture of sisal/coir fibers and matrix was discharged into the female die cavity. The male die was laid down on the female die and pressed to 410.4 kg/cm2 from hydraulic pressing machine for 3 hours. After drying, the hybrid composite sheet was cut to specimens to conduct vertical and horizontal burning rates and moisture absorption test according to UL standard (125*13.3*3 mm) and ASTM D-570 respectively.

3. Results and discussion

3.1. Flammability behavior of hybrid composites

Fire retardant of sisal/coir fiber reinforced hybrid composite samples of size 125*13*3 mm was characterized for flammability behavior according to UL-94 standard for getting vertical and horizontal burning rates. For horizontal test, a 25mm Bunsen flame was applied to the leading edge of the sample. 6.35mm length of samples is introduced to Bunsen flame for 30 sec without altering place of Bunsen flame and sample. Then the sample is withdrawn from the flame. When the sample burns to the 25.4 mm mark before 30 seconds then the flame is removed. If the sample constantly burns after displacement of the flame, the time for the fire front to travel from the 25.4 mm to the 101.6 mm mark from the free end is noted to calculate the rate of burning. In vertical rate of burning test, a 19.05 mm Bunsen flame is applied to the bottom of the sample for 10 sec, replaced and then reapplied for a supplementary 10 seconds. The time of flaming and glowing is recorded as soon as the sample is replaced. The experimental setup for burning tests is as shown in figure 1.

(a) Vertical burning set-up (b) Horizontal Burning set-up

Figure 1. Flammability burning setup of hybrid composites

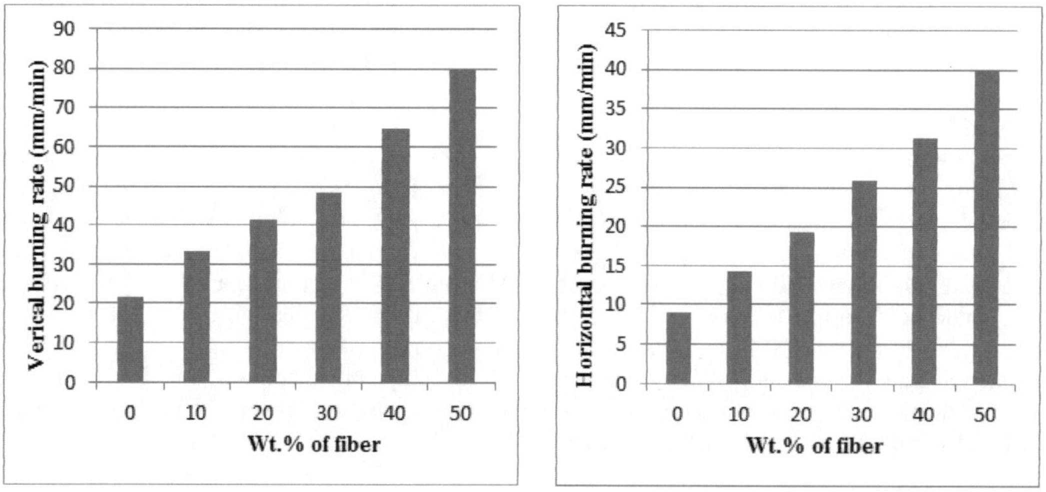

Figure 2. Burning rate of hybrid composites with different wt.% of sisal and coir fiber loading

The figure 2 represents the graphical representation of burning rate of vertical and horizontal burning for various wt. % of sisal/coir fiber reinforcement. The results indicate that addition of natural fiber increases the flammability. Hence natural fiber supports fire and is a bad flame retardant due to the generation of a surface layer during pyrolysis of the natural fiber composites which discloses poor fire retardancy. This layer behaves as supporter of fire, which spreads the heat to the un-pyrolised substance. In vertical burning test, the speed of flame is much faster than in horizontal position due to preheating of the specimen.

3.2. Water absorption of hybrid composites

Water absorption property of sisal/coir-epoxy resin hybrid composites was evaluated according to ISO 62:1999 standard procedure. Fabricated hybrid composite specimens were dried in oven at 500C and cooled to room temperature. After weighing the dried specimen to an accuracy of 0.1mg, they were immersed in distilled water in a plastic tub at room temperature. Once in 24 hours, the specimens were taken out from plastic tub, all the water from the surface was removed with clean cloth and the specimens were weighed. Same procedure was repeated for 28 days. The percentage of water absorption of sisal and coir fiber reinforced hybrid composites increases with increasing the fiber content (4). The percentage of water absorption was plotted against the time (days) as shown in Fig 3 which shows that the water absorption becomes stable after 25 days. Fig 4 shows the variation in the amount of moisture absorption with increase in fiber content. This shows that the moisture absorption increases with the increase in fiber content, thus composite sample shrinks, expands and de-bonding between fibers and matrix. As the composite sample shrinks and expands, de-bonding between fibers and matrix creating a voids which act as a reservoir for the water, so increasing its overall saturation level. Indication of this model is suggested by drying out the samples and gauging the post-test weight. With recognition that, moisture plays an important role influencing the mechanical behavior, the long term durability of PMC's. Water diffuses through the matrix and reaches the inter phase region and the reinforcement. Dissolution of polymer matrix, de-bonding of fiber matrix inter phase and degradation of fibers during aging in water, may all lead to the reduction in mechanical properties of the PMC's. Shrinkage and swelling of fibers may also happen, causing the decoupling at interphase, which leads to the de-bonding of the fibers. It is seen that the samples are immersing in water leads to rapid de-bonding, delamination and loss of structural integrity. Thus, the strength of PMC's is affected by immersion in water. The figure also shows that the moisture content increases with the increase in the soaking duration. Higher water absorption rates can be attributed to the hydrophilic property of cellulosic fiber due to the hydroxyl group (-OH) on the cellulosic molecules. Hydrogen bond forms between water molecules and hydroxyl group (8).

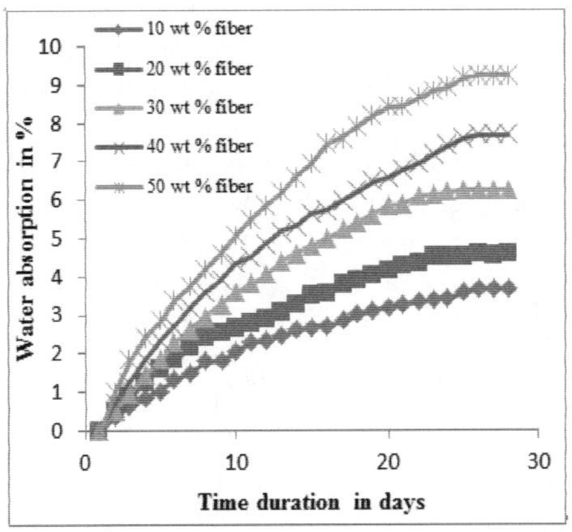

Figure 3. Wt.% of moisture absorption of different sisal/coir content of hybrid composites

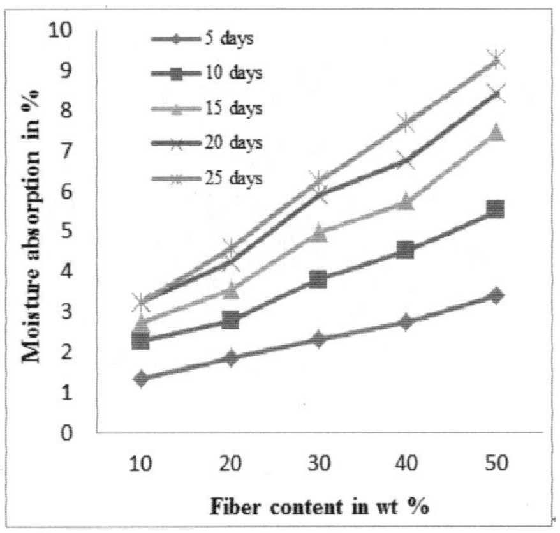

Figure 4. Variation in the amount of moisture absorption with increase in fiber content.

4. Conclusions

Experiments were conducted to establish the fire retardant behavior and moisture absorption behavior of different wt. % of sisal and coir fiber reinforced hybrid bio-composites. The experimental results lead to the following conclusions.

- Increasing the weight percentages of fibers in composites increases flammability and discloses poor fire retardant.
- Among all fabricated hybrid composites, 50 wt. % fibers reinforced hybrid composites burst into flames earlier, liberates more quantity of heat, and deforms in shape notably during combustion.
- Moisture absorption (%) behavior reveals that the absorption of water content of hybrid composites increases with increase in the wt.% of fiber content and also increases with the increase in the soaking duration.0
- Sisal/coir fiber reinforced polymer matrix composites can used in many applications like domestic (window panels, tables, etc.,), automobiles (car dash-boards), etc., but they are not suitable where the fire response is serious consideration.

References

[1] B.C. Mitra, "Environment Friendly Composite Materials: Biocomposites and Green Composites", Defence Science Journal, Vol. 64, No. 3, pages: 244-261, May 2014.

[2] Libo Yan, Nawawi Chouw, Liang Huang, Bohumil Kasal, "Effect of alkali treatment on microstructure and mechanical properties of coir fibres, coir fibre reinforced-polymer composites and reinforced-cementitious composites", Construction and Building Materials 112; pages:168-182, 2016.

[3] P. Noorunnisa Khanam, H. P. S. Abdul Khalil, G. Ramachandra Reddy, S. Venkata Naidu, "Tensile, Flexural and Chemical Resistance Properties of Sisal Fibre Reinforced Polymer Composites: Effect of Fibre Surface Treatment," J Polym Environ Vol. 19; pages: 115-119, 2011.

[4] K. Murali Mohan Rao, K. Mohana Rao, A.V. Ratna Prasad, "Fabrication and testing of natural fibre composites: Vakka, sisal, bamboo and banana", Materials and Design Vol. 31; pages: 508-513, 2010.

[5] Akash, K V Sreenivasa Rao, N S Venkatesha Gupta, Arun kumar D S, "Mechanical Properties of Sisal/Coir Fiber Reinforced Hybrid Composites Fabricated by Cold Pressing Method", Materials Science and Engineering 149, 2016.

[6] M.W. Chai, S. Bickerton, D. Bhattacharyya, R. Das, Influence of natural fibre reinforcements on the flammability of bio-derived composite materials, Composites: Part B 43; pages 2867-2874, 2012.

[7] K. Ramanaiah, A.V.Ratna Prasad, K.Hema Chandra Reddy, Mechanical, thermophysical and fire properties of sansevieria fiber-reinforced polyester composites, Materials and Design 49 pages: 986-991, 2013.

[8] Satyanarayana K G, "Recent Developments in Green Composites based on Plant Fibers-Preparation, Structure Property Studies", Bioprocessing & Biotechniques, Vol. 5; pages 1-12, 2015.

IOP Conf. Series: Materials Science and Engineering **191** (2017) 012004 doi:10.1088/1757-899X/191/1/012004

Application of Analytic Hierarchy Process (AHP) in the analysis of the fuel efficiency in the automobile industry with the utilization of Natural Fiber Polymer Composites (NFPC)

E Jayamani[1*], D S Perera[1], K H Soon[1] and M K B Bakri[1]

[1]Faculty of Engineering, Computing and Science, Swinburne University of Technology Sarawak Campus, Jalan Simpang Tiga, 93350, Kuching, Sarawak, Malaysia

*E-mail: elammaranj@gmail.com

Abstract. A systematic method of material analysis aiming for fuel efficiency improvement with the utilization of natural fiber reinforced polymer matrix composites in the automobile industry is proposed. A multi-factor based decision criteria with Analytical Hierarchy Process (AHP) was used and executed through MATLAB to achieve improved fuel efficiency through the weight reduction of vehicular components by effective comparison between two engine hood designs. The reduction was simulated by utilizing natural fiber polymer composites with thermoplastic polypropylene (PP) as the matrix polymer and benchmarked against a synthetic based composite component. Results showed that PP with 35% of flax fiber loading achieved a 0.4% improvement in fuel efficiency, and it was the highest among the 27 candidate fibers.

1. Introduction

Currently, natural fuel resources are running low and this phenomenon receiving attention globally. The escalating fuel consumption for transportation is one of the major reasons of natural fuel resource depletion. However, this issue can be alleviated if weight reduction can be achieved on vehicles by utilizing natural fiber polymer composites (NFPCs) as replacements to conventional materials such as metals and synthetic fiber polymer composites for certain vehicle components [1, 2].

Synthetic-fiber polymer composites such as embedding glass or carbon fibers in a plastic matrix as alternatives to the steel or aluminium components may avert the fuel consumption issue. They possesses strengths, stiffness, and other mechanical properties, that are similar to a plain steel or aluminium component, with the additional benefit of being a much lighter material. Thus, started the easiest solution to increasing the efficiency, as reducing weight and maintaining the strength is very much desired in this industry.

However, the problem now lies within replacing these synthetic fiber polymer composites with those NFPCs to ascertain same or stronger properties than the synthetic counterparts and help the automobile industry "go green" [3]. Therefore, this study aims to perform statistical analysis using Analytic Hierarchy Process (AHP) to assess the benefits of utilizing various NFPCs to obtain multiple options of candidate fibers to justify the grounds of replacing synthetic fibers in the automobile industry to achieve better fuel efficiency.

IOP Conf. Series: Materials Science and Engineering **191** (2017) 012004 doi:10.1088/1757-899X/191/1/012004

1.1. The rule of composites mixtures

Composites composed of two or more materials (i.e. matrix and reinforcement phase). The "rule of mixtures" is a theory that involves predicting the final composites properties based on the weighted average of the constituents' properties. Example of this rule would apply to the density (ρ) and among other properties, and thus the following equation may be written:

$$m_c = m_m + m_r \tag{1}$$

Where, m_c is the mass of the composite (kg), m_m is the mass of the matrix phase (kg), m_r is the mass of the reinforcement phase (this case the fibers) (kg).

Given the relation that volume, mass and density have in common, i.e. *density is the ratio of mass over volume.* Therefore, the following equation is formed.

$$\rho_c = f_m \rho_m + f_r \rho_r \tag{2}$$

Where, f_m and f_r are the volume fractions of matrix and reinforcing phase with respect to the composite and final phase. For this research, it is assumed that the composite's fibers are aligned in the longitudinal direction [4]. Therefore, by using the equation (2) for density and similar models for the other properties, the predicted properties are computed using MATLAB and the results were recorded in the database. The input data is based from various sources [5, 6, 7, 8] and a sample of the data at 60% natural fiber loading based on the rule of mixtures is shown in Table 1. It is also noted that the rule of mixtures is an estimation method as it is a general rule for a weighted mean calculation of a composite which is made up of continuous and unidirectional fibers [4, 9].

1.2. The analytic hierarchy process

AHP has been used successfully by many applications such as business field, government, military, and industry [10]. It was adopted by Farag [11] and used as a decision-making process for selecting the right composite for automobile industry. The key elements of AHP described by Bodin and Gass [10] have been laid out to understand the theory and practical situations application. Few of the theories relate directly to this study are the AHP Fundamental Pairwise comparison scale, ratio scales, and ratings model. AHP fundamental aspects will accurately analyze the data in a multi-factor manner. Noted that this process follows a similar pattern of the weighted objective method, in which the decision maker (DM) assigns weights/scale ratings from 1 – 9 and estimates the unknown weights based on research, insights, or even experience and assign them relative to the multi-criteria factors [10].

The pairwise comparison question deals with assigning accurate weights to certain criterion, are based on the DM's approximation (1 – 9) by asking whether one entity (say for example, Entity A) (i.e. criteria/alternative) is more important that the other (Entity B) in question. Bodin and Gass [10] have mentioned that the pairwise comparison must be interpreted by taking the ratio of A's assigned weight to B's and should be normalized, so that the denominator is always equal to 1. An example would clarify the above mentioned by using the verbal scale and associating it with the fundamental 1- 9 numeric scales. Thus, by saying Entity A is "more important" than Entity B, the ratio may be written as 5/1. What this means is if the DM assigned A's weight to be 40 and B's to 8, the relative ratio thereby is 5/1 [10]. Thus, AHP's resultant weights are ratio-scale numbers and consequently form a ratio scale that is units free.

Therefore, the resultant weights are normalized ratio-scale numbers. This allows the DM to compare alternatives. Subsequently, this method would allow for comparing alternative NRPCs, seeing as there are a number of combinations to select from (24 volume fraction variants; 20%, 35%, and 60% of NFPCs – with PP, with 3 combinations of PP with E-glass with respect to each volume fraction as a benchmark). Since the resulting weights are units free, it would allow easy comparisons based on numeric values alone. For example, if A acquires a resultant weight of 0.680 and B with a weight of 0.340; a ratio of 4/2, it would ultimately conclude that A is preferred twice as much as B [10]. PP has been selected as the candidate matrix due to the fact that it has been used as a polymer matrix previously [11] and it achieve higher rankings and physical properties in the final outcome [12].

For evaluation of the criteria, the eigenvector of the pairwise matrix must be calculated. This is to ensure that the DM's scores are related to each of the other weighted scores and gives clarity to the values. Quantitative values, such as preferences, are evaluated using the eigenvector method, while quantitative values, like data on mechanical properties, need to be normalized in the matrix.

Table 1. Sample of input data of NFPC at 60% fiber loading (value based on [5, 6, 7, 8])

Composite combos	Density (kg/m3)	Tensile strength (MPa)	Young's modulus (GPa)	Elongation at break (%)	Mass (kg)
PP-Flax	1,364.00	916.00	24.04	161.92	5.45
PP-Hemp	1,346.00	556.00	42.64	162.40	5.38
PP-Jute	1,358.00	496.00	16.54	160.90	5.43
PP-Kenaf	1,424.00	574.00	32.44	160.96	5.69
PP-Sisal	1,334.00	436.00	13.84	164.20	5.33
PP-Coir	1,184.00	121.00	36.64	178.00	4.73
PP-Curaua	1,304.00	676.00	18.64	162.58	5.21
PP-Abaca	1,364.00	503.80	20.80	161.74	5.45
PP-E-glass	1,994.00	2,116.00	44.44	162.22	7.97

2. Methodology

2.1. Initialization of the hierarchy

The overall method to setup the AHP is in accordance to the published work by Saaty [13]. The decision hierarchy of AHP is shown in Figure 1.

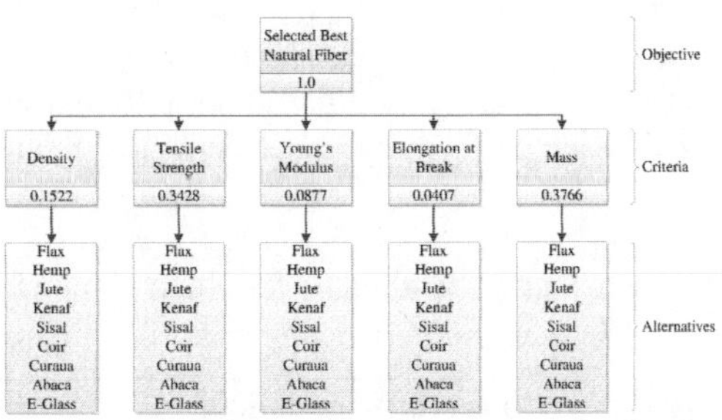

Figure 1. The decision hierarchy of AHP for the research. The values shown above correspond to the weighted scores from the pairwise comparison calculations which will be illustrated in Figure 2.

2.2. AHP parameters

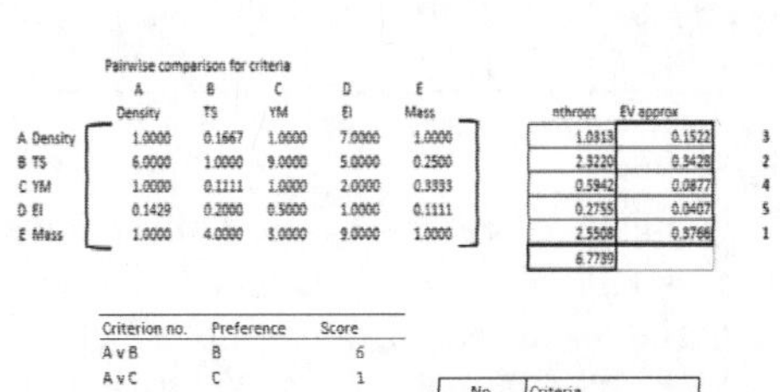

Figure 2. Pairwise comparison (top) shows the evaluated weighted scores with respective eigenvector values. Preference weights against each criterion (bottom) with legend.

IOP Conf. Series: Materials Science and Engineering **191** (2017) 012004 doi:10.1088/1757-899X/191/1/012004

The criteria were assessed according to the scaled weights set out by Bodin and Gass [10] and utilized by Farag [11]. This research explores the fuel efficiency improvement through mass (weight) reduction. The mass criterion would need to hold the highest weighting. The pairwise comparison matrix, as seen in Figure 2, shows the evaluated version of the ratio-scale numbers. Also depicted are the scores attributed to each criterion against each other, and the eigenvector evaluation.

2.3. MATLAB computation

Raw data were fed into MATLAB codes to generate the AHP results. The computation was repeated for each fiber's weight percentile group and then again for a variant design of the car hood, to observe if any improvements were found. MATLAB coding was performed to ease the computation for large matrices of data, i.e. to obtain the eigenvectors from multiples (27x27) matrices of fiber data, necessary for the AHP.

2.4. Solid modelling proposed engine hoods

Two engine hood designs were considered in this study. Solid modelling of the proposed engine hood designs were completed using SolidWorks. The mass property of the two hood models was captured and it was used for subsequent interpretation for fuel efficiency.

The two hood dimensions are as follows: Hood 1: L = 1404.25 mm, w = 1702.27 mm, and t = 1.6677 mm, and Hood 2: L = 1410.65 mm, w = 1268.10 mm, and t = 1.2673 mm.

3. Results and discussions

3.1. Simulation output

3.1.1. Hood 1 results and Hood 2 results. Hood 1 showed promising results, and the top 5 ranks are displayed in Table 2. Majority of the values resulted from the NFPC with 60% fiber loading, while Flax at 35% ranked 5th out of the 27 candidate fibers. Hood 2 results, showed that the mass reduction achieved were greater than Hood 1, and produced similar ranking to that of the Hood 1 results with the exception of the Curaua fiber at 60% outperforming kenaf fiber at the same percentage for 3rd place, as shown in Table 2.

Table 2. Top 5 ranked candidates for Hood 1 and Hood 2

	Hood 1			Hood 2		
Rank	Score	Fiber	Mass (kg)	Score	Fiber	Mass (kg)
1	0.546006	H1 Flax 60%	5.45	0.398046	H2 Flax 60%	3.09
2	0.54465	H1 Hemp 60%	5.38	0.391702	H2 Hemp 60%	3.05
3	0.541365	H1 Kenaf 60%	5.69	0.390966	H2 Curaua 60%	2.96
4	0.541301	H1 Curaua 60%	5.21	0.389881	H2 Kenaf 60%	3.23
5	0.539208	H1 Flax 35%	5.11	0.387631	H2 Flax 35%	2.90

3.2. Fuel efficiency improvement

By comparing both top 5 NFPC candidates of Hood 1 and 2 and benchmarking it with the synthetic PP + E-glass (35% and 60%), the mass reduction was calculated, and consequently the fuel efficiency through the reduction was derived. The results showed that fuel efficiency can be achieved by utilizing the NFPC with kenaf fiber (60% loading), Curaua fiber (60% loading) and flax fiber (35% loading) in Hood 1, while Table 3 also shows that by using NFPC with flax fiber (60% loading) and hemp fiber (60% loading) with design Hood 2, poor fuel efficiency was resulted. NFPC with 60% flax fiber is common for both designs. However, the highest improvement of fuel efficiency was achieved in the case of PP with 35% flax loading at 0.4%.

Table 3. Fuel efficiency improvements due to mass reductions for two Hood designs.

Fiber	Hood 1 Mass	Hood 2 Mass	Mass reduction H1	Mass reduction H2	Fuel efficiency improvement
PP-Flax 60%	5.45	3.09	31.62%	31.64%	0.06%
PP-Hemp 60%	5.38	3.05	32.50%	32.52%	0.08%
PP-Kenaf 60%	5.69	3.23	28.61%	28.54%	0.24%
PP-Curaua 60%	5.21	2.96	34.63%	34.51%	0.34%
PP-Flax 35%	5.11	2.9	22.34%	22.25%	0.40%
PP-E-glass 60%	7.97	4.52	-	-	-
PP-E-glass 35%	6.58	3.73	-	-	-

4. Conclusions

It can be concluded that fuel efficiency were achieved through mass reduction by utilizing NFPC, and statistically analyze the mechanical properties by adopting the AHP method of decision making. The results achieved by this simulation are consistent with the current findings, and show that the optimum candidate material is PP - flax at 35% fiber loading. The cost analysis shows many other fibers that come close to being optimal, which opens the path into more extensive research into other possible fibers in the chart with more criteria, focused on environmental factors and then can be compared to these findings.

Acknowledgments

We would like to thank the Ministry of Higher Education Malaysia (MOHE) for supporting this work through the FRGS grant scheme (FRGS/2/2013/TK04/SWIN/03/1).

References

[1] Karus M and Kaup M 2002 Natural Fibres in the European Automotive Industry *Journal of Industrial Hemp* 7 119-131.

[2] Kandelaars P P A A H and Van Dam J D 1998 An analysis of variables influencing the material composition of automobiles *Resources, Conservation and Recycling* 24 323-333.

[3] Alves C, Ferrão P M C, Silva A J, Reis L G, Freitas M, Rodrigues L B, and Alves D E 2010 Ecodesign of automotive components making use of natural jute fiber composites *Journal of Cleaner Production* 18 313-327.

[4] Holbery J and Houston D 2006 Natural-fiber-reinforced polymer composites in automotive applications *JOM* 58, 80-86.

[5] Groover M P 2002.Fundamentals of modern manufacturing: materials, processes and systems (New York: Wiley–Interscience)

[6] Ashori A 2008 Wood–plastic composites as promising green-composites for automotive industries! *Bioresource Technology* 99 4661-4667.

[7] Koronis G Silva A and Fontul M 2013 Green composites: A review of adequate materials for automotive applications *Composites Part B: Engineering* 44 120-127.

[8] Saba N, Paridah, M T and Jawaid M 2015 Mechanical properties of kenaf fibre reinforced polymer composite: A review', *Construction and Building Materials* 76 87-96.

[9] Timings R 2005.Mechanical Engineer's Pocket Book,(Oxford: Newnes-Jordan Hill).

[10] Bodin L and Gass S I 2003 On teaching the analytic hierarchy process *Computers & Operations Research* 30 1487-1497.

[11] Farag M M 2008 Quantitative methods of materials substitution: Application to automotive

components *Materials & Design* 29 374-380 .
[12] AL-Oqla F M, Sapuan S M, Ishak M R and Nuraini A A 2015 A model for evaluating and determining the most appropriate polymer matrix type for natural fiber composites. *International Journal of Polymer Analysis and Characterization* 20 191-205.
[13] Saaty T L 1983 Introduction to a modeling of social decision processes *Mathematics and Computers in Simulation* 25 105-107.

IOP Conf. Series: Materials Science and Engineering **191** (2017) 012005 doi:10.1088/1757-899X/191/1/012005

Mechanical and thermal properties of melt processed PLA/organoclay nanocomposites

Norazura Ibrahim[1], Margaret Jollands[2] and Rajarathinam Parthasarathy[2]

[1]Faculty of Applied Sciences, Universiti Teknologi MARA (UiTM), 40450 Shah Alam, Selangor, Malaysia.
[2]Rheology and Materials Processing Centre, School of Civil, Environmental and Chemical Engineering, RMIT University, Melbourne, Australia.

E-mail: norazura.ibrahim@gmail.com

Abstract. Polylactic acid (PLA) and organically modified layered silicates (organoclay) with concentrations of 2-10 wt% were prepared by melt intercalation technique. The effects of organoclay on the mechanical and thermal properties of PLA were studied. Tensile properties were evaluated using an Instron Universal Tester. Modulated differential scanning calorimetry (MDSC) and Thermogravimetric analyses (TGA) were performed to study the thermal behaviour of the prepared composites. The nanocomposites exhibited superior improvement of practical materials properties such as Young's modulus and thermal stability, as compared to the neat PLA. The Young's modulus drastically increased, whereas tensile strength and elongation at break decreased. The maximum degradation temperature of the hybrid increased linearly with an increasing amount of organoclay. However, MDSC has determined that the glass transition, cold crystallisation, and melting point temperatures were not significantly influenced by the presence of organoclay.

1. Introduction

Polylactic acid (PLA) has received increased interest in the last decade because of emergence of more environmentally aware consumers and increased price of crude oil. PLA is mainly used in applications such as plastic bags for household waste, disposable cups, and plates, as its degradability is a marketing advantage. PLA has higher modulus and strength [1], but lower toughness and thermal stability [2], as well as higher price, which limits its current commercial usage. Additives are commonly used to improve performance of plastics. This study focuses on the use of fillers to improve PLA properties.

A promising approach to improving properties is to prepare nanocomposites by adding montmorillonite (MMT) clay in the polymer matrix. Advantages include: (1) MMT can improve physical as well as mechanical properties; (2) the filler loadings are relatively low so the melt processing conditions do not need to be changed and (3) the final product cost may be reduced [1]. However, improvement in the nanocomposite's properties depends on several factors, including MMT concentration as well as degree of its dispersion [3]. Dispersion is controlled by the processing conditions as well as compatibility of MMT and polymer.

Several studies of PLA/organoclay nanocomposites prepared by melt intercalation technique have been reported [4-6]. However, these reports did not include how the melt processing conditions were chosen. In our recent paper [7], we described how the optimum processing conditions for processing PLA/MMT nanocomposites by melt intercalation can be identified using experimental design. In this paper, properties of PLA/MMT nanocomposites prepared at the optimum processing conditions were reported, and the effects of filler level on mechanical and thermal properties were examined. The level of filler was varied to determine if there is an optimum level.

2. Experiment

2.1. Materials

NatureWorks® PLA polymer 7000D was used in this study. The MMT nanoclay Cloisite® 30B from Southern Clay Products was used as filler. This is a natural montmorillonite modified with a quaternary ammonium salt. PLA pellets and clays were dried to remove volatiles in a vacuum oven at 90°C for 24 hours prior to processing.

2.2. Preparation of the nanocomposites

PLA pellets and various amounts of MMT were melt blended in a counter rotating Haake Rheomix 600. The processing conditions were set at the following temperature, rotor speed, and mixing time: 175 °C and 100 rpm for 7 min. Samples were made with seven different clay loadings (from 2 to 10 wt%). The pure PLA sample was also processed in a mixer. Specimens for property testing were obtained by compression molding. These were produced from warm melt from the mixer placed in a suitable mould in a hot-plate hydraulic press. The press platens were set at 190 °C and samples were pressed for 3 minutes and allowed to cool to ambient temperature under pressure (10 kN). All samples were then stored in a desiccator prior to testing.

2.3. Characterisation

An Instron model 4465 Universal Testing Frame and a load cell of 10 kN was used to measure tensile properties. The test was conducted at a constant rate crosshead speed of 5 mm/min at room temperature per ASTM D638 standard.

Thermal analysis was performed using a modulated differential scanning calorimeter (MDSC, TA2920) with an RCS unit, from TA Instruments. Sample and reference chamber with an empty pan were heated from -20 to 220 °C at heating rate of 2 °C/min with a heating/cooling cycle of the modulation period of 40 s and the amplitude of ±0.5 °C. After heating up to 220 °C, the PLA samples were isothermally held for 5 min to eliminate the previous heat history and subsequently cooled down to -20 °C at a cooling rate of 2 °C/min. A second heating scan at 2 °C/min was also performed. The data analysis reported here is from the second heating cycle. The degree of crystallinity was calculated by considering a melting enthalpy of 93.6 J/g for 100 % crystalline PLA. All the data were recorded using TA Instrument Control software installed with the equipment.

The thermal stability was investigated using a Thermogravimetric Analyzer (Perkin Elmer STA6000). Samples of 10-20 mg were heated with a heating rate of 10 °C/min from room temperature to 600 °C both in oxidizing (air) and inert (nitrogen) atmosphere. The resulting thermograms were processed using Pyris Version 11. Each sample was run in duplicates.

3. Results and discussion

3.1. Mechanical characterisation

The tensile results of neat PLA and PLA nanocomposites are presented in Figure 1 and summarised in Table 1. Values reported here are the averages of four specimens for each sample. The modulus of the nanocomposites increased monotonically with increasing clay content, by as much as 54 % for the 10 wt% sample, which is a very significant improvement.

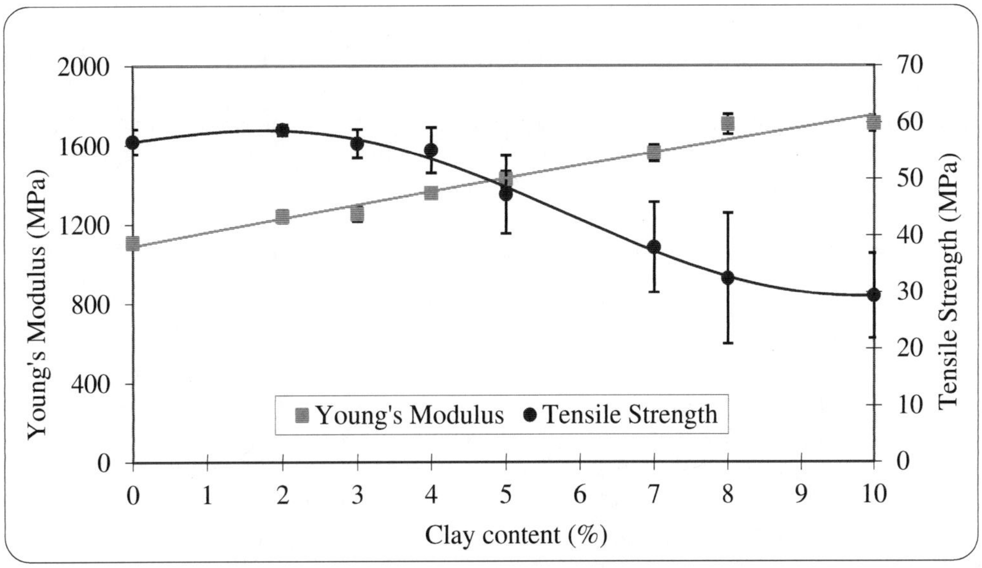

Figure 1. Tensile properties of PLA nanocomposites as a function of clay loading

Table 1. Tensile properties of PLA nanocomposites at various clay loadings

Samples	**Young's Modulus (MPa)**	**Tensile Strength (MPa)**	**Elongation at break (%)**
PLA-30B-0	1110	57	7.0
PLA-30B-2.00	1240 (+12 %)	59 (4 %)	9.6
PLA-30B-3.25	1250 (+13 %)	56 (-1 %)	7.0
PLA-30B-4.50	1360 (+23 %)	55 (-3 %)	6.3
PLA-30B-5.75	1430 (+29 %)	47 (-17 %)	4.5
PLA-30B-7.00	1560 (+41 %)	38 (-35 %)	3.5
PLA-30B-8.25	1710 (+54 %)	32 (-42 %)	2.5
PLA-30B-10.00	1710 (+54 %)	29 (-47 %)	2.3

There are only a few published results on tensile (tension) properties of PLA/organoclay nanocomposites [8]. Most studies on the mechanical properties of PLA/organoclay nanocomposites were reported using dynamic mechanical analysis [9, 10]. Nevertheless, two studies did report improved tensile properties of the PLA nanocomposites. Thellen et al. [8] reported the increase in the modulus of blown film PLA nanocomposites with 5 wt% MMT was 30 to 40 %. Li et al. [11] also reported that adding 5 wt% MMT, Cloisite® 20A, or Cloisite® 30B, to PLA increased the modulus by 15% and 21 %, respectively. The nanoclay acts as a mechanical reinforcement of the polymer at a nanoscale molecular level and thus gives a higher modulus value [12].

The tensile strength and elongation of the nanocomposite increased slightly at low levels of clay content (between 2 and 4 wt%), and decreased almost linearly with increases in the clay content. Alexandre and Dubois [13] reported increased Young's modulus, but lowered tensile strength and elongation at break. Increased tensile strength was reported only for systems where strong polymer/organoclay interaction occurred [14, 15]. In this study this indicates the clay is dispersed uniformly at low concentrations (between 2 and 4 wt%) with a high surface attraction between the clay and the polymer matrix. Above this concentration, the attractive force between the clay and the polymer matrix was disturbed by the tactoid form of the clay. This corresponds to passing through the percolation threshold region, estimated at 4.2 wt% clay loading [16]. According to Chang et al., [17], strength and elongation decrease if agglomerates are present.

3.2. Thermal characterisation

Table 2 shows that the T_g of PLA decreased by 1-2 °C with the addition of the organoclay, which is not significant compared with the measured SD (0.7 °C). The T_g depends primarily on chain flexibility, molecular weight, branching/crosslinking, intermolecular attraction, and steric effects [18, 19]. The drop in T_g could be due to a number of reasons such as clay aggregates, interphase regions, and adhesion problems at the clay-matrix interface at elevated temperatures [20].

Table 2. Thermal properties of PLA nanocomposites at various clay loadings

Samples	T_g (°C)	T_{cc} (°C)	ΔH_c (J/g)	χ_c (%)	T_m (°C)	ΔH_m (J/g)
PLA-30B-0	61.9	120.6	25.2	26.9	151.0	21.5
PLA-30B-2.00	61.2	121.0	27.1	29.5	151.2	23.6
PLA-30B-3.25	61.0	114.6	28.3	31.3	155.1	23.3
PLA-30B-4.50	60.6	114.2	28.4	31.8	154.9	23.1
PLA-30B-5.75	60.8	116.5	28.2	32.0	155.0	23.2
PLA-30B-7.00	60.2	116.7	28.5	32.7	155.1	23.1
PLA-30B-8.25	60.0	115.2	28.3	33.0	155.0	23.1
PLA-30B-10.00	59.9	113.6	26.8	31.8	155.0	22.7

T_g, glass transition temperature; T_{cc}, cold crystallisation temperature; ΔH_c, enthalpy of crystallisation; χ_c, degree of crystallinity; T_m, melting temperature; ΔH_m, enthalpy of fusion. T_{cc} and T_m are taken at the peak maximum of the crystallisation and melting peaks.

Paul et al. [21] reported increasing filler content did not significantly influence the T_g or T_m of PLA Cloisite® 30B nanocomposites. Because of the creation of aggregates, the increase of the nanofiller content did not lead to an extra increase of the total nanofiller surface area, and hence there was no effect on T_g and T_m. MMT nanoparticles have a great tendency to create aggregates because of their larger surface area [22].

PLA sample had a melting point at about 151 °C and a heat of fusion of 21.5 J/g. The incorporation of MMT clays led to an increase by 4 °C at 3.25 wt% of clay loading. Thereafter the T_m did not change significantly. Most studies [4, 21] reported that the incorporation of MMT clays did not significantly change the melting point of the materials.

Table 2 also shows that the cold crystallisation temperature, T_{cc}, of the PLA nanocomposites was significantly reduced by the addition of organoclay. The T_{cc} of PLA-30B-10.00 nanocomposites was 7 °C lower than neat PLA. Similar observations were reported by Lee et al. [12] and Day et al. [23]. Both stated that clay nanoparticles can act as effective nucleating sites for crystallisation in PLA. This was attributed to the nucleating effect of the nanosized clay. The large surface area facilitates the PLA crystallisation process, thus, T_{cc} decreases with the addition of nanoclays [12].

From Table 2, it can also be seen that the χ_c of PLA nanocomposites increased significantly (by ±2.2 SD) in the presence of organoclay up to around 8.25 wt%. Above 8.25 wt%, χ_c decreased. This indicates that the clay had a nucleation effect at lower clay loadings [24]. The nucleation efficiency will be the highest below the percolation threshold. Thereafter the rate of increase of nucleation due to the presence of fillers reduces, finally becoming very low at very high clay loadings [25]. This behaviour is attributed to the creation of larger agglomerates above the percolation threshold, and hence a reduction in the number of available nucleation sites. It can be said that a small amount of nucleating agent helps form polymer crystals while a larger amount hinders the ordered arrangement of molecular chains resulting in low crystallinity [26].

Of these thermal properties, crystallisation is the most important from both scientific and commercial viewpoints, because many mechanical properties of PLA are significantly influenced by its crystallinity [27]. The higher modulus at higher clay loadings is consistent with the higher crystallinity. The higher modulus of the nanocomposites is generally attributed to the reinforcement effect of the filler, but may also indicate that the crystallinity is higher [28].

3.3. Thermogravimetric analysis
In general, the addition of a filler can improve the thermal stability of the filled polymer composite systems. As inorganic species have good thermal stabilities, it is generally believed that the introduction of inorganic components into organic materials can improve their thermal stabilities. Many studies have reported that polymer/MMT composites have higher thermal stability than that of the neat polymer matrix [22, 29].

Table 3 summarises the maximum degradation temperatures. The degradation temperature was higher in all nanocomposites than in PLA. The shift was more pronounced for the nanocomposites with higher filler content. This suggests that the introduction of nanofillers increases the thermal stability of PLA [29]. The increase in thermal stability could be caused by an ablative reassembling of the silicate layers, which may occur on the surface of the nanocomposites, creating a protective barrier. The increase in the thermal stability has also been attributed to the high thermal stability of clay and to the interaction between the clay particles and the polymer matrix [30]. This is a common behaviour observed for nanocomposites and is partly responsible for their flame-retardant properties that are usually improved as well as their thermal stability [13].

Table 3. Thermal degradation behaviour of PLA nanocomposites at various clay loadings

System	Onset degradation temperature (°C)	Maximum degradation temperature (°C)	Residue (wt %) at 600 °C
PLA-30B-0	341	357	1.0
PLA-30B-2	342	359	3.4
PLA-30B-3.25	340	362	3.5
PLA-30B-4.50	341	363	5.2
PLA-30B-5.75	343	365	5.7
PLA-30B-7.00	343	369	7.2
PLA-30B-8.25	342	368	8.7
PLA-30B-10.00	342	370	9.1

The thermal properties of PLA are significantly influenced by the addition of the organoclay Cloisite® 30B. The results suggest the nanocomposites have a degree of dispersion, but the dispersion decreases as the filler level increases. PLA nanocomposites containing a few percent of Cloisite® 30B are more thermally stable than neat PLA. The role of the organoclay in the nanocomposite acts as a heat barrier, which enhances the overall thermal stability of the system. The superior performance in this study of nanocomposites samples compared with PLA suggests that the optimised processing condition meets this requirement.

4. Conclusions
PLA/organoclay nanocomposites were prepared at optimum processing conditions with varying levels of organically modified MMT (from 2 to 10 wt%). Mechanical properties showed a large increase of Young's modulus compared with the unfilled control PLA (up to 54 %). However, the tensile strength was reduced, by up to 50 %. As for thermal analysis, increasing clay loading decreased the glass transition slightly and increased crystallinity: this suggests that modified MMT is a nucleating agent for PLA. Thermal analysis showed that at higher clay loading thermal stability improved significantly. Investigations of the nanocomposite morphology (transmission electron microscopy and small-angle X-ray scattering) will be reported in a forthcoming article.

IOP Conf. Series: Materials Science and Engineering **191** (2017) 012005 doi:10.1088/1757-899X/191/1/012005

Acknowledgments
The first author gratefully acknowledges the Ministry of Higher Education Malaysia (MOHE) and Research Management Centre (RMC) of Universiti Teknologi MARA, Malaysia, for financial support through Grant No: 600-RMI/DANA 5/3/REI (7/2015).

References
[1] Carrasco, F., et al., *Processing of poly(lactic acid)/organomontmorillonite nanocomposites: Microstructure, thermal stability and kinetics of the thermal decomposition.* Chemical Engineering Journal, 2011. 178(0): p. 451-460.
[2] Mathew, A.P., K. Oksman, and M. Sain, *Mechanical properties of biodegradable composites from poly lactic acid (PLA) and microcrystalline cellulose (MCC).* Journal of Applied Polymer Science, 2005. 97(5): p. 2014-2025.
[3] Pluta, M., *Melt compounding of polylactide/organoclay: Structure and properties of nanocomposites.* Journal of Polymer Science Part B-Polymer Physics, 2006. 44(23): p. 3392-3405.
[4] Di, Y., et al., *Poly(lactic acid)/organoclay nanocomposites: Thermal, rheological properties and foam processing.* Journal of Polymer Science Part B: Polymer Physics, 2005. 43(6): p. 689-698.
[5] Paul, M.-A.l., et al., *New nanocomposite materials based on plasticized poly(L-lactide) and organo-modified montmorillonites: thermal and morphological study.* Polymer, 2003. 44: p. 443-450.
[6] Pluta, M., et al., *Polylactide/montmorillonite nanocomposites and microcomposites prepared by melt blending: Structure and some physical properties.* Journal of Applied Polymer Science, 2002. 86(6): p. 1497-1506.
[7] Ibrahim, N., M. Jollands, and R. Parthasarathy. *Optimising Processing Conditions of PLA Nanocomposites Using Response Surface Methodology.* in *3rd International Conference on Science and Social Research (CSSR2016).* 2016. Everly Hotel, Putrajaya.
[8] Thellen, C., et al., *Influence of montmorillonite layered silicate on plasticized poly(l-lactide) blown films.* Polymer, 2005. 46(25): p. 11716-11727.
[9] Sinha Ray, S., et al., *New polylactide-layered silicate nanocomposites. 2. Concurrent improvements of material properties, biodegradability and melt rheology.* Polymer, 2003. 44(3): p. 857-866.
[10] Pluta, M., J.K. Jeszka, and G. Boiteux, *Polylactide/montmorillonite nanocomposites: Structure, dielectric, viscoelastic and thermal properties* European Polymer Journal, 2007. 43(7): p. 2819-2835
[11] Li, T., et al., *Polylactide, nanoclay, and core–shell rubber composites.* Polymer Engineering & Science, 2006. 46(10): p. 1419-1427.
[12] Lee, J.H., et al., *Thermal and mechanical characteristics of poly(l-lactic acid) nanocomposite scaffold.* Biomaterials, 2003. 24(16): p. 2773-2778.
[13] Alexandre, M. and P. Dubois, *Polymer-layered silicate nanocomposites: preparation, properties and uses of a new class of materials.* Materials Science and Engineering, 2000. 28: p. 1-63.
[14] Lan, T. and T.J. Pinnavaia, *Clay-Reinforced Epoxy Nanocomposites.* Chemistry of Materials, 1994. 6(12): p. 2216-2219.
[15] Fornes, T.D., et al., *Nylon 6 nanocomposites: the effect of matrix molecular weight.* Polymer, 2001. 42(25): p. 09929-09940.
[16] Ibrahim, N., *Structure-property relationship of polylactide/organoclay nanocomposites*, 2013, RMIT University.
[17] Chang, J.-H., Y.U. An, and G.S. Sur, *Poly(lactic acid) nanocomposites with various organoclays. I. Thermomechanical properties, morphology, and gas permeability.* Journal of Polymer Science Part B: Polymer Physics, 2003. 41(1): p. 94-103.

IOP Conf. Series: Materials Science and Engineering **191** (2017) 012005 doi:10.1088/1757-899X/191/1/012005

[18] Clegg, D.W. and A.A. Collyer, *Mechanical properties of reinforced thermoplastics*. 1986: Elsevier Applied Science Publishers.
[19] Cowie, J.M.G. and V. Arrighi, *Polymers : chemistry and physics of modern materials*. 2007, Boca Raton: Taylor & Francis.
[20] Yasmin, A., et al., *Mechanical and thermal behavior of clay/epoxy nanocomposites*. Composites Science and Technology, 2006. 66(14): p. 2415-2422.
[21] Paul, M.-A., et al., *New nanocomposite materials based on plasticized poly(l-lactide) and organo-modified montmorillonites: thermal and morphological study*. Polymer, 2003. 44(2): p. 443-450.
[22] Kontou, E., M. Niaounakis, and P. Georgiopoulos, *Comparative study of PLA nanocomposites reinforced with clay and silica nanofillers and their mixtures*. Journal of Applied Polymer Science, 2011. 122(3): p. 1519-1529.
[23] Day, M., A. Nawaby, and X. Liao, *A DSC study of the crystallization behaviour of polylactic acid and its nanocomposites*. Journal of Thermal Analysis and Calorimetry, 2006. 86(3): p. 623-629.
[24] Lewitus, D., et al., *The Effect of Nanoclays on the Properties of PLLA-modified Polymers Part 1: Mechanical and Thermal Properties*. Journal of Polymers and the Environment, 2006. 14(2): p. 171-177.
[25] Bhattacharya, S., et al., *Foaming behavior of high-melt strength polypropylene/clay nanocomposites*. Polymer Engineering & Science, 2009. 49(10): p. 2070-2084.
[26] Liao, R., et al., *Isothermal cold crystallization kinetics of polylactide/nucleating agents*. Journal of Applied Polymer Science, 2007. 104(1): p. 310-317.
[27] Kulinski, Z. and E. Piorkowska, *Crystallization, structure and properties of plasticized poly(l-lactide)*. Polymer, 2005. 46(23): p. 10290-10300.
[28] Tanniru, M., Q. Yuan, and R.D.K. Misra, *On significant retention of impact strength in clay–reinforced high-density polyethylene (HDPE) nanocomposites*. Polymer, 2006. 47(6): p. 2133-2146.
[29] Krishnamachari, P., et al., *Thermal Characterization of Biodegradable Poly (Lactic Acid)/Clay Nanocomposites*, in *Proceedings of the 2007 National Conference on Environmental Science and Technology*, E. Nzewi, et al., Editors. 2009, Springer New York. p. 219-225.
[30] Fischer, H.R., L.H. Gielgens, and T.P.M. Koster, *Nanocomposites from polymers and layered minerals*. Acta Polymerica, 1999. 50(4): p. 122-126.

IOP Conf. Series: Materials Science and Engineering **191** (2017) 012006 doi:10.1088/1757-899X/191/1/012006

From the experience of development of composite materials with desired properties

I A Garkina and A M Danilov

Department of mathematics and mathematical modeling, Penza State University of Architecture and Construction, 28 Herman Titov street, Penza, Russia

E-mail: i.a.naum@mail.ru

Abstract. Using the experience in the development of composite materials with desired properties is given the algorithm of construction materials synthesis on the basis of their representation in the form of a complex system. The possibility of creation of a composite and implementation of the technical task originally are defined at a stage of cognitive modeling. On the basis of development of the cognitive map hierarchical structures of criteria of quality are defined; according to them for each allocated large-scale level the corresponding block diagrams of system are specified. On the basis of the solution of problems of one-criteria optimization with use of the found optimum values formalization of a multi-criteria task and its decision is carried out (the optimum organization and properties of system are defined). The emphasis is on methodological aspects of mathematical modeling (construction of a generalized and partial models to optimize the properties and structure of materials, including those based on the concept of systemic homeostasis).

1. Introduction

Due to the complexity and diversity of the functioning processes in real systems can not get them absolutely adequate mathematical models. The developed mathematical model describing the process of formal operation of the system, is able to cover only the basic characteristic patterns. You can not specify the formal rules to select the characteristics of the studied states and parameters of real systems. The researcher has to be guided only by their own intuition, relying on the statement of the problem and applied understanding of the processes of functioning of the system. Analysis of each complex system - this is a unique problem that requires not only diverse cultures, but also the inventiveness and talent. System studies are a symbiosis of theoretical model simulations with observations, empirical studies, full-scale (laboratory) experiments. One of the main problems is to determine the general laws governing the functioning of the system based on individual private observations. This also applies to the problems of building materials [1-4].

System operation conditions are determined by the input variables $x_1, x_2, \ldots, x_n$ (factors); some of them amenable to regulation or partial management. The behavior, the result, the effectiveness of functioning of the system are characterized by output variables $y_1, y_2, \ldots, y_m$ (responses, resulting variables). Random latent variables $\varepsilon_1, \varepsilon_2, \ldots, \varepsilon_m$ (not amenable to direct measurement) reflects the impact on the value of unrecorded at the input factors (random measurement error of the analyzed indicators, residues). So at winter concreting electro-conductive concrete which durability depends on its structure is applied. The factors are: the percentage of Portland cement - x_1, sand - x_2, graphite - x_3, water - x_4; strength conductive concrete - y. The main objective is to construct a function

$$\mathbf{f}(x_1, x_2, \ldots, x_p) = (f_1(x_1, x_2, \ldots, x_p), \ldots, f_m(x_1, x_2, \ldots, x_p))^{\mathrm{T}},$$

adopted in some sense makes the best restore the value of the $\mathbf{y} = (y_1, y_2, \ldots, y_m)^{\mathrm{T}}$ by $\mathbf{x} = (x_1, x_2, \ldots, x_p)^{\mathrm{T}}$, T - transposition symbol.

2. Formulation of the problem

The main objectives of the statistical analysis in building materials are:

- establish a statistically significant association between the variables $\mathbf{y}$ and $\mathbf{x}$ (components $\mathbf{x}$ and $\mathbf{y}$ can be: *quantitative* (for example, the percentage of component in material), *ordinal* (ordering of objects analyzed according to the degree of manifestation being studied properties; for example, education worker, discharge working, etc.), *classification* (divide objects into homogeneous classes, not amenable to streamline; for example, methods for the treatment of concrete products, the availability of competing additives or fillers et al., employee profession, industry, the motive of family migration, etc.);
- outlook (recovery) $\mathbf{y}$ to $\mathbf{x}$ (determined the values of the function $\mathbf{f}(\mathbf{x})$, but not its structure);
- identification of causal relationships between $\mathbf{x}$ and $\mathbf{y}$, partial control $\mathbf{y}$ by regulation $\mathbf{x}$.

The solution of these tasks to determine the physical meaning of the statistical relationships (for example, depending on the properties of the materials studied on factors). *Standard techniques and methods to determine the structure of the model (the general form of the function $\mathbf{f}(\mathbf{x})$) does not exist.* In the staging phase of statistical analysis each elementary object of research is associated with a set of analyzed parameters $\{x_1, x_2, \ldots, x_p, y_1, y_2, \ldots y_m\}$; determined the final purpose of the study, the type of study dependencies, the shape and the degree of accuracy of statistical conclusions. Information step is to collect the results of measurement of statistical information

$$\left\{x_1^{(i)}, x_2^{(i)}, \ldots, x_p^{(i)}, y_1^{(i)}, y_2^{(i)}, \ldots, y_m^{(i)}\right\},\ i = \overline{1, n}.$$

The structure of connections and their tightness (class $\mathbf{F}$ of feasible solutions; view, the desired connection between the structure $\mathbf{x}$ and $\mathbf{y}$ the (family of functions $\mathbf{f}$) between the study variables) are determined based on the correlation analysis. Calculates an estimate of the parameters in the test equation the statistical relationship. If the class $\mathbf{F}$ is defined as a parametric family of functions $\{\mathbf{f}(\mathbf{x}, \mathbf{\Theta})\}$, the problem boils down to the selection of (statistical estimation) Θ values of the conditions for an extremum of a functional (for example, the method of least squares). Of course, it provides an analysis of the accuracy of the constraint equations.

In a dynamic system, there is an interaction between its heterogeneous parts, and formed observed (output) signals. All other signals are considered as perturbations (input signals, measured and is not measured noise). *The totality of the alleged links between the observed signals and is a model (in the broad sense).* The model can take different forms and degrees of mathematical detail (or at all without using the language of mathematics). Ambiguous choice of identification method (the incompleteness of knowledge about the object, limitations in the observations of the object in time, the inaccuracy of the measurement signals at the input and output of the object, and so on.).

The adequacy of the system model and the object of study can not prove: model can be accepted or rejected on the grounds that obtained the assessments and conclusions contradict the observed facts. System model will always be different from the original; we can talk about the asymptotic approximation to it under certain conditions for each practical task. Asymptotic convergence is ensured by the fact that the object of the simulation is a concrete system with inherent only to her autonomous operation of law (peculiar and unique to this system). The adequacy of the system model is achieved by narrowing the scope of its use *(limited practical applicability of the model). Model experiments provide an intuitive knowledge of nature, which can be used to adjust the model taking into account the properties of the object being studied.* The simulation allows a deeper insight into the essence of the original object, and modeling studies led to the discovery of new properties and the laws of the functioning of the system under study. Today is no unified theory of constructing system models, to fully implement the concept of systemic homeostasis. System modeling studies are always iterative. In assessing the effectiveness and reliability of the model be used quantitative criteria;

IOP Conf. Series: Materials Science and Engineering **191** (2017) 012006 doi:10.1088/1757-899X/191/1/012006

qualitative criteria allow us to estimate a model in terms of its clarity, usability, testability, applications in other areas of concern (an intermediate position between the quantitative and qualitative criteria is the criteria with point scale). Model experiments precede full-scale and direction information collection point in the process of observation (the leading role modeling). At the same time, the results of modeling studies need to be the actual source of data and require experimental verification. A systematic approach can reduce or even eliminate the uncertainty inherent in the problem at hand; reconstruct it in the models that meet the objectives of the study; identify objects, properties and relationships of the studied system, taking into account the mutual influence of the external environment. It allows you to understand the connections between individual facts and on a higher level to carry out the study. The theoretical value of the systems approach is that it allows you to define general patterns that are isomorphic for different classes; the most important of them is isomorphic backbone factor.

3. Synthesis of materials for special purposes as the complex systems

The approach the authors used effectively in the development of materials for special purposes as the complex systems. The role and place of the identification in the synthesis follow from Fig. 1.

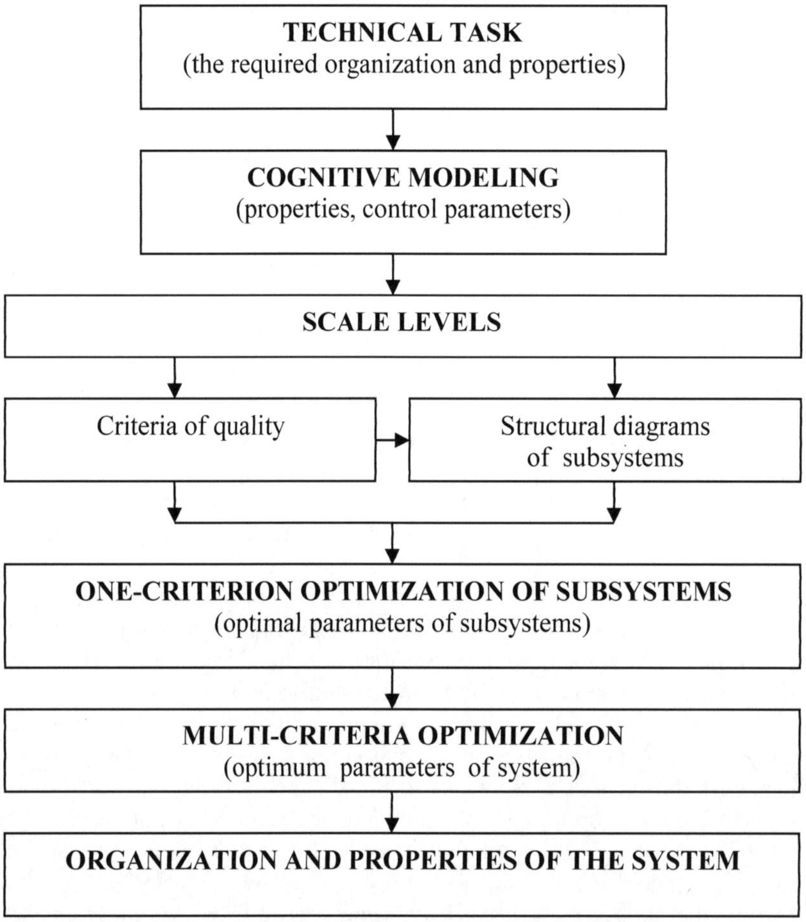

Figure 1. Synthesis of complex system

The synthesis began with a cognitive modeling. Identification is interdisciplinary. Therefore, in the composition of the expert group to develop terms of reference included experts of various fields (materials scientists, physicists, chemists, mathematicians, managers, experts in the field of system analysis, etc.). Each of the representatives played a dominant role in determining the individual pieces of the system; inter-element connections are established by the whole group. The cognitive model has allowed to produce material decomposition (synthesis of an integrated system reduces to solving the problems of synthesis of subsystems). Class of mathematical models of radiation-protective composites was selected; later they were used to optimize the structure and properties (fragments were seen as a subsystem; identify ways of measuring their properties). In drawing up the cognitive map is important not to miss the connection, which is determined by the considered property; «... quite insignificant reason escapes us by its smallness, it is a significant action that we can not foresee, and then we say that this phenomenon is a result of the case» (A.Poincare; a violation of the principle of transitivity). Linked to this is the difficulty of determining the number of partial criteria that are determined by the required performance of the material.

If the material is multi-layered structure, the construction of a mathematical model is almost impossible. Therefore, postulated the existence of a mathematical model to replace the identified object. Material properties were seen as particular criteria; structure of the material defined as the totality of their.

According to the results of cognitive modeling (using simulated principle) a complex system (composite) was represented as a finite set of models that reflect the essence of a particular facet of the system. Each of the properties of the material investigated by one or more simplified (narrowly focused) models. Building the set of simplified models revealed new properties (often without building a model generalizing). In other words, the composite material is determined by the interaction of simplified models [5, 6].

Structural identification is carried out (selection-end models) the solution of the general problem of identification in the selected class of models, as well as the decision of private problems of identification (parametric). Accordance with the basic requirements for the model: ease of interpretation of the physical meaning of the coefficients included in the model. Thus, the regression models are illustrative and are not used for the prediction and purposeful change factor values.

Each of the operating characteristics of the material was considered as an asymptotic value of the solution of the differential equation (kinetic process of the formation of physical and mechanical characteristics). Such representation is not the sole (test class may be replaced if necessary). Unfortunately, as a rule, there is a problem of correctness. We need the correct formulation of the problem with the involvement of additional information on the desired solution (or should use the classic control algorithms for ill-posed problem). The main kinetic processes (set of strength, changes in the modulus of elasticity, contraction and shrinkage, the increase of internal stress, heat resistance, chemical resistance, water absorption and water resistance) were modeled in a class of ordinary fourth-order differential equations with constant coefficients and real roots of the characteristic polynomial [5-7]. Concentrations and granulometric characteristics were determined by the material ingredients depending on the requirements (critical concentration, dependence of the critical volume concentration of the mean aggregate size distribution, the averaged effective strength of the material).

Proceeding from the above, we propose the following synthesis scheme:

- modeling of individual properties;
- selection of main parameters of models;
- determining of model parameters from composition, technological characteristics;
- determining a plurality of partial criteria;
- ranking the properties of materials;
- minimizing of dimension of the criterions space;
- multi-criteria optimization of the quality of the material and the corresponding prescription and process parameters.

Multi-criteria optimization was to optimize the vector of the objective function $\mathbf{q}(\mathbf{x})=(q_1(\mathbf{x}),q_2(\mathbf{x}),\ldots,q_m(\mathbf{x}))$, if the criteria adopted are equally significant. In some cases, multi-criteria problem of minimization of the vector $\mathbf{q}(\mathbf{x})$ (using weighted sums of the strategy) was reduced to a scalar problem (accepted $q(x)=\sum_{i=1}^{n}\alpha_i\cdot(q_i(x))^2$). Definition of integrative properties of systems at a qualitative level was made on the basis of independent studies of separate subsystems (with a certain degree of autonomy). Combining separate subsystems into a single system carried with allowance for interconnections. Each element in the hierarchical structure of the quality system was defined autonomously (without all the interconnections). Interconnections eliminated by introducing customizable reference models with a simultaneous decentralization of modules for inputs. Selection of parameters was done using the experimental samples.

4. Conclusions

Given the experience in the development of a number of composites developed a scheme of synthesis of building materials with predetermined properties and structure. It is proposed methodological principles of creation of materials (as complex systems) using a system approach.

References

[1] Korolev E V, Smirnov V A 2013 Using Particle Systems to Model the Building Materials (Advanced Materials Research Vol 746) pp 277-280
[2] Grishina A N, Korolev E V 2016 Influence of Nanoscale Barium Hydrosilicates on Composition of Cement Stone (Key Engineering Materials Vol 683) pp 90-94
[3] Loganina V I, Makarova L V, Tarasov R V 2016 The Effect of Filler Based on Calcium Silicates on Strength of Lime Composites (Applied Mechanics and Materials Vol 835) pp 531-534
[4] Kiselev D G, Korolev E V, Smirnov V A 2014 Structure Formation of Sulfur-Based Composite: The Model (Advanced Materials Research Vol 1040) pp 592-595
[5] Danilov A M and Garkina I A 2015 Systems approach to the modeling and synthesis of building materials (Contemporary Engineering Sciences Vol 8) pp 219-225
[6] Budylina E A, Garkina I A, Danilov A M 2015 Control of multi-objective complex systems (Contemporary Engineering Sciences Vol 8) pp 441-445
[7] Garkina I A 2015 Modeling of kinetic processes in composite materials (Contemporary Engineering Sciences Vol 8) pp 421-425

IOP Conf. Series: Materials Science and Engineering **191** (2017) 012007 doi:10.1088/1757-899X/191/1/012007

The influence of alkaline treatment on thermal stability of flax fibres

J CHAISHOME and S RATTANAPASKORN

Department of Food Engineering, Faculty of Engineering, King Mongkut's Institute of Technology Ladkrabang, Chalongkrung Rd. Ladkrabang, Bangkok 10520, Thailand

jedsada.ch@kmitl.ac.th

Abstract. The paper pursued the hypothesis that alkaline treatment removed hemicellulose and pectin from the flax fibres, the fibre thermal stability will be improved. The use of plant fibres as reinforcement in thermoplastic composites comes with the detrimental effect of thermal degradation on the tensile performance of fibres and composites, particularly during long consolidation times at high temperature. SEM was used to observe the treated fibre surface. The micrographs show that the treatment with a higher concentration of NaOH solution results in a more obviously rough fibre surface. Moreover, FTIR and TGA were used to examine the chemical decomposition and thermal stability, respectively. The spectra of treated materials indicate that both hemicellulose and pectin were dissolved from the fibre surface following treatment. Finally, TGA results revealed that the loss of mass belonging to hemicellulose and pectin in treated fibres results in a shift of the main degradation temperature to higher temperature.

1. Introduction

There has been a high sustainability agenda toward industrial products made from biodegradable or renewable materials. Consequently, it has been shown that there is a growing interest in the application of plant fibre thermoplastic composites [1-3]. Furthermore, research and development (R&D) in this area has resulted in the improvement of the performance of plant fibre composites in industries such as the automotive sector. Natural fibre composites offer several advantages such as eco-friendliness, safety from hazards and light weighting [4, 5].

Nevertheless, the low thermal stability of plant fibres is a limiting factor affecting the composite performance. Plant fibres are combustible materials. The natural properties and the various constituents of plant fibres can generate poor composite performance due to thermal degradation [6-8]. Thermal degradation of the plant fibres at high processing temperatures, typical of thermoplastic systems, is an important issue. The thermal degradation of plant fibre depends on a number of factors namely, chemical composition of materials, temperature and heating rate. The tenacity decreases and the degree of depolymerisation increases with an increase in temperature and/or dwell time at high temperature due to chain scission [7]. For example, the flax fibre retains its strength at 170°C for 120 minutes and at 200°C for 30 minutes, while it shows a 50% drop at 210°C for 120 minutes [7, 9]. Furthermore, the thermal decomposition temperature of fibres limits the choice of polymer as a matrix material due to the melted temperature of the polymer could be less than the thermal decomposition temperature of fibres. Low processing temperatures leads to low impregnation with high viscosity thermoplastic resins.In addition, volatile products from their chemical constituents may be generated during heating and settle in the laminates. These lead to low impregnation and to the formation of voids during moulding processes [10].

The chemical composition of plant fibres comprises different compounds with various concentrations. For instance, flax fibre consists of 64.1 – 71.9 ($\%_{wt}$) of cellulose, 16.7 – 20.6 ($\%_{wt}$) of hemicellulose, 2.0 – 20.6 ($\%_{wt}$) of lignin, 8 – 12 ($\%_{wt}$) of moisture content, 1.8 - 2.3 ($\%_{wt}$) of pectin and 1.5- 1.7 ($\%_{wt}$) of wax [6, 11-13]. Hemicellulose and pectin seem to dominate the degraded fibre property due to these material's low thermal stability and low tensile strength. Hemicellulose in the fibres degraded with higher dwelling time at high temperature. Therefore, a decrease of hemicellulose and pectin content of the fibres should enhance their mechanical properties *[14]*.

An attempt is to be made to improve the thermal stability of the fibres and consequently improve composite performance by using alkaline treatment methods. Alkaline treatment is a chemical treatment commonly used to prepare good quality fibres [15]. The removal of hemicellulose and pectin after the mercerisation process also improves the fibre strength and stiffness [10, 16, 17]. Moreover, this treatment increases the fibre rough surface, resulting in improved interfacial adhesion and mechanical performance of the composites.

2. Materials and experimental method

2.1. Materials

The flax fibres were soaked in NaOH solution of 1, 2, 3, 4, and 5% by weight of concentrations for 20 minutes at room temperature. Subsequently, these fibres were then followed by washing in running water several times to clean the fibre surface. The fibres were then dried in a hot air oven at 80°C for 8 hours.

2.2. Fourier transform infrared spectrometry

The Fourier Transform Infrared spectrometry (FTIR) technique was used to establish the chemical composition of treated and untreated flax fibres. In each measurement, five repeats of the treatments were tested at room temperature. The spectrum of the fibres was measured using a Bruker FTIR Tensor 27 spectrometer in attenuated total reflectance (ATR) mode, between 600 and 4000 cm^{-1} frequency range. Alcohol was used for cleaning the spectrometry after every test.

2.3. Thermogravimetric analysis

Thermal stability of treated and untreated flax fibres was studied by thermogravimetric analysis with three repeats for each material. All materials were stored in a plastic zip lock bag before testing. The fibres were cut to weights between 5 and 7 mg and were tested in a platinum pan. Thermogravimetric (TG) and differential thermogravimetric (DTG) curves were obtained. The analysis was conducted using a thermogravimetric analyzer/DSC TA SDT Q600, under nitrogen atmosphere. Nitrogen is commonly used as an inert gas which does not react with plant fibres. A nitrogen flow rate of 150 ml/min was used with a pre-heating process for 15 minutes and temperature range from room temperature to 600°C at heating rates of 10°C/min.

3. Results and discussion

3.1. Effect of alkaline treatment on the fibre surface

Fig.1(a) shows scanning electron microscopy (SEM) images of untreated flax fibre surface. It can be seen that the fibres are not uniform in shape and the diameter varies along the fibre. Figs. 1(b)-(f) show fibre treated by the concentrations of NaOH solution of 1, 2, 3, 4 and 5% by weight, respectively. These micrographs indicate that the treatment with a higher concentration of NaOH solution results in a more obviously rough fibre surface. The SEM micrographs reveal that alkaline treatment changes the fibre surface characteristics due to the treated fibres having lost some of their constituents, with a

IOP Conf. Series: Materials Science and Engineering **191** (2017) 012007 doi:10.1088/1757-899X/191/1/012007

relatively high roughness, particularly apparent in the treated fibres with 4 and 5%$_{wt}$ NaOH solution, as shown in Figs. 1(e) and 1(f), respectively.

Furthermore, it is postulated that the rough surface of treated fibres could enhance mechanical keying at the interface region of the fibres and the matrix material in the composite. The advantages of improved fibre-matrix interfacial adhesion and improved consolidation resulting in the improved tensile properties of the composite could be achieved.

3.2. Effect of alkaline treatment on the chemical compositions of fibres

The chemical composition changes of the fibre surface after treatment were measured by FTIR. Infrared spectra of treated fibres of various treatment conditions are shown in Fig. 2. The spectra present the absorption bands of chemical group characteristics namely, cellulose, hemicellulose, pectin, lignin and water components. The absorption bands of treated and untreated fibres show both similar and different peaks. The presence of similar peaks observed confirms the existence of both hemicellulose and pectin before and after treatment. More important, very small peaks at 1243 and 1731 cm^{-1}, which exist in the untreated fibre (see the arrows in Fig. 2), weaken in the treated fibre. These peaks indicate that both hemicellulose and pectin were dissolved, leaving cellulose and lignin components after treatment.

3.3. Effect of alkaline treatment on the thermal stability of fibres

The effect of alkaline treatment on the differential thermogravimetric (DTG) peak position is shown in Fig. 3. An initial low peak, in the range of 40°C to 150°C is attributed to a mass loss of water from the moist fibres. After this peak, the DTG curve of untreated fibre shows a small shoulder peak between 230°C and 260°C, which is attributed to thermal degradation of the hemicellulose and pectin. Moreover, the curve also shows the main peak at 330°C, which is attributed to cellulose degradation. It can be seen from Fig. 3 that the alkaline treatment changes the DTG main peak position, shifting it to higher temperature, compared with untreated fibres. This shifting depends significantly on the concentration of NaOH solution. The main degradation temperature increases from 330°C for untreated fibre to 355°C for 5% NaOH mercerized fibres.

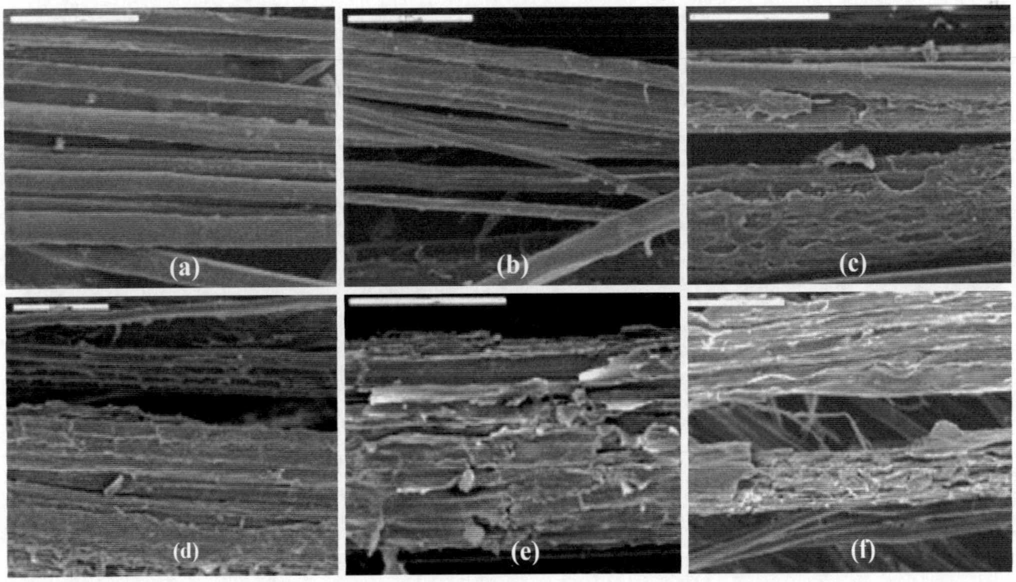

Figure 1. (a) SEM micrograph of untreated flax fibres (b) flax fibres treated with 1%$_{wt}$ NaOH (c) flax fibres treated with 2%$_{wt}$ NaOH (d) flax fibres treated with 3%$_{wt}$ NaOH (e) flax fibres treated with 4%$_{wt}$ NaOH (f) flax fibres treated with 5%$_{wt}$ NaOH.

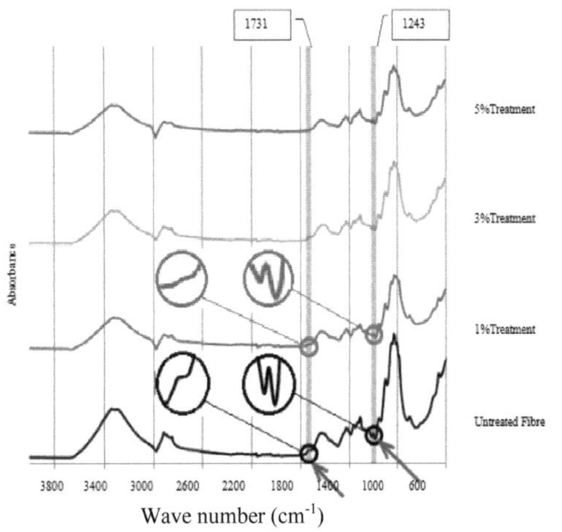

Figure 2. FTIR spectra of treated and untreated flax fibres.

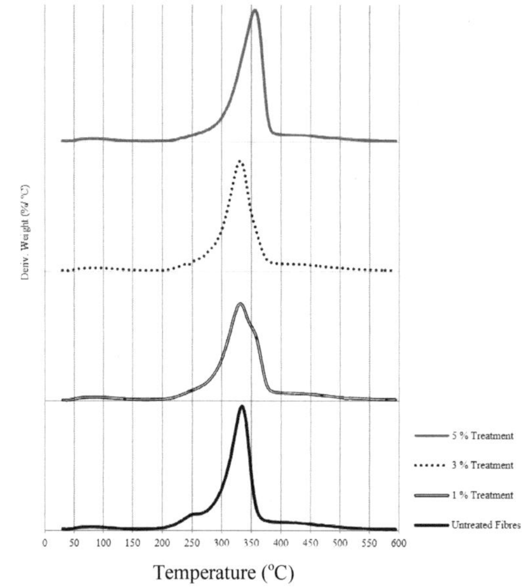

Figure 3. The differential thermogravimetric (DTG) curves of treated and untreated flax fibres.

The shoulder peak of the DTG curve, at about 230°C to 260°C, disappears after treatment with a higher concentration of NaOH solution, supporting the hemicellulose and pectin loss. Hemicellulose and pectin have low thermal stability, are removed from the fibre surface by the alkali. The long chain polymers, i.e., cellulose and lignin, are retained. Thus, the hypothesis is that treated fibres have prolonged thermal stability at higher temperature due to removal of the layers susceptible at lower temperatures.

3.4. The void content and the fractured failure surface of composites

Fig 4 shows the void content of 3%$_{wt}$ NaOH treated flax fibre composites with the fibre volume fraction (V_f) of 30% manufactured at 200°C for 30, 60 and 90 minutes of consolidation time in an isothermal vacuum forming process [error bars are 95% confidence limit]. The consolidation time affects the void content in a similar way to that of the untreated composite [14]. The void content of 0.96% is achieved with 90 minutes of consolidation time.

The tensile fractured surfaces of 3%$_{wt}$ NaOH treated fibre composites, manufactured at 200°C for 30 and 90 minutes, are shown in Fig. 5. Voidage presence in Fig. 5(a) reveals the poor consolidation of the composites due to insufficient consolidation time. The fibres were easily pulled out from their bundle and the matrix, leaving gaps and holes on their fractured surfaces. By contrast, the fractured surface of the composite, manufactured with 90 minutes consolidation time, shown in the micrographs of Fig. 5(b). It can be seen that the improved bonding is due to the enhanced surface roughness of the fibres as a result of treatment. The contribution of high fibre surface roughness resulted in good mechanical interlocking mechanism of the fibres and the matrix.

4. Summary

The treatment with a higher concentration of NaOH solution results in a more obviously rough fibre surface.

The treatments reduce the hemicellulose and pectin from the fibre surface following treatment.

The loss of mass belonging to hemicellulose and pectin in treated fibres results in a shift of the main degradation temperature to higher temperature.

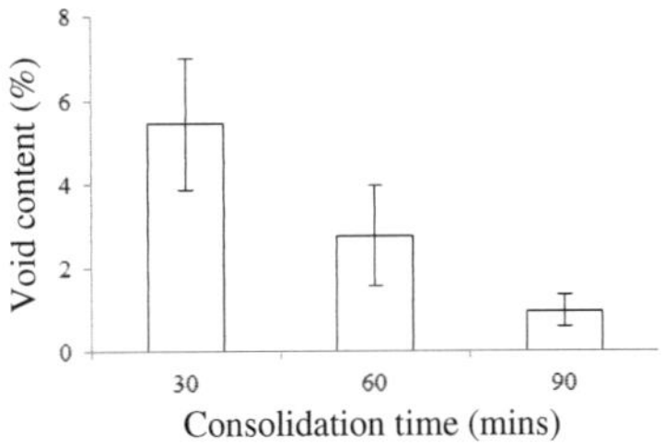

Figure 4. Void content of treated fibre composites (V_f% of 30%) [error bars are 95% confidence limit].

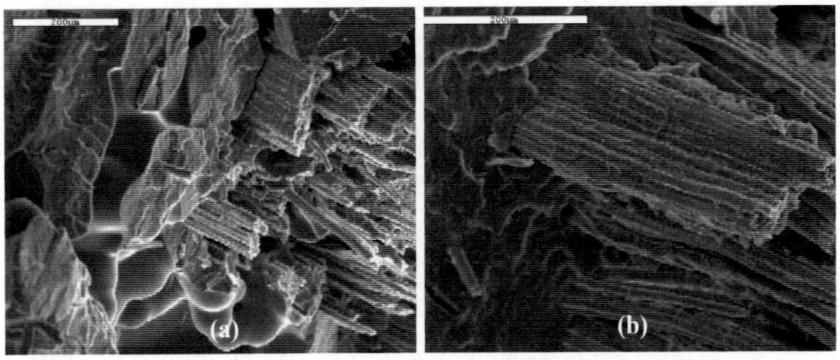

Figure 5. SEM images of tensile fractured surfaces of 3%$_{wt}$ NaOH treated flax/PP composite, manufactured at 200°C (a) a consolidation time of 30 minutes (b) a consolidation time of 90 minutes.

References

[1] Agopyan, V., et al., Developments on vegetable fibre-cement based materials in Sao Paulo, Brazil: an overview. Cement and Concrete Composites, 2005. 27(5): p. 527-536.

[2] Alavudeen, A., et al., Review of natural fiber woven composite. Reviews on Advanced Materials Science, 2011. 27(2): p. 146-150

[3] Baillie, C., Green composites Polymer composites and the environment. 1 ed. 2004, Boca Raton New York Washington, DC: Woodhead Publishing Ltd and CRC Press LLC. 308.

[4] Bodig, J. and B.A. Jayne, Mechanics of wood and wood composities. Reprint ed. with corrections. ed. 1993, Malabar, Fla.: Krieger Pub.

[5] John, M.J. and S. Thomas, Biofibres and biocomposites. Carbohydrate Polymers, 2008. 71(3): p. 343-364.

[6] De Rosa, I.M., et al., Morphological, thermal and mechanical characterization of okra (Abelmoschus esculentus) fibres as potential reinforcement in polymer composites. Composites Science and Technology, 2010. 70(1): p. 116-122.

[7] Gassan, J. and A.K. Bledzki, Thermal degradation of flax and jute fibers. Journal of Applied Polymer Science, 2001. 82(6): p. 1417-1422.

[8] Ouagne, P., et al., Analysis of the film stacking processing parameters for PLLA/flax fibre biocomposites. Journal of Composite Materials, 2010. 44(10): p. 1201-1215.

[9] Wielage, B., et al., Thermogravimetric and differential scanning calorimetric analysis of natural fibres and polypropylene. Thermochimica Acta, 1999. 337(1â€"2): p. 169-177.

[10] Pielichowski, K.N., James Rapra Technology Limited Staff, Thermal Degradation of Polymeric Materials. 2005, Shrewsbury: Smithers Rapra. 320.

[11] Bledzki, A.K. and J. Gassan, Composites reinforced with cellulose based fibres. Progress in Polymer Science, 1999. 24(2): p. 221-274.

[12] Faruk, O., et al., Biocomposites reinforced with natural fibers: 2000-2010. Progress in Polymer Science, 2012. 37(11): p. 1552-1596.

[13] De Rosa, I.M., C. Santulli, and F. Sarasini, Mechanical and thermal characterization of epoxy composites reinforced with random and quasi-unidirectional untreated Phormium tenax leaf fibers. Materials & Design, 2010. 31(5): p. 2397-2405.

[14] Chaishome, J., et al., Thermal Degradation of Flax Fibres as Potential Reinforcement in Thermoplastic Composites. Advanced Materials Research, 2014. 894: p. 32-36.

[15] Ray, D., et al., The mechanical properties of vinylester resin matrix composites reinforced with alkali-treated jute fibres. Composites Part A: Applied Science and Manufacturing, 2001. 32(1): p. 119-127.

[16] Summerscales, J., et al., A review of bast fibres and their composites. Part 1 - Fibres as reinforcements. Composites Part A: Applied Science and Manufacturing, 2010. 41(10): p.

IOP Conf. Series: Materials Science and Engineering **191** (2017) 012007 doi:10.1088/1757-899X/191/1/012007

1329-1335.
[17] Thomason, J.L. Why are Natural Fibres Failing to Deliver on Composite Performance. in Proceedings of ICCM17. 2009. Edinburgh, Scotland.

IOP Conf. Series: Materials Science and Engineering **191** (2017) 012008 doi:10.1088/1757-899X/191/1/012008

Water treatment: Chitosan associated with electrochemical methods

Priscila Tamiasso-Martinhon[1,*], João Marques Teixeira de Souza[2], Silvia Maria Cruzeiro da Silva[2], Fernando Luiz Pellegrini Pessoa[2] and Célia Sousa[1]

[1]Federal University of Rio de Janeiro (UFRJ/IQ/DFQ), Av. Athos da Silveira Ramos, 149 - CT - Bloco A Lab. 411, Cidade Universitária, 21941-909, Rio de Janeiro, RJ, Brazil.
[2]Federal University of Rio de Janeiro (UFRJ), Escola de Química, Programa de Pós-Graduação em Tecnologia de Processos Químicos e Bioquímicos (TPQB), Av. Horácio Macedo, 2030 – CT – Bloco E – Sala 200, Cidade Universitária, 21941-909, Rio de Janeiro, RJ, Brazil.

[*]E-mails: pris-martinhon@hotmail.com; pris@iq.ufrj.br

Abstract. Pollution of water bodies due to the presence of toxic metals and organic compounds, bring out a series of environmental problems of public, government and social character. In addition, water pollution, has become the target and source of concern in many industrial sectors. Therefore, it is essential to develop technologies for treatment and purification of water. Chitosan is a natural product derived from chitin, extracted mainly from the shells of crustaceans. It is a low cost, renewable and biodegradable biopolymer of great socioeconomic and environmental importance. The classic treatment of wastewater containing metals involves physical chemistry processes of precipitation, ion exchange and electrochemistry. Electrochemical technology has been presented as the most promising methods for treating wastewater polluted with metals, colloids, dyes or oil in water emulsions; besides being used in removing organic compounds. Alternative methods like adsorption with biosorbents have been investigated. The great advantage of this latter over other techniques is the low generation of residues, easy recovery of metals and the possibility of reuse of the adsorbent. This article aimed to carry out an exploratory study, of bibliographical nature, on the use of chitosan in electrochemical methods for water treatment.

1. Introduction

Chitosan is a copolymer obtained by deacetylation of chitin, only differing in functional groups. It is a natural product of low cost, renewable, biodegradable and of great economic and environmental importance. Chitin, the main biopolymer used to extract chitosan, it is found in the exoskeletons of arthropods like crustaceans - crab, shrimp; and lobster; insects – ants, bees and butterflies; molluscs, as well as in the cell walls of fungus. Due to your wide range of use and application, technological feasibility has been increasingly studied [1-5].

There are those define chitosan as a linear chain polymer, cationic with protonation of the amino group (NH3+) [6, 7]. It can be obtained by deacetylation of chitin, differentiating from this in terms of proportion of amino groups and the solubility. This characteristic of chitosan to provide free amino groups increases their ability to react with multiple molecules, making the biopolymer with increased availability of pendant groups, it can be employed as powder, beads or films [8].

There are a variety of studies on the properties, features and applications of chitosan, which is currently used in various fields such as medicine, agriculture, biotechnology, cosmetics industry products, food and as adsorbents in removing dyes and metal species [9-15]. Due to the simplicity of obtaining a polymer in different physical forms, in various countries, the chitosan is produced on a large scale, with high industrial application [16].

Pollution of water bodies due to the presence of toxic metals and organic compounds, bring out a series of environmental problems of public, government and social character. In addition, water pollution, has become the target and source of concern in many industrial sectors Due to population growth and the unsustainable use of water resources, water scarcity has become a major challenge in the world. Therefore, it is essential to develop efficient technologies for treatment and purification of water [17].

Electrochemical technology can be a powerful tool in addressing these challenges, because it has the following advantages: relative availability of electricity, reduced energy reaction conditions (cold processes), reproducible systems and easily controllable allowing the automatization and ease of assembly of relatively compact plants. The electrolytic process takes place when a potential is applied from an external current source causing a non-spontaneous chemical reaction in an aqueous medium. This process has been presented as the most promising methods for treating wastewater polluted with colloids, dyes or oil in water emulsions; besides being used in removing BTEX (benzene, toluene, ethylbenzene and xylene) [18].

A literature review on the subject and some of its various applications has been performed, since this is a source of renewable raw material, economically viable. It was possible to verify that there are very few studies using chitosan specifically in electrochemical methods for the treatment of effluent, confirming the importance of conducting studies in this area. This study is focused on the use of chitosan in electrochemical methods as a strategy for the treatment of effluent (water).

2. Theoretical foundations

Chitosan was industrially produced around 1971 in Japan and the US, leading countries of the world chitosan market. By 1993, Brazil also started marketing chitin and chitosan on a large scale. The chitosan is a biodegradable copolymer consisting of D-glucosamine units containing a free amino group [19]. The deacetylation of chitin leads to the production of copolymers with varying composition and molecular weight, depending on the conditions and extent of reaction.

Thus, the reproducibility of characteristics and properties is not always the case, even if similar reaction conditions are employed in the deacetylation of chitin, aiming at preparing chitosan. It is soluble in dilute acid, forming a cationic polymer by protonation of amino group (NH_3^+), which gives special properties different with respect to vegetable fibers, as cellulose [2].

Chitosan is a linear cationic polysaccharide, naturally occurring or can be obtained by deacetylation of chitin. While there is no definitive nomenclature for establishing a difference between chitin and chitosan, the term chitosan is generally copolymers of 2-amino-2-deoxy-D-glucose [8]. The deacetylation of chitin leads to chitosan obtained, its most important derivative whose primary structure is identical to that of chitin, but in chitosan predominate 2-amino-2-deoxy-D-glucopyranose units. The complete deacetylation of chitin is rarely performed, because many consecutive reactions are needed, which also favor the progressive depolymerization [8]. Thus, the term chitosan includes all polymers containing at least 50-60% of units of 2-amino-2-deoxy-D-glucopyranose. There are no well-defined limits in terms of content units of 2-acetamido-2-deoxy-D-glucopyranose and 2-amino-2-deoxy-D-glucopyranose for chitin and chitosan distinction, but due to their different compositions, polymers exhibit very different properties.

From a practical point of view, it is the solubility which allows simple and quick distinction, since chitosan is soluble in several dilute acidic aqueous solutions (the acid and hydrochloric acid solutions are the most commonly used), but chitin is not soluble in these media and it is dissolved in only a few solvent systems. According Campana-Filho (2007), deacetylation of chitin may be effected by chemical or enzymatic processes, but the latter are not employed on an industrial scale due to the high costs of deacetylases extraction, as well as low productivity [19]. The chemical processes of deacetylation of chitin can be achieved by the homogeneous route, which is made alkali-chitin, or heterogeneous, being the most extensively used and studied.

Generally, chitin is suspended in concentrated aqueous solutions of NaOH or KOH (40-60%) for varying times (0.5 to 24 h) and at relatively high temperatures (50-130°C). The characteristics of the obtained chitosan are determined by the concentration of the alkaline solution and ratio chitin/alkaline solution, the particle size of chitin, temperature, time and atmosphere of the reaction. The most severe conditions are generally employed to favor the efficiency of deacetylation, but it results in marked depolymerization via alkaline hydrolysis.

Electrochemical methods have great potential in the treatment of industrial effluents because they are fast, low cost and can be used as a powerful tool in the identification, quantification and degradation of organic and inorganic substances present in these media.

A procedure widely used in electrochemical studies is the modification of the surface of the electrodes. The development and application of modified electrodes, has received great interest in science and technology, with application in various fields of research. The modification of the surface of an electrode can be performed through physical or chemical processes.

An example of physical surface modification is the construction of a partially blocked surface using an insulating resin that will render some areas of the electrode surface inactive while other areas will remain available for electrochemical reactions to occur. This type of procedure allows studies on the influence of parameters such as the diameter of the active sites, the active surface area and the density of active sites, on mass transport. Thus, allowing important information to be obtained from the kinetic point of view, which are applicable to the study of processes such as electrodissolution of metals, the mechanism of action of corrosion inhibitors and adsorption.

The chemical modification of the surface of an electrode is a means of controlling its properties because it allows the characteristics of interest of the modifier to be transferred to it. This procedure allows different conditions of selectivity and reactivity to be achieved in electrochemical systems. The manufacture of chemically modified electrodes depends on the ability to immobilize chemical species on the surface of the electrode.

In general, we can relate the following methods to accomplish this immobilization: Adsorption of the reagent on the surface of the electrode; Covalent bond formation between the electrode and the electroactive reagent; Coating the electrode with polymers.

Adsorption is an older and simpler technique for anchoring redox centers on the electrode surface. Although the first studies were carried out with platinum, most of the adsorption works were performed with pyrolytic graphite and vitreous carbon, directed to electrocatalytic applications, especially molecular oxygen reduction. This method is very simple. However, it is subject to interference from impurities that can adsorb more strongly than adsorbate. In addition, it does not allow you to create structures that are thicker than monolayers.

The first surface modification works employing covalent bonding were based on the use of silane reactants, which could react with hydroxyl groups found on metal oxide electrodes, to give species covalently bound via siloxane bonds. Tin oxide electrodes were initially used, which in analogy to the glass have hydroxyl groups reactive on the surface, with which the silanes could bind. However, it has been found that many metal electrodes, when oxidized in acidic media, are coated with a thin layer of oxide, which is reactive to silane reactants, thus allowing even the platinum to be modified with covalently bound silanes.

Although silane reactants offer the simplest means for covalent anchoring, other approaches are advantageous, including cyanuric chloride linkages and direct addition reactions to the graphite surface. The great interest in organic synthesis was responsible for the appearance of a line of modification of carbon surfaces with the use of pyrolytic graphite electrodes, vitreous carbon and carbon paste electrodes. The carbon paste electrodes present a particularity as to the modification process, since in this case the material of interest is incorporated into the pulp through a simple mixing process, which does not characterize a covalent attachment procedure. The advantage of this electrode is the ease of surface renewal.

Polymer-modified electrodes have a greatly enhanced signal when compared to those obtained with monolayer-modified electrodes. In this way, the realization of the measures is easier, besides having a greater stability. The preparation of the electrodes can be performed in different ways: (i) employing a preformed polymer containing the redox center; (ii) by binding a redox center to a previously prepared polymer; (iii) by in-situ generation, by electropolymerization, of active redox monomers. This type of study has been carried out employing mainly voltammetric electrochemical techniques. More recently we have used impedance techniques for the kinetic study of modified electrodes [20-21].

The development and application of modified electrodes, has received great interest in science and technology, with application in various fields of research. The modified electrode is constructed by adding chemically active substances immobilized on the electrode substrate, in order to pre-establish and control the physical chemistry nature of the electrode/solution interface. The electrode modification usually gives the properties (reactivity, selectivity, etc.) of the modifier, thereby enabling the development of devices with adequate response for various purposes and applications [22, 23].

3. State of the art

Different methodologies and strategies for applications of chitosan have been proposed in the literature. The majority of studies showed the chitosan applications in the coating of edible films [24-27]; effluent treatment due to the interaction of this with metal ions [28-30] and fat reduction [2]. Worry and concern with environmental issues have increased over the decades. Scientists around the world are developing research studies to assess and reduce pollution impacts on the biosphere, and to develop clean technologies that generate less waste. Among the many pollutants, metals have received special attention because some are very toxic to a wide variety of organisms, even at extremely low concentrations [29].

The classic treatment of wastewater containing metals involves physical chemistry processes of precipitation, ion exchange and electrochemistry. However, these traditional techniques are unsuitable for decontamination of large volumes of wastewater containing metals in low concentrations due to low operational efficiency and high costs of this process. Therefore, alternative methods have been investigated as, for example, electrodialysis, reverse osmosis, ultrafiltration and adsorption with biosorbents. The great advantage of this latter over other techniques is the low generation of residues, easy recovery of metals and the possibility of reuse of the adsorbent [29].

Among the natural materials, chitosan, the plus abundant and low cost is a good adsorbent for heavy metals, since it has the ability to form complexes with transition metal ions in aqueous solution and cause the neutralization of acidic solutions. Chitosan can be physically modified, one of the most interesting advantages being its great versatility in being prepared in different forms, such as powders, flakes, microspheres, nanoparticles, membranes, sponges, hives, fibers and hollow fibers [9, 17].

The versatility of chitosan is attributed to free amino groups, which are exposed after the deacetylation of chitin [31, 32]. The preparation of chitosan microspheres is a strategy for increasing their adsorption capacity, once the microspheres have a surface area greater than the chitosan flake. Moreover, microspheres have faster absorption kinetics and greater ease of handling and operation [2].

Several recent studies show that chitosan can be used as microspheres for different purposes. Laus et al. (2006) proposed a method using chitosan microspheres crosslinked with tripolyphosphates used for removal of acidity, Fe3+ and Mn2+ water contaminated with coal mining [30]. Among the environmental impacts related to the mining of coal, the most aggravating is the pollution of water sources in the regions close to the mines, where the mineral is exploited.

Water pollution, observed in mining, is due to the generation of acid effluents from mines and coal wastes. This characteristic of acidic mining waste is given mainly by the presence of pyrite (FeS2), a very common sulfide mineral in the Brazilian coals. Pyrite is rapidly oxidized and dissociated when exposed to air and water, releasing Fe2+. In this study, chitosan crosslinked with tripolyphosphate were used in the form of microspheres used in remediation of contaminated water by coal mining, being shown

The dissolution of sodium tripolyphosphate (Na5P3O10) in a chitosan solution induces an ionic crosslinking among the ions tripolyphosphate (TPF) and the protonated amino group of chitosan. Crosslinking of chitosan prevents the polymer is dissolved in an acidic medium and introduces phosphate groups, which have basic sites that might interact with H3O+ and metal ions in aqueous solution can increase its remediation capacity of aquatic environments contaminated by coal mining.

According to Ma (2016), due to population growth and the unsustainable use of water resources, water scarcity has become a major challenge in the world. Therefore, it is necessary to construct new technological alternatives for treatment and purification of water [33].

Carbon materials such as activated carbon and carbon nanotubes have been used electrochemically modified for the treatment of water and industrial effluents. The adsorption process is recognized as an effective method for the removal of environmental contaminants from aqueous solutions. Researchers have studied the use of carbon nanotubes to remove electrochemically pollutants. The adsorption capacity of carbon nanotubes can be significantly improved by electrochemical methods (34-37].

The electrosorption process, which is an adsorption induced by the potential of undesirable ions and contaminants on the surface of charged electrodes, has attracted much attention due to its compatibility with the environment and its low energy consumption [38]. The development of analytical and electroanalytical techniques to determine low concentrations of metals in natural water and wastewaters have been of extreme interest. Janegitz et al. (2007) developed a carbon paste electrode modified with chitosan, for the determination of Cu2+ in wastewater by voltammetry [31, 32].

The use of chitosan as a modifying agent for metal ions sensors has been little explored, despite the high affinity of this material by metal cations. In this experiment, they used a system with three electrodes, wherein the working electrode is a chemically modified electrode with chitosan, the reference electrode, an Ag/AgCl electrode and the auxiliary a platinum electrode.

In the treatment of industrial effluents the modified electrodes are increasingly being used to detect trace metals. However, electrodes modified with chitosan nanotubes are still little explored. Martínez-Huitle et al. (2010) developed a research modifying vitreous carbon electrode with chitosan [39]. According to the proposed study, it was used to determine trace metals (Cu, Pb, Cd, Co, As and Pt) in aqueous solutions. The modified surface of the electrode exhibited an affinity with metal ions in solution, forming complexes.

4. Conclusion

Among the natural materials, chitosan, the plus abundant and low cost, is a good adsorbent for heavy metals, since it has the ability to form complexes with transition metal ions in aqueous solution and cause the neutralization of acidic solutions. The application of electrochemical techniques to determine and remediate low concentrations of metals in natural water and wastewaters it is an advantageous and cost-effective alternative. The use of chitosan as a modifying agent for metal ions sensors has been little explored, despite the high affinity of this material by metal cation, therefore it appears as a promising alternative for the treatment of wastewater.

Acknowledgments

The authors acknowledge the financial support of CAPES, CNPq, FAPERJ and GIEESAA.

References

[1] R. Yang, H. Li, M. Huang, H. Yang, A. Li. Water Research 95, 59 (2016).
[2] V. V. C. Azevedo, S. A. Chaves, D. C. Bezerra, M. V. L. Fook, A. C. F. M. Costa. Rev. Eletr. Mater. Proc. 2, 27 (2007).
[3] M. A. Vigilato, M. M. Horn, V. C. A. Martins, A. M. G. Plepis. *In:* SIICUSP, USP, SP (2014).
[4] K. Shang, Z. Qiao, B. Sun, X. Fan, S. Ai. J. Solid State Electrochem. 17, 1685 (2013).
[5] A. Pavinatto, L. A. Mercante, C. S. Leandro, L. H. C. Mattoso, D. S. Correa. J. Electroanal. Chem. 755, 215 (2015).
[6] B. Moura, P. Muszinski, C. Schmidt, J. Almeida, L. Pinto. Rev. Cienc. Ex. Eng. 15, 7 (2005).
[7] J. Varna, S. V. Deshpande and J. F. Kennedy. Carbohydrate Polym. 55, 227 (2004).
[8] Airoldi. Quim. Nova 31, 144 (2008).
[9] M. C. M. Laranjeira, V. T. Fávere. Quim. Nova 32, 672 (2009).
[10] A. Aranaz, R. Harris, A. Heras. Current Org. Chem. 14, 308 (2010).
[11] Huang, Q. Zeng, L. Wang. Electrochim. Acta 206, 176 (2016).
[12] M. Z. Elsabee, E. S. Abdou. Mater. Sci. Engineering C. 33, 1819 (2013).

[13] S. K. Shukla, A. K. Mishra, O. A. Arotiba, B. B. Mamba. Int. J. Biol. Macromol., 59, 46 (2013).
[14] Y. C. Yoon, K.B. Kim, D.H. Lee, C. S. Park, Y. H. Kim. *In:* SMAE-2016. MATEC Web of Conferences 67, 06054 (2016).
[15] R. H. Ramli, C. F. Soon, A. Z. M. RUS. *In:* IConGDM-2016. MATEC Web of Conferences 78, 01031 (2016).
[16] V. H. Prashanth, R. N. Tharanathan. Trends Food Sci. Technol. 18, 117 (2007).
[17] R. Yao, F. Meng, L. Zhang, D. Ma, M. Wang. J. Hazardous Mater. 165, 454 (2009).
[18] P. Canizares, C. Martínez, C. Jiménez, C. Sáez, M. A. Rodrigo. J. Hazardous Mater. 151, 446 (2008).
[19] S. P. Campana-Filho, D. Britto, C. Curti, F. R. Abreu, M. B. Cardoso, M. V. Battisti, P. C. Sim, R. C. Goy, R. Signini, R. L. Lavall. Quim. Nova 30, 644 (2007).
[20] P. Tamiasso-Martinhon, H. Cachet, C. Debiemme-Chouvy and C. Deslouis. Electrochim. Acta 53, 5752 (2008)
[21] P. Tamiasso-Martinhon, J. Carreño, C. Sousa, O. E. Barcia and O. R. Mattos. Electrochim. Acta 51, 3022 (2006).
[22] C. Pereira, L. T. Kubota. Quim. Nova 27, 725 (2004).
[23] C. Pereira, A. D. SANTOS, L. T. Kubota. Quim. Nova 25, 1012 (2002).
[24] B. G. Assis, H. C. Alves. Rev. Biotecnol. Cienc. Desenvolv. 4, 33 (2003).
[25] P. S. Tanada-Palmu, P. S. P. Proença, P. E. Trani, F. A. Passos, C. R. F. Grosso. Rev. Cienc. Agron. Bragantia 64, 291 (2005).
[26] F. Borgognoni, B.Polakiewicz, R. N. M. Pitombo. Cienc. Tecnol. Alim. 26, 502 (2006).
[27] A. Botrel, N. F. F. Soares, R. M. Geraldine, R. M. Pereira, E. A. F. Fontes. Cienc. Tecnol. Alimentar 27, 32, (2007).
[28] V. A. Spinelli, V. T. Fávere, M. C. M. Laranjeira. React. Function. Polym. 61, 347 (2004).
[29] V. A. Spinelli, M. C. M. Laranjeira, V. T. Fávere, I. Y. Kimura. Rev. Polim. 15, 218, (2005).
[30] R. Laus, M. C. M. Laranjeira, A. O. Martins, V. T. Fávere. Quim. Nova 29, 34 (2006).
[31] C. Janegitz, L. H. M. Júnior, O. Fatibello-Filho. Quim. Nova 30, 1673, (2007).
[32] B. C. Janegitz, B.C. Lourenção, K. O. E. Lupetti, O. F. Filho. Quim. Nova 30, 879 (2007).
[33] Y. Ma, S. C. Huang, P. H Chou, W. Den, C. H. Hou. Chemosphere 146, 113 (2016).
[34] X. N. Li, S. Chen, X. Quan, Y. B. Zhang. Environ. Sci. Technol. 45, 8498 (2011).
[35] X. N. Li, S. Chen, L. Y. Li, X. Quan, H. M. Zhao. J. Colloid Interface Sci. 415, 159 (2014).
[36] H. Liu, K. C. Zuo, C. D. Vecitis. Environ. Sci. Technol. 48, 13871 (2014).
[37] X. F. Fan, H. M. Zhao, Y. M. Liu, X. Quan, H. T. Yu, S. Chen. Environ. Sci. Technol. 49, 2293 (2015).
[38] S. Y. Huang, C. S. Fan, C. H. Hou. J. Hazard. Mater. 278, 8 (2014).
[39] A. Martínez-Huitle, N. S. Fernandes, M. Cerro-Lopez, M. A. Quiroz. Port. Electrochim. Acta 28, 39 (2010).

IOP Conf. Series: Materials Science and Engineering **191** (2017) 012009 doi:10.1088/1757-899X/191/1/012009

Green way genesis of silver nanoparticles using multiple fruit peels waste and its antimicrobial, anti-oxidant and anti-tumor cell line studies

Kiruthika Naganathan[1] and Somanathan Thirunavukkarasu[1]

[1]Department of Nanoscience, Vels University, Chennai, Tamil Nadu, India

E-mail: krithi.karthick05@gmail.com; soma_nano@yahoo.co.in

Abstract. Green synthesis of silver nanoparticles (SNP) opens a new path to kill and prevent various infectious diseases and also tumor. In this study, we have synthesized silver nanoparticles using multiple fruit peel waste (pomegranate, orange, banana and apple (POBA)). The primarily nanoparticles formation has been confirmed by the color change. The synthesized SNP were analyzed by various physicochemical techniques such as UV- Visible spectroscopy, x-ray diffraction (XRD), fourier transform infra red (FT-IR) spectroscopy and transmission electron microscope (TEM). The formation of SNP was confirmed by its absorbance peak observed at 430 nm in UV-Visible spectrum. Further, the obtained SNP were identified by XRD and TEM, respectively to know the crystalline nature and size and shape of the particles. The activities of SNP were checked with human pathogens (Salmonella, E.coli and Pseudomonas), plant pathogen (Fusarium) and marine pathogen (Aeromonas hydrophila) and also studied the scavenging effect and anticancer properties against MCF-7 cell lines. This studies proves that the SNP prepared from fruit waste peel extract approach appears extremely fast, cost efficient, eco friendly and alternative for conventional methods of SNP synthesis to promote the usage of these nanoparticles in medicinal application.

1. Introduction

Nanotechnology is an important field of modern research dealing with design, synthesis, and manipulation of particles structure ranging from approximately 1-100 nm. Metal nanoparticles have a high specific surface area to volume ratio. Because of the unique properties of nanoparticles, including optical properties, electronic properties, medicinal properties and magnetic properties [1-5]. These metal nanoparticles pulls in an expanding interest as a result of novel elements and alluring application in the therapeutic field [6]. Several processes physical and chemical [7] were developed including chemical reduction using a variety of organic and inorganic reducing agents, electrochemical techniques, physicochemical reduction, and radiolysis are widely used for the synthesis of silver nanoparticles and considering the real life application of nanoparticles in the area of medicine [8], catalysis [9], detection [10], etc. Recently, nanoparticle synthesis is most interesting scientific areas for attention to produce nanoparticles using environmentally friendly methods (green chemistry) [11] for the search of benign methods for the development nanoparticles and searching antibacterial, antioxidant, and antitumor activity of natural products.

Most of the reported research concentrate on single fruit peel waste extract for the preparation of SNP and they have also used thermal decomposition method [12]. But in this study, we have proven that SNP prepared from multiple fruit waste peel extract approach appears extremely fast, cost efficient, eco-friendly and alternative for conventional methods. Fruits peel wastes of POBA are assembling more consideration for choosing a full range of pharmacological activities. It is experienced for its general tonic, anticancer, hostile toleprotic, against hyperglycemias, hostile to unfavorably susceptible against diabetic properties [13]. It enhances the phagocytic and bactericidal limit of polymorphs, ensures against gastric mucosal harm and searches free radicals. Since this plant has in like manner been depicted to have against fibrotic, hostile to oxidant, mitigating, resistant module as nontoxic [14]. Among numerous metal nanoparticles, SNP plays crucial role as a drug as it is profoundly had antimicrobial [15, 16] anti-inflammatory [17] and hostile to tumor activity. In this study, we have synthesized silver nanoparticles using multiple fruit peel waste of POBA. The SNP formation were confirmed and analyzed by various physicochemical techniques such as UV-Visible spectroscopy, x-ray diffraction (XRD), fourier transform infra-red (FT- IR) spectroscopy and transmission electron microscope (TEM). The activities of SNP were checked with human pathogens (Salmonella, E.coli and Pseudomonas), plant pathogen (Fusarium) and marine pathogen (Aeromonas hydrophila) and also studied the scavenging effect and anticancer properties against MCF-7 cell lines.

2. Materials and method

2.1. Preparation of fruit peel extract

The multiple fruit peel waste of POBA used in this experiment was collected from a local market in Chennai Koyambedu. Each fruit peel waste of 2g were cut into smaller pieces and dried under direct sunlight for 48 hrs. Then the dried materials were grid into powder form and collected in a sample bottle. About 0.1g of the POBA powder dissolved in 100 ml distilled water and stirred for 30 min to obtain the POBA peel solution. This aqueous peel solution is used for the preparation of SNP.

2.2. Preparation of SNP

According to our previous investigation [18], we have synthesized SNP using about 3ml of POBA was added to 47 ml of 0.1mM solution of silver nitrate and mix the solution well, rapidly the color of the solution has turned which implies that the reduction was completed within a short period (2 min) at room temperature with the appearance of yellowish brown color which confirms the formation of SNP (Scheme 1). The obtained SNPs were analysed by using UV-Visible spectroscopy, XRD, FT-IR and TEM.

2.3. Antimicrobial activity

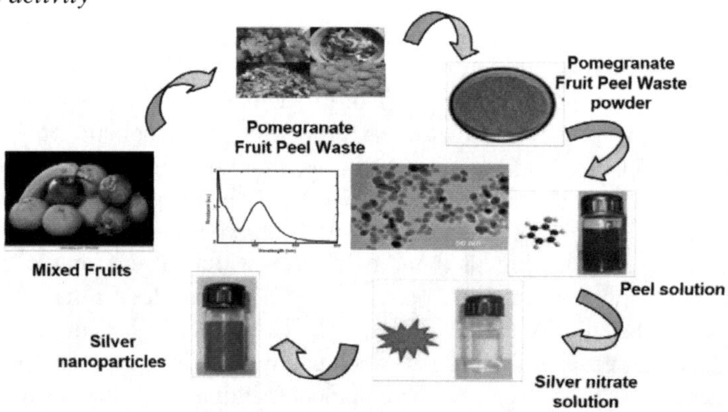

Figure 1. Schematic representation of SNP from fruit peel waste.

The antimicrobial activity has tested by using the agar well diffusion method. The isolated human pathogens (***Salmonella, Pseudomonas, E*. coli)**, plant pathogen **(Fusarium *soaps*)**, marine pathogen (***Aeromonas hydrophila***). A pure culture has sub cultured in respective broth and the strain is spread on the plates. Four circular wells were made and loaded with the SNP as well as control. Then the plates are incubated to observe zone of inhibition.

2.4. Antioxidant activity based on DPPH assay

The scavenging activity towards 1, 1-diphenyl-2-picrylhydrazyl (DPPH) radical was studied with slight modification an aliquot (1000 - 100μL) of SNP or control and of (1000 - 1600μL) H2O was mixed with 1ml of10μm in absolute methanol. The mixture was shaken vigorously and left to stand for 30 min in the dark before measuring the absorbance at 517 nm against a control and the free radical scavenging activity was calculated using following equation.

% scavenging = OD control – OD test / OD control x 100

2.5. Cell viability test (cell line study)

The assay is based on the reduction of soluble yellow tetrazolium salt to insoluble purple formazan crystals by metabolically active cells. Only live cells are able to take up the tetrazolium salt. The enzyme (mitochondrial dehydrogenase) present in the mitochondria of the liver cells is able to convert internalized tetrazolium salt to formazan crystals, which are purple in color. Then the cells are lysed and dissolved in DMSO solution. The color development is then determined in an ELISA reader at 570 nm. The cell lines were plated separately in 96 well plates at a concentration of 1 × 104 cells/well. After 24 h, cells were washed twice with 100 μl of serum-free medium and starved for an hour at 37 °C. After starvation, cells were treated with different concentrations of test compound (5-20μg/ml) for 24 h. At the end of the treatment period the medium was aspirated and serum free medium containing MTT (0.5 mg/ml) was added and incubated for 4 h at 37°C in a CO2 incubator. The 50% inhibitory concentration value (IC50) of the treated drug was identified in normal (Untreated) cell line. The MTT containing medium was then discarded and the cells were washed with PBS (200 μl). The crystals were then dissolved by adding 100 μl of DMSO and this was mixed properly by putting up and down. Spectrophotometrical observance of the purple, blue formazan dye was measured in a microplate reader at 570 nm (Beograd 680). Cytotoxicity was determined using Graph pad prism5 software.

3. Results and discussion

3.1. UV and FTIR

Figure 2 shows the UV–vis spectra of SNP formation using POBA at room temperature. The surface plasmon resonance (SPR) band appeared at 430 nm which clearly indicates the formation of SNP and revealed that the size of 25 nm [19]. Figure 3 illustrates the FTIR spectrum of SNP formed using POBA. The bands located at 3433.41 cm^{-1} were related to O-H extending and 2067.76 cm^{-1} was allocated to aromatic stretch, individually 1633.76 cm^{-1} has considered as basic and amine bunches extending.

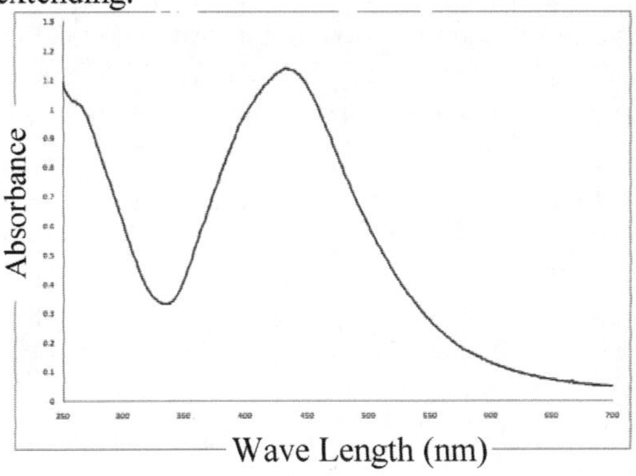

Figure 2. Visible Spectra of SNP at 430nm.

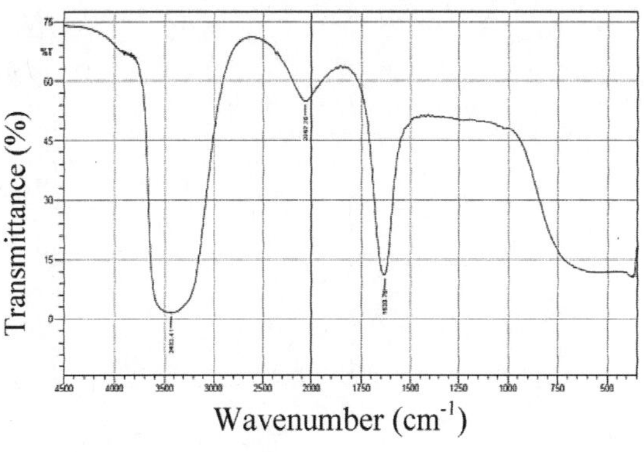

Figure 3. FT-IR of SNP from POBA.

3.2. XRD Pattern and TEM of SNP from POBA

Figure 4 illustrate the XRD pattern of biosynthesized AgNPs, shows the diffraction peaks at 2θ = 38.13°, 44.21°, 64.47° and 77.37° corresponding to the (111), (200), (220) and (311) can be assigned to face-centered cubic (fcc) lattice of metallic silver. Similar results were reported by previous investigator [20, 21]. Figure 5 shows the size and shape of SNP by using TEM analysis, the results have observed nanoparticles size in 25 nm which indicates well dispersion and spherical in shape.

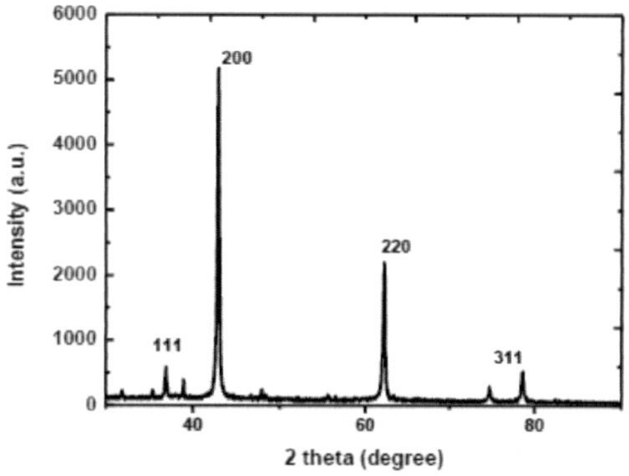

Figure 4. XRD pattern of the SNP.

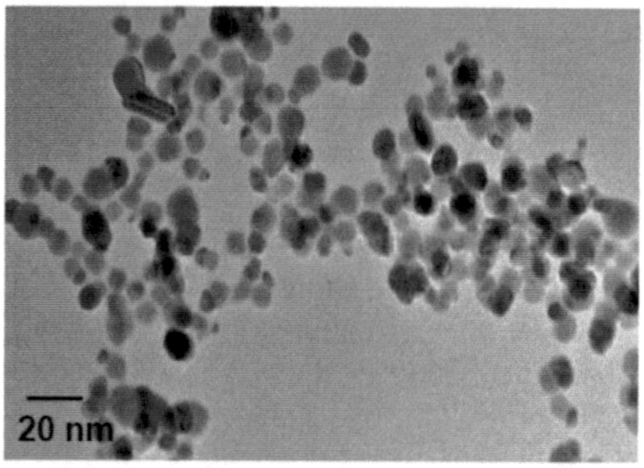

Figure 5. TEM Image of the SNP.

3.3. General Mechanism of SNP

Fruit extract is primarily composed of alkaloids and polyphenols. The active constituent that appears to be responsible for its multiple health benefits is Ellagic acid and it is naturally occurring phenolic compound found in several fruits and nuts and also it effectively protects cells from damaging free radicals. Phenolics possess hydroxyl and carboxyl groups, able to bind to heavy metals. They may inactivate metal ions by chelating. When metal salts come in contact with the ellagic acid present in the peel extract the ester oxygen atom and the ortho- phenolic hydroxyl of the ellagic acid form p track conjugation effect. The esterification of the carboxyl and hydroxyl groups of ellagic acid make the ortho-phenolic hydroxyl loose the hydrogen much easily, forming a steadier semi-quinone structure. Thus, ellagic acid has an easy electron loosing capacity which results in the formation of H+ radical, which reduces the size of silver to nanosize Punicalagins, tannin, along with Ellagic acid is a major antioxidant component of POBA and have free radical scavenging capacity. They are water soluble and have high bioavailability [11].

3.4. Antimicrobial activity

The fruits peel waste interceded SNP has demonstrated better antibacterial movement against human pathogen. This outcome may get cell demise brought on by the SNP in the particular pathogens [22]. Figure 6(a) obviously shows that the SNP has solid antimicrobial movement towards human pathogens contrasted with plant and marine pathogens. Refer Figure 6(b) Fusarium and Figure 6(c) Aeromonas Hydrophila respectively. Minimal zone of restraint has thought about separately in Table 1.

Table 1. Zone of Inhibition

Organisms	5µl	10µl	15µl	Control
Salmonella	8	8.2	9	9
E.coli	6.5	6.8	7	7.2
Pseudmonas	6.4	6.8	6.9	7
Fusarium	6.3	6.5	6.7	7.3
	5.2	5.7	6	6.2

IOP Conf. Series: Materials Science and Engineering **191** (2017) 012009 doi:10.1088/1757-899X/191/1/012009

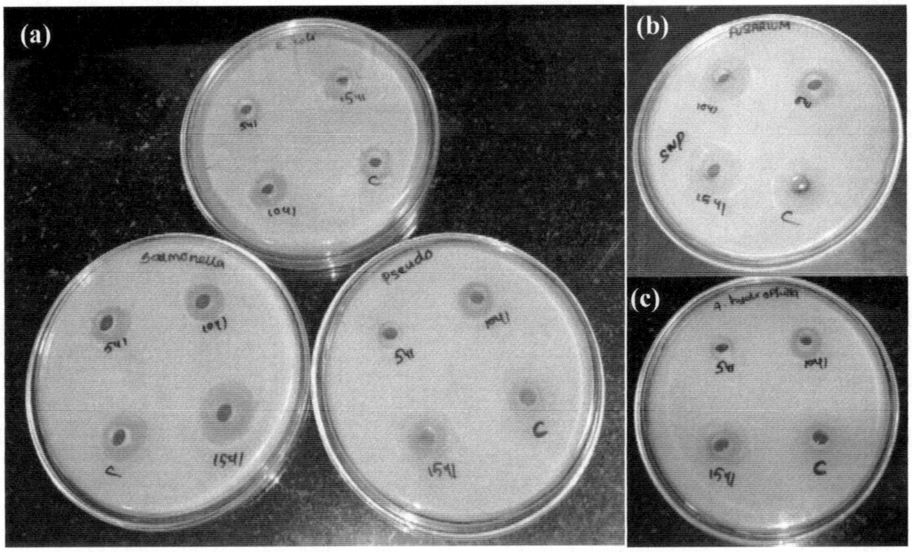

Figure 6. Antimicrobial activity.

3.5. Antioxidant activity

DPPH scavenging activity was studied using POBA mediated SNP by varying the concentrations 100, 200, 300, 400, 500 µg/ml was found in 25%, 37%, 40%, 60% and 80% at the different concentrations (Figure 7). The 1, 1-diphenyl-2-picryl hydrazyl (DPPH) radical was widely used as the model system to investigate the scavenging activities of several natural compounds. DPPH radical is scavenged by antioxidants through the donation of proton forming the reduced DPPH which can be quantified by its decrease of absorbance at wavelength 517 nm. Radical scavenging activity increased with increasing percentage of the free radical inhibition. The assay is based on the measurement of the scavenging ability of antioxidants towards the stable radical DPPH which reacts with a suitable reducing agent.

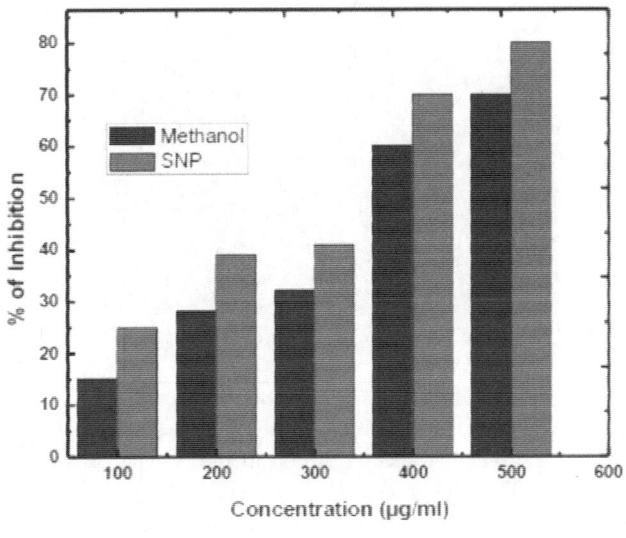

Figure 7. DDPH scavenging activity.

IOP Conf. Series: Materials Science and Engineering **191** (2017) 012009 doi:10.1088/1757-899X/191/1/012009

3.6. Cell line study

The cytotoxicity of cell line study has done with human breast cancer cells MCF-7. The effect of silver nanoparticle was studied at different concentrations (10 μg, 25 μg, 50 μg, 100 μg, 250 μg, 500 μg) against MCF7 cell lines by MTT assay which was shown in Figure 8. The IC50 value was calculated for silver nanoparticles and it was found to be 500 μg/ml. Figure 8 has illustrated that the bright field microscopic images of morphological characterization of MCF7 cell lines for normal cell (a) and treated cells with different concentrations of SNP (b) 10 μg, (c) 25 μg, (d) 50 μg, (e) 100 μg, (f) 250 μg, (g) 500 μg. Absorbance values that are lower than the control cells indicate a reduction in the rate of cell proliferation. An increase in proliferation may be offset by cell death; evidence of cell death may be inferred from morphological changes. This study shows that the cytotoxicity has increased by dose dependant while the high concentration of the silver nanoparticle has better result comparing with minimal concentration. This is the first study to report the cytotoxicity of AgNPs synthesized using multiple fruit peel waste extract against MCF7 cell lines.

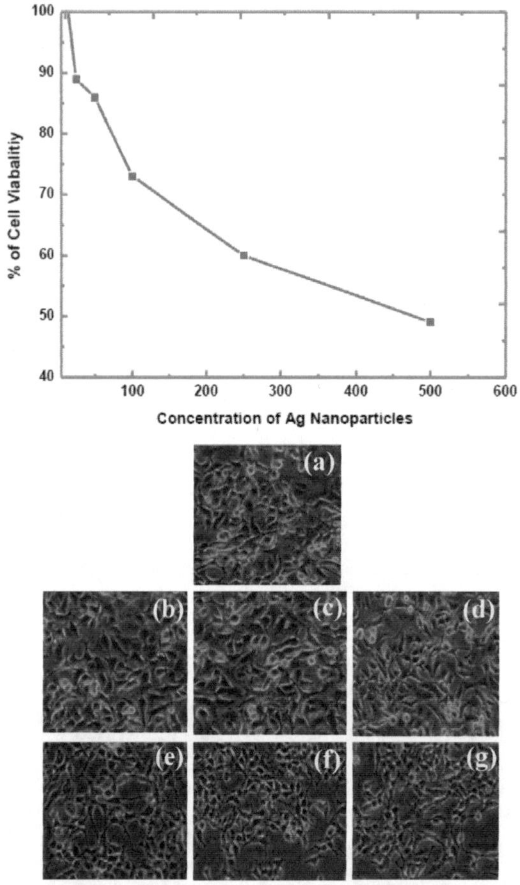

Figure 8. Cytotoxicity effect of SNP from POBA on MCF-7 Cell Lines.

4. Conclusion

The present study has shown the "green" synthesis route combination of silver nanoparticles from fruit peel waste materials has received increasing attention due to the development of eco-friendly technologies in material science. The characterization with UV- Visible spectroscopy, XRD, FT-IR and TEM is the evidence for the formation of nanoparticles. This study shows the successful synthesis of SNP with excellent morphological properties by a very facile, cost effective and green method. We have concluded that the SNP synthesized using fruit waste possess high antimicrobial activity against human, plant and marine pathogens and also the study show great attention of anticancer activity against human breast cancer cell (MCF-7) lines which further suggest the potential use of this kind nanoparticles in various pharmacology field.

References

[1] Granqvist C, Buhrman R, Wyns J and Sievers A 1976 *Phys. Rev. Lett.* 37 625.
[2] Catauro M R M, De Gaaetano F D and Marotta A 2005 *J. Mater. Sci. Mater.* Med. 16 261.
[3] Crabtree J H, Siddiqi Ra B R, Huen I T, Handott L L and Fishman A 2003 *Perit. Dial. Int.* 23 368.
[4] Krolikowska A K A, Michota A and Bukowska J 2003 *Surf. Sci.* 532 227.
[5] Zhao G S J 1998 *Biometals* 11 27.
[6] Priyadarshini S, Gopinath V, Priyadharshini N M, MubarakAli D and Velusamy P 2013 *Colloids Surfaces B: Biointerfaces* 102 232.
[7] Choi S M, Seo M H, Kim H J and Kim W B 2011 *Syn. Metals* 161 2405.
[8] Chandran S P, Chaudhary M, Pasricha R, Ahmad A and Sastry M 2006 *Biotechnol. Prog.* 22 577.
[9] Amkamwar B, Damle C, Ahmad A and Sastry M 2005 *J. Nanosci. Nanotechnol* 5 1665.
[10] Li S, Shen Y, Xie A, Yu X, Qiu L, Zhang L and Zhang Q 2007 *Green Chem.* 9 852.
[11] Ahmad N, Sharma S and Rai R 2012 *Adv. Mat. Lett.* 3 376.
[12] Goudarzi M, Mir N, Kamazani M M, Bagheri S and Niasari M S 2016 *Scientific Reports.* 6 32539.
[13] Asthana J G, Jain S, Mishra A and Vijaykanth M S 2001 *Indian Drugs* 38 82.
[14] Goel H C, Prasad J, Singh S, Sagar R K, Agrawala P K, Bala M, Sinha A K and Dogra R 2004 *J. Radiation research.* 45 61.
[15] Kurek A, Grudniak A M, Dowjat A K and Wolska K I 2011 *Polish J. Microbio.* 60 3.
[16] Morons J R, Elechiguerra J L, Camacho A and Ramirez J T 2005 *Nanotechnology* 16 2346.
[17] Percival Steven L, Bowler P G and Dolman J 2007 *International Wound Journal* 4 186.
[18] Kiruthika N and Somanathan T 2016 *International Journal of Pharma and Bio Sciences* 7 7.
[19] Aravinthan A, Govarthanan M, Selvam K, Praburaman L, Selvankumar T, Balamurugan R, Kamalakannan S and Kim J H 2015 *International Journal of Nanomedicine* 10 1977.
[20] Bar H, Bhui D K, Sahoo G P, Sarkar P, Pyne S and Misra A 2009 *Colloids Surf. A.* 348 212.
[21] Yang N and Li W H 2013 *Industrial Crops and Products* 48 81.
[22] Duraipandiyan V, Ignacimuthu S, Balakrishna K and Al-Harbi N A 2012 *Asian Journal of Traditional Medicine* 7 1.

IOP Conf. Series: Materials Science and Engineering **191** (2017) 012010 doi:10.1088/1757-899X/191/1/012010

Fabrication of nanoparticles using *Annona squamosa* leaf and assessment of its effect on liver (Hep G2) cancer cell line

V. Vanitha[1*], S. Hemalatha[1], N. Pushpabharathi[1], P. Amudha[1], M. Jayalakshmi[1]

[1]Department of Biochemistry, Vels University, Velan Nagar, P.V. Vaithiyalingam Road, Pallavaram, Chennai - 600 117, Tamil Nadu, India

E-mail: vanitha.sls@velsuniv.ac.in

Abstract. *Annona squamosa* is a fruit bearing plant possesses potent bioactive compounds in all its part. In this present investigation iron oxide nanoparticle was synthesized from hydroethanol extract of *Annona squamosa* leaves at 60^0 c temperature. Production of iron oxide nanoparticles in extraction is detected by UV–V spectrophotometer, Scanning electron microscopy was employed to analyse the structure of nanoparticles. Fourier transform infrared spectroscopy (FT-IR) analysis were performed, in order to determine the functional groups on *Annona squamosa* leaves extract. The synthesized Fe_3O_4 NPs shows potential cytotoxicity against liver carcinoma cell line (HepG2), and there is no toxicity on the normal liver cell line. Our reports confirmed that the *Annona squamosa* leaf is a very good eco-friendly and nontoxic bioreductant for the synthesis of Iron oxide nanoparticle and opens up further opportunities for fabrication of drugs towards cancer therapy.

1. Introduction

From our ancient history, our ancestors used various plant as a drugs for many diseases, many of our traditional plants has healing properties. Cancer treatments do not have potent medicine as the currently available drugs are causing side effects in some instances.The main obstacle in treating cancer is delivering of drug to a specific carcinoma cell. Thus besides from plants, effort has also been made to discover natural products from iron oxide nanoparticles to provide clinic drugs directly for the synthetic. More specifically for drug delivery purposes, the use of nanoparticles is attracting, increasing attention due to their unique capabilities and their negligible side effects not only in cancer therapy but also in the treatment of other ailments. Among all types of nanoparticles, biocompatible superparamagnetic iron oxide nanoparticles (SPIONs) with proper surface architecture and conjugated targeting ligands/proteins have attracted a great deal of attention for drug delivery applications [1]. *Annona squamosa* plant, a angiosperm fruit bearing plant which exhibit medicinal property in all their parts such as leaves of *Annona squamosa* have strong antibacterial activity [2], Hypoglycemic [3] and Anticancer activity [4]. In the present study, we report the use of *Annona squamosa* leaf in the biosynthesis of iron oxide nanoparticle and its cytotoxic effect on liver (HepG2) cancer cell lines.

2. Materials and methods

2.1. Chemicals

The chemicals were purchased from Sigma, USA. All other chemicals used are of analytical grade.

2.2. Collection of sample and authentication

Fresh leaves of *Annona squamosa* plant was collected locally during the month of November to January and authenticated by the Taxonomist Prof. P. Jayaraman, Director, Plant Anatomy Research Centre, Chennai, India. (PARC/2009/456).

2.3. Preparation of leaf extract

Extract were prepared by cold maceration method using 30:70 hydro ethanol and air dried *Annona squamosa* leaf powder. The concentrated crude extracts were lyophilized into paste (5 and 15 g respectively) and were taken for further investigation.

2.4. Synthesis of iron oxide nanoparticles

Iron oxide nanoparticles were synthesized by a modified co-precipitation method [5]. Ferric chloride ($FeCl_3$, 0.074 g) and ferrous chloride ($FeCl_2$, 0.190 g) at a ratio of 2:1 were dissolved in 20 ml deionized water, which was then stirred with 1:1, 1:2, 1:3, 1:4 and 1:5 ratios of plant extract and heated to 60°C. The nanoparticles were then removed from solution by magnetic separation.

2.5. Determination of synthesized iron oxide nano particles

2.5.1 UV-Visible spectroscopy: The Surface Plasmon Resonances (SPR) of synthesized iron oxide nanoparticles have been studied by UV-Visible double-beam bio-spectrophotometer ELICO-BL-198 using the software SPECTRAL TREATS VERSION 2.37.4 REL-1.

2.5.2. SEM analysis (scanning electron microscopy): SEM was used to verify uniformity of nanoparticle shape and size as previously described. The fabricated nanoparticles were dropped onto black carbon tape with a double-side. After that, they were vacuum-coated with a platinum mixture for 45 s and morphologically analyzed with a FE-SEM (CamScan Apollo 300 Field-Emission SEM, UK) at 20 kV.

2.5.3. Fourier transform infrared spectrophotometer (FTIR): Fourier Transform Infrared Spectrophotometer (FTIR) is perhaps the most powerful tool for identifying the types of chemical bonds (functional groups) present in compounds. The Liquid extract of plant specimen was loaded in FTIR spectroscope (Shimadzu, IR Affinity 1, Japan), with a scan range from 400 to 4000cm^{-1} with resolution of 4 cm^{-1}.

2.6. Antiproliferation assay

2.6.1 Cell lines and culture: Human hepatocellular carcinoma cell lines, HepG2 cells (American Type Culture Collection [ATCC] CRL 8024) were obtained from National center for cell science, Pune, India was maintained in monolayer culture at 37°C and 5% CO_2 in DMEM supplemented with 10% FBS, 100 U/ml of penicillin, 50 µg/ml of streptomycin.

2.6.2. Cytotoxicity assay: The cells were plated in 96 well flat bottom tissue culture plates at a density of approximately 1.2 X 10^4 cells/well and allowed to attach overnight at 37°C. The medium was then discarded and cells were incubated with different concentrations of the iron oxide nanoparticles sythesized from leaf extract of *Annona squamosa* (25, 50 and100µg) for 24 hours. After the incubation, medium was discarded and 100µl fresh medium was added with 10µl of MTT (5mg/ml)[14]. After 4 hours, the medium was discarded and 100µl of DMSO was added to dissolve the formazan crystals. Then, the absorbance was read at 570nm in a microtitre plate reader. Cyclophosphamide was used as a positive control. Cell survival was calculated using standard formula.

3. Result and discussion

In the present study, we report on rapid biosynthesis of Iron oxide nanoparticle using *A. squamosa* leaf extract. The plant mediated iron oxide nanoparticles was prepared at room temperature.

IOP Conf. Series: Materials Science and Engineering **191** (2017) 012010 doi:10.1088/1757-899X/191/1/012010

3.1. Synthesis of iron oxide nanoparticles

The addition of Ferric chloride solution to the *Annona squamosa* leaves extract the reduction reaction takes place of Fe^{3+} to Fe_3O_4. A possible reaction beyond the formation of iron oxide nanoparticles is Ferric chloride ($FeCl_3$ $.6H_2O$) and *Annona squamosa* leaf extract are involved in the reaction of aqueous phase medium. Figure 1 represents the formation of iron oxide complex by dark brown colour appearance. Initially, the C=O of aldehyde group in *Annona squamosa* leaf extract chelated with Fe^{3+} ions to form ferric protein chains HO^- .... Fe^{3+}...... bonds and as result in the formation of suspended ferric hydroxide $Fe(OH)^3$. Subsequently on slow evaporation, ferric hydroxide in a core is dehydrated ($-H_2O$) to form a black colored magnetite (Fe_3O_4) nanoparticle as a crystals. The protein chain in *Annona squamosa* leaf extract covered on Fe_3O_4 surface through chelation of COO- ......Fe^{3+} .

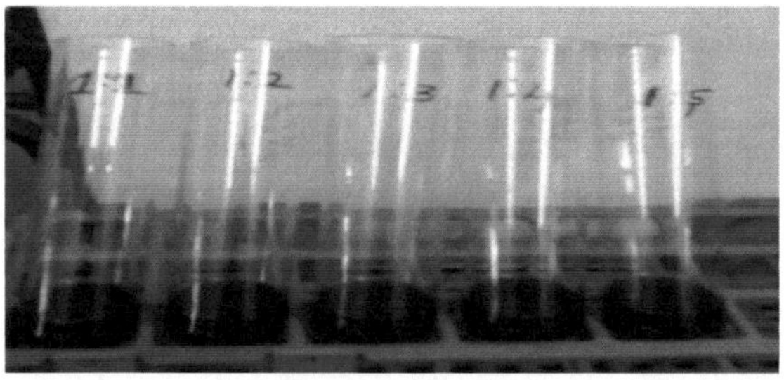

Figure 1. Photograph of visible color change of Annona squamosa leaf extract.

3.2. UV-Visible spectroscopy

After the addition the *Annona squamosa* leaf extract into the aqueous solution of $FeCl_3$, the solution was filled in glass cuvette of path length 10mm and UV-Vis spectral analysis has been done in the range of 300 to 700 nm. DI water was used as blank. Figure 2 represents the absorbance obtained by UV-spectroscopy on various concentration gradient of sample.

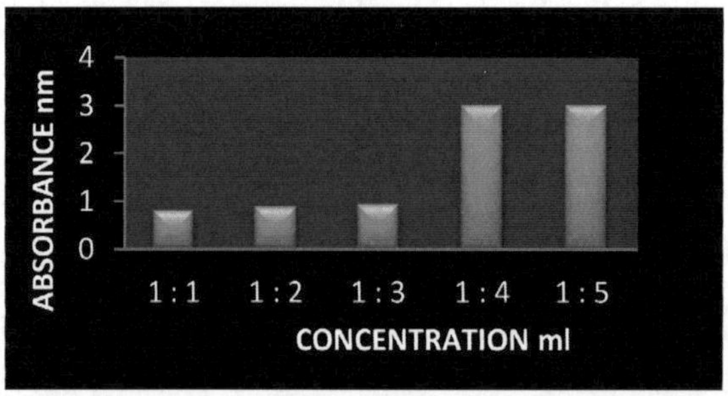

Figure 2. Absorbance obtained by UV-spectroscopy on various concentration of sample.

3.3. SEM analysis (scanning electron microscopy)

The surface morphology of the iron oxide and silica particles has been studied by scanning electron microscopy method. As shown in Figure 3a-3c represents the SEM images of Fe_2O_3 nanoparticles grafted with *Annona squamosa* with small spot by SEM analysis. Figure 3a-3c suggests the aggregation of the particles and small grains are present at the surface. A scanning electron microscopy was employed to analyse the structure of nanoparticles that were formed. Figure.3a shows the low magnification 25k x 27,000 SEM image of Fe_3O_4 powder; it can be seen that the particles are agglomerated. The size distribution and morphology is irregular. The particles are plate like structure with coarsened grains, whereas the magnification at 25k x17,000 in Figure3b, the Fe_3O_4 nanoparticles showed uniformly distributed small spherical shaped particles. The magnification at 25kx 10,000 (Figure3c) under the same conditions, which showed that large number of homogeneous nano capsule like morphology of iron oxide nanoparticles.

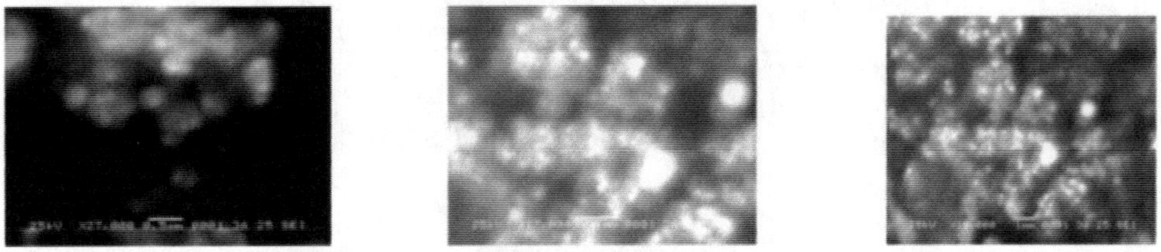

Figure 3a: Typical electron energies are 1-25 kV with a spot size of 0.5nm.

Figure 3b: Typical electron energies are 1-17 kV with a spot size of 0001nm a

Figure 3c: Typical electron energies are 1-17 kV with a spot size of 1nm.

Figure 3. The SEM images of Fe_2O_3 nanoparticles grafted with *Annona squamosa.*

3.4. Fourier transform infrared spectrophotometer (FTIR)

FTIR analysis was performed, in order to determine the functional groups on *Annona squamosa* leaves extract and predict their role in the synthesis of iron oxide nanoparticles. Figure 4 shows the FT-IR spectrum of *Annona squamosa* leaf extract. The strong absorption peak at 3340 cm-1 is assigned to O-H stretching of alcohol and phenolic compounds or stretching of the –NH band of amino group. The presence of peak at 2980 cm-1 are assigned to aliphatic C-H stretching in methyl and methylene groups. The peak at 1640 cm-1 is due to stretching vibration of CO groups in the ketones, aldehydes and carboxylic acids. The peak belonging to 1387 cm-1 is due to COO stretching vibration. The peak at 1640 cm-1 is attributed to the presence of carboxylate ions (COO-), which is responsible for the formation of iron oxide nanoparticles. The peaks at 1083 and 1043 cm-1 indicate the presence of CO groups. This peak was absent in plant extract which indicate the formation of iron oxide nanoparticles. The absorbance band at 872 cm-1 might be assigned to the existence of some amount of oxidised iron oxide on the surface. FT-IR analysis confirmed that the bio reduction of ferric chloride into iron oxide nanoparticles is due to the reduction by capping material of *Annona squamosa leaf* extract.

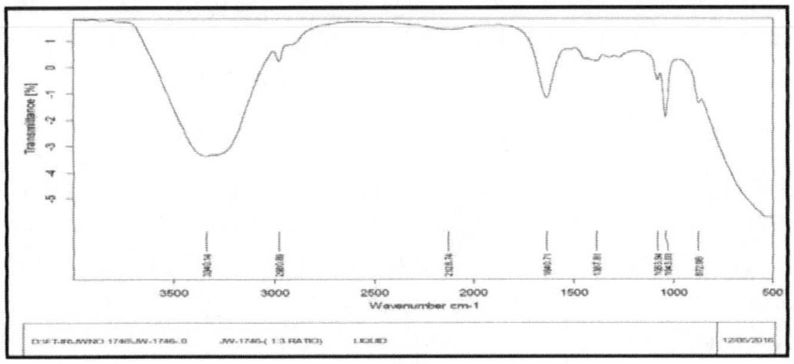

Figure 4. Representation of wave number cm^{-1} obtained by FT-IR analysis.

3.5. Anti proliferative assay- MTT against Hep G2 cells

The synthesized iron oxide nanoparticle obtained was subjected to MTT assay against Hep G2 cell line. Observation of morphological changes in cells indicated that Fe_3O_4 NPs inhibited proliferation of the various cancer cell lines in a dose-dependent and time-dependent manner. No toxicity was seen in the normal liver cell line. Figure 5 represent the percentage of cell viability of HepG2 Cell line after treated with synthesized Fe_3O_4 NPs and figure 6 represent the percentage of cytotoxicity exhibited by synthesized Fe_3O_4 NPs on HepG2 Cell line.

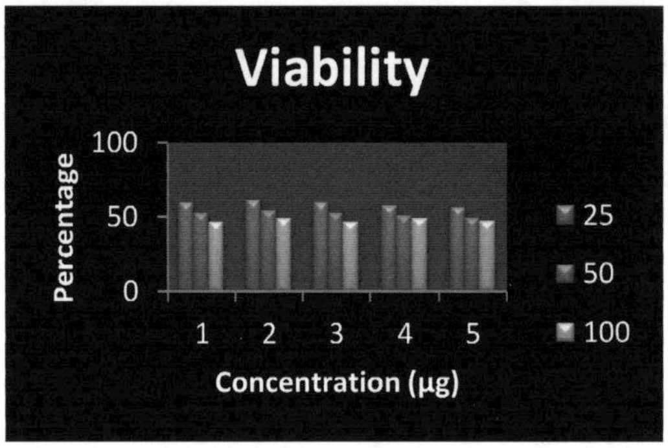

Figure 5. Percentage of cell viability by synthesized Fe_3O_4 NPs on HepG2 Cell line.

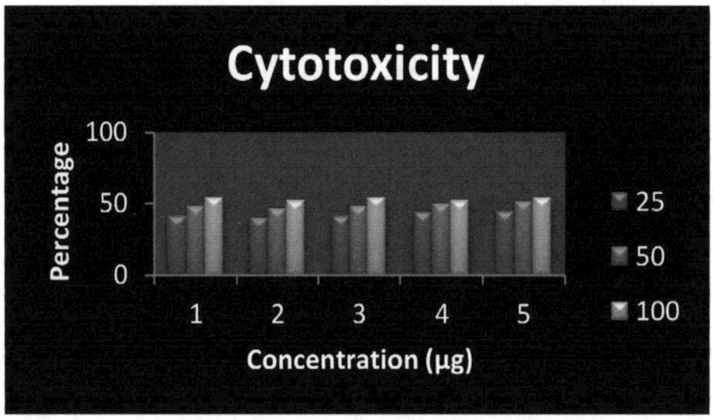

Figure 6. Percentage of cell cytotoxicity by synthesized Fe_3O_4 NPs on Hep G2 Cells.

The cyotoxicity of the iron oxide nanoparticle was studied against the Hep G2 cell line by MTT assay. The cytotoxicity effect of synthesized iron oxide nanoparticle from the ratio 1:1 to 1:5 was studied at different concentrations 25 μg, 50 μg, and 100 μg against Hep G2 cell line. Nanoparticles synthesized from this ratio also had significantly higher yield. The cytotoxic activity of the nanoparticles synthesized from various ratios had almost the cytotoxicity which indicates that the activity arises from compounds of extract rather than ferrous sulphate.

4. Conclusion

This study shows that the *Annona squamosa* iron oxide nanoparticle exhibits cytotoxic activity at very less dose. The synthesized iron oxide nanoparticles showed promising anticancer activity against human liver cancer cell line. From the study, it can be concluded that the iron oxide nanoparticles synthesized using *Annona squamosa* possess high anticancer activity against cell lines which further suggest the potential therapeutic use of these nanoparticles.

References

[1] Mahmoudi M, Sant S, Wang B, Laurent S, and Sen T 2011 Superparamagnetic iron oxide nanoparticles (SPIONs): Development, surface modification and applications in chemotherapy *Advanced Drug Delivery Reviews* 63 24-46.

[2] Padhi L P, Panda S K, Satapathy S N and Dutta S K 2011 In vitro evaluation of antibacterial potential of *Annona squamosa* L. and Annona reticulata L. from Similipal Biosphere Reserve, Orissa, India *Journal Agricultural Technology* 7(1) 133-142.

[3] Ranveer S, Tomar and Siddharaj S 2012 Antidiabetic activity of *Annona squamosa* L. in experimental induced diabetic rats *International Journal of Pharmaceutical & Biological Archives* 3(6) 1492-1495.

[4] Khar A, Pardhasaradhi B, Reddy M, Ali M and Kumari L 2005 Differential cytotoxic effects of *Annona squamosa* seed extracts on human tumour cell lines: Role of reactive oxygen species and glutathione *Journal of Bioscience* 30(2) 237-244.

[5] Mulvaney P 1996 Surface plasmon spectroscopy of nanosized metal particles advanced mineral products (Research Centre, School of Chemistry, University of Melbourne, Parkville, Victoria, 3052, Australia) 788–800.

IOP Conf. Series: Materials Science and Engineering **191** (2017) 012011 doi:10.1088/1757-899X/191/1/012011

Characteristic of Starch-Poly(N-Vinyl-Pyrrolidone) for an encapsulation material in floating drug delivery system

N Rositaningsih[1] and E Budianto[1]

[1]Chemistry Department of Mathematics and Natural Sciences Faculty of Universitas Indonesia, Depok, West Java, Indonesia 16424

E-mail: nungki_rositaningsih@yahoo.com

Abstract. Modified starch has been widely used in many studies in the field of drug delivery for gastrointestinal drug release. In this study, the starch was modified with poly(N-vinyl-pyrrolidone) by interpenetrating polymer network (IPN) method. Characterizations were done to observe its characteristics in floating drug delivery applications. Three modified starch-based hydrogels were synthesized, i.e. crosslinked starch, semi-IPN, and full-IPN hydrogels. Non-modified starch hydrogel was also synthesized for benchmark purpose. All materials were characterized and analysed by Swelling Test, Buoyancy Test, Differential Scanning Calorimetry (DSC), Thermal Gravimetric Analysis (TGA), and Fourier Transform Infrared Spectroscopy (FTIR). Swelling test showed that the crosslinked-starch hydrogel has the lowest swelling percentage compared to other hydrogels, whereas non-modified hydrogels tend to have similar swelling percentage to Semi-IPN. Morphology and visual analysis results showed that non-modified hydrogels were physically the most fragile, followed by a crosslinked-starch, semi-IPN-starch, and full-IPN-starch hydrogel. Therefore, full-IPN-starch hydrogel had the highest elasticity. From IR spectrum result, it can be seen that a wavenumbers shift was observed for the modified starch compared with the IR spectrum of pure starch powder. TGA and DSC characterizations showed the degradation temperature for non-modified hydrogel was at 250°C. The degradation temperature for both crosslinked and semi-IPN starch were at 275°C. For full-IPN, however, the degradation temperature was at 225°C. On the other hand, the degradation level could be observed from DSC and TGA results as well. Full-IPN appeared to be the slowest, while non-modified hydrogel seem to be the fastest to degrade. $CaCO_3$ was used as the pore forming agent (PFA) in this research for buoyancy characterization. Buoyancy test showed that the full-IPN had the fastest floating lag and longest floating time followed by semi-IPN-starch, crosslinked-starch, and non-modified starch. From all characterizations done in this research, it can be suggested that the full-IPN provided the most suitable characteristics as an encapsulation material candidate in a floating drug delivery system.

1. Introduction

Biomaterial for encapsulation in drug delivery system is increasingly developed this days, i.e. carbohydrate based material, biopolymer, and modified carbohydrate based hydrogel and many more. Biopolymer hydrogel is developed as an encapsulant in the drug delivery system to delay the metabolism process, therefore, the interaction with water in the body can be slowed. It means that the drug will be slowly released and can be absorbed optimally in the body [1].

Hydrogel expands as it absorbs water or solvent, and retains their mechanical stability. Even though the levels of water or solvent absorbed in the hydrogel reached 99%, it will keep its shape and characteristics as a solid material. Therefore, hydrogel can maintain its structure with low or high water or solvent content in it. Chemically crosslinked methods is a versatile route to create a hydrogel with a better degree of mechanical stability [2]. Interpenetrating polymer network hydrogel (semi and full) provides additional strength and new properties for drug delivery system [3]. These hydrogels are widely used for drug delivery on a specific target (site-specific) [4], such as colon [5], and intestine [6]. Floating drug delivery system, in principle, is used to deliver drugs that are absorbed in the stomach. This carrier will absorb the water content in the stomach, therefore, it will inflate and float in the stomach. During this absorption, the drug in hydrogel will be released gradually in to the body. This method is used, so that the drug will be absorbed optimally and effectively over a time interval of metabolism in the stomach before entering the intestinal system [7]. With this system, the function of the drug could be optimized even in low doses as it increases the absorption effectiveness in the body [8].

2. Experimental

2.1. Preparation of starch hydrogels

For this experiment, several floating hydrogels were synthesized i.e. non-modified, crosslinked, semi-IPN, and full-IPN starch hydrogels. Non-modified starch hydrogel was synthesized by adding 2 grams of starch powder into 70 ml hot distillate water (80 – 90°C), then cooled down to room temperature. To synthesize the crosslinked-starch hydrogel, 2% of acetaldehyde 0.1 M was added to hydrogel solution and stirred for 3 hours. Crosslinked-starch hydrogel was added 90:10 % w/w of poly(N-vinyl-pyrrolidone) and stirred in 1 hour to synthesized Semi-IPN hydrogel. Full-IPN had the most complex synthesis steps. The first step is similar as the process for crosslinked starch hydrogel, and then put in polymerization reactor with nitrogen gas condition at 70°C. The process was then continued by adding 90:10% w/w of N-vinyl- pyrrolidone in crosslinked starch hydrogel, ammonium persulfate, and N,N'-N,N'-methylene-bis-acrylamide under reflux. This solution was stirred for 1 hour. To synthesize the floating hydrogel, 1.00% calcium carbonate was added to each hydrogel solutions as pore forming agent. The solutions was stirred to ensure no air bubbles formed in solutions. All hydrogel solutions were dried in an oven at 60°C for 18 hours. [9]

2.2. Swelling ratio measurements

Swelling ratio measurements were done for floating and non-floating hydrogels. Dried hydrogels were immersed in pH 1.2 buffer solution at room temperature. For non-floating hydrogels, measurement were done in 30, 60, and 180 minutes [10]. On the other hand, floating hydrogels were measured only in 60 minutes. Filter paper is used to remove excess water on the surface of hydrogels. The swelling ratio percentage was the determined using equation 1.

$$Swelling\ ratio\ (\%) = \left[\frac{W_t - W_0}{W_n}\right] x\ 100\ \% \quad (1)$$

Where W_0 is the initial weight of samples and W_t is the weight of swollen hydrogel at predetermined time (t).

2.3. In vitro buoyancy observations

The floating properties was analysed with 50 ml of pH 1.2 buffer solution at 37°C. The time required for hydrogels to rise to the surface and float (floating lag time), and the time required for hydrogel started to sink back (floating time) was measured by visual observation [11].

2.4. Physical characterizations

Differential Scanning Calorimetry (DSC) and Thermal Gravimetric Analysis (TGA) were done for physical characterizations of the hydrogels on PerkinElmer STA 6000. The test was done under temperature ranging from 50 – 500°C. Fourier Transform Infrared Spectroscopy (FTIR) was used within the range 400 – 4000 cm^{-1} on IRPrestige-21 Shimadzu. Morphology of all hydrogel was observed by Stereo-inverter microscope from Olympus SZX16.

3. Results and discussion

Morphology of hydrogel as shown in Figure 1 indicated that floating hydrogels had more pores in their surface than non-floating hydrogels. It's also indicated non-modified starch hydrogel had the roughest surface, followed by a crosslinked starch, semi-IPN starch, and full-IPN starch hydrogel. The rough surface of hydrogels can cause fragility to the material [12]. Therefore, full-IPN-starch hydrogel had the highest elasticity.

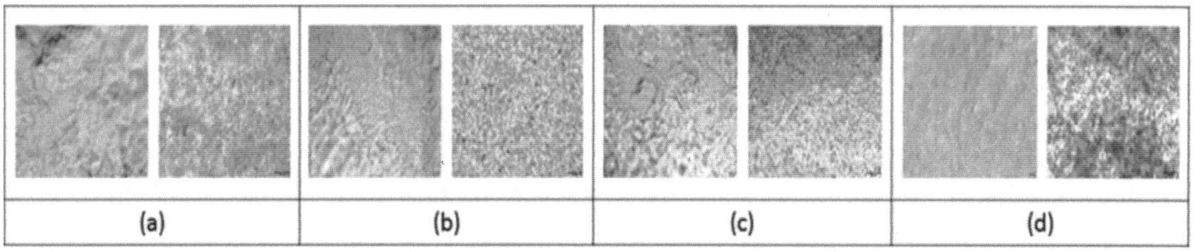

Figure 1. Microscope stereo-inverter scanning of non-floating hydrogels (left) and floating hydrogel (right): (a) Non-modified starch hydrogel (b) Crosslinked starch hydrogel (c) Semi-IPN starch hydrogel (d) Full-IPN starch hydrogel.

The result of swelling study of hydrogels showed that, in general, swelling value increased proportionally with the duration of immersion for non-floating hydrogels (Figure 2a). The highest swelling ratio was achieved at 30, 60, and 180 minutes immersion for full-IPN, non-modified, and semi-IPN starch hydrogen respectively. Floating hydrogels, however, it was observed that floating method affect the swelling ratio as shown in Figure 2b. The swelling percentage of floating hydrogels were significantly lower than non-floating hydrogel. It might be because of pores formation in the floating hydrogel which reduce the ability of the hydrogel to absorb water.

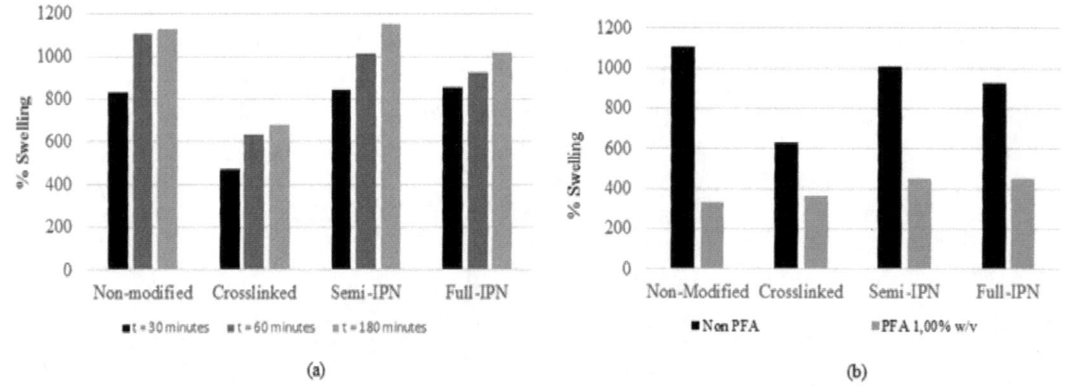

Figure 2. Results on swelling properties observation of hydrogels: (a). Non floating hydrogel in various time (b). Comparison between non floating hydrogel to floating hydrogel with 1.00% $CaCO_3$.

The buoyancy tests showed that pores formation gave the buoyancy characteristics for the hydrogels. Non-floating starch hydrogels were not floating due to the absence of the pore. Full-IPN starch hydrogel had the fastest floating lag time followed by semi-IPN starch, crosslinked starch and non-modified starch hydrogel. The floating properties were related to gas content of polymer network [13], showed in table 1.

Table 1. Floating lag and floating time for non-floating hydrogel ($CaCO_3$ 0%) and floating hydrogel ($CaCO_3$ 1.00%)

Hydrogel – PFA	Floating Lag Time	Floating Time
NK-$CaCO_3$ 0%	No Floating activity	No Floating activity
NK-$CaCO_3$ 1,00%	42 minutes 56 seconds	> 3 hours (180 minutes)
Crosslinked-$CaCO_3$ 0%	No Floating activity	No Floating activity
Crosslinked-$CaCO_3$ 1,00%	32 minutes 39 seconds	> 3 hours (180 minutes)
Semi-IPN-$CaCO_3$ 0%	No Floating activity	No Floating activity
Semi-IPN-$CaCO_3$ 1,00%	21 minutes 53 seconds	> 3 hours (180 minutes)
Full-IPN-$CaCO_3$ 0%	No Floating activity	No Floating activity
Full-IPN-$CaCO_3$ 1,00%	18 minutes 21 seconds	> 3 hours (180 minutes)

FT-IR spectrum of non-floating hydrogels (Figure 3a) shown that there were interaction between starch and crosslinking agent, and also PVP. This was indicated by band shifting if compared to starch powder spectrum. Starch spectrum showed O-H band at 3550 – 3000 cm^{-1}, C-H band at 2750 – 3000 cm^{-1} and C-O band at 1250 – 1000 cm^{-1}. Floating hydrogel spectrum (Figure 3b) has shifting band compare to non-floating hydrogel. O-H band from non-floating hydrogel spectrum at 3629 cm^{-1} to 3566 cm^{-1}. The peak at around 2900 cm^{-1} of C-H band for floating hydrogels were stretching compared to non-floating hydrogels. The absorption band of non-floating hydrogel at around 1700 cm^{-1} has shifted to at around 1600 cm^{-1}. This indicated that the C-OH, might be caused by formation of bonds with carbonyl groups [14,15].

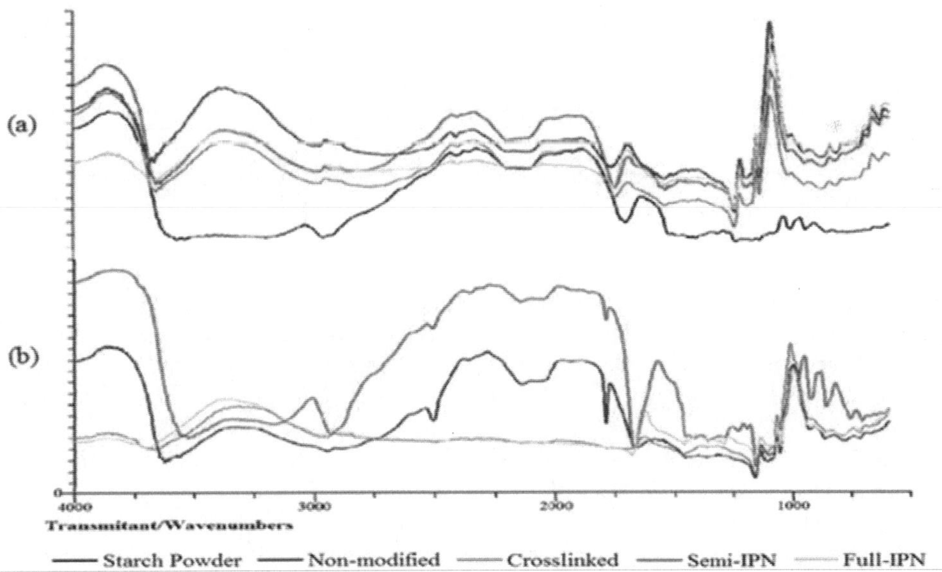

Figure 3. FT-IR Spectrum of Hydrogels: (a) Non Floating Hydrogels compared to Starch powder (b) Floating Hydrogels.

The thermal spectrum of DSC and TGA (figure. 4) showed the thermal properties of non-floating starch hydrogels. The degradation temperature for non-modified, crosslinked, semi-IPN, and full-IPN hydrogel were roughly at 250°C, 275°C, 275°C, and 225°C respectively. Furthermore, the degradation level could be observed from DSC and TGA results as well. Full-IPN appeared to be the slowest, while non-modified hydrogel seem to be the fastest to degrade. The decomposition temperature for non-modified starch hydrogel was at 300°C [16]. Crosslinked starch hydrogel had similar decomposition properties to non-modified starch hydrogel, but it has higher heat flow and lower decomposition temperature at 290°C, followed by semi-IPN and full-IPN starch hydrogel.

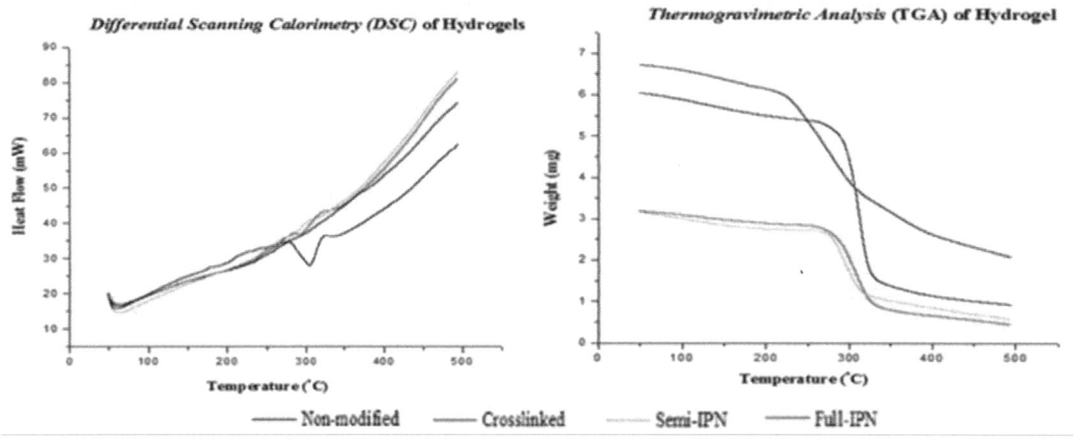

Figure 4. Thermal Curves (DSC and TGA) for Non Floating Hydrogels.

4. Conclusion

Non floating hydrogels showed better structure than floating hydrogels. However, for drug delivery purposes, floating hydrogels characteristics were more preferable. From all characterizations done in this research, it could be suggested that the full-IPN provided the most suitable characteristics as an encapsulation material candidate in a floating drug delivery system. Therefore, full-IPN floating hydrogel could be the most suitable material for drug release applications.

Acknowledgment

We would like to thank the Department of Chemistry, Faculty of Mathematics and Life Science, Universitas Indonesia, and Directorate of Research and Community Services Universitas Indonesia for the support for this research.

References

[1] Babu K, Bera A, Kumari K, Mandal A and Saxena V K 2015 *J.Polym. Eng.* 35(1) 79.
[2] Changez M, Koul V, Krishna B, Dinda A K and Choundray V 2004 *Biomaterials* 25 139.
[3] Choi B Y, Park H J, Hwang S J and Park J B 2002 *International Journal of Pharmaceutics* 239 81.
[4] Fakhouri F M, Daryne C, Yamashita F, Martelli S M, Jesus R C, Alganer K, Collares-Queiroz F P, Innocentini-Mei L H 2013 *Carbohydrate Polymers* 93 681.
[5] Geever L M, Cooney C C, Lyons J G, Kennedy J E, Nugent M J D, Devery S and Higginbotham C L 2008 *European Journal of Pharmaceutics and Biopharmaceutics* 69 1147
[6] Grabowskaand B and Holtzer M 2009 *Archives of Metallurgy and Materials* 54 427.
[7] Hennink W E and Van Nostrum C F 2002 *Advanced Drug Delivery Review* 54 13
[8] Kulkarni R V, Shreedar V, Mutali S, Setty M and Sa B 2010 *International Journal of Biological Macromolecular* 47 520.
[9] Kumar N, Niranjan S K, Irchhaiya R, Vishwakarma K S and Akhtar A 2012 *International Research Journal of Pharmacy* 3(8) 29.
[10] Musa M B, Yoo M J, Kang T J, Kolawole E G, Ishiaku U S, Yakubu M K and Whang D J 2013 *Research and Reviews: Journal of Engineering and Technology* 2(4) 9.

[11] Narang N 2011 *International Journal of Applied Pharmaceutics* 3(1) 1.
[12] Risbud M V, Hardikar A A, Bhat S V, Bhonde R R 2000 Journal of Controlled Releas, 68 23.
[13] Rosa M, Zia H and Rhodes T 1994 *International Journal of Pharmaceutics* 105 65.
[14] Sandeep C, Harikumar S L and Kanupriya 2012 *International Journal of Research in Pharmacy and Chemistry* ISSN: 2231-2781
[15] Selvakumaran S, Muhammad I I and Razak S I A 2010 *Carbohydrate Polymers* 135 207.
[16] Vaghani S S, Pathel M M, Satish C S 2012 *Carbohydrate Research* 347 76.

Parametric study of nanofiber fabrication by biocompatible polymers

S G Park[1] and H W Choi[2]

1 Graduate School of Mechanical Design, Keimyung University, Daegu, Dalgubeol-daero 1095, Republic of Korea
2 Department of Mechanical and Automotive Design, Keimyung University, Daegu, Dalgubeol-daero 1095, Republic of Korea

E-mail: psg5655@kmu.ac.kr

Abstract. This paper introduces a parametric study of nanofiber fabrication. Poly-Caprolactone(PCL) and Chloroform are basic material for nanofiber fabrication with electrospinning method. The device for electrospinning was composed of a syringe pump, syringe, high voltage generator and 32gauge plastic nozzle. PCL concentration, applied voltage, solution flow rate and the tip to collector distance (TCD) were parameters in electrospinning method. As a result of electrospinning experiment, it was found that as TCD and applied voltage increased, the diameter of fibres was decreased. In feed rate experiment, the size of fibres decreased when feed rate decreased. The optimized PCL concentration was found at 6w/v% for the minimum of fibres.

1. Introduction

Nanofibers can provide a highly porous mesh for the biomaterials. Their high surface area per unit volume is advantageous for the various applications. A porous structure made with nanofibers is useful platform where the pore size and form can be easily changed unlike regular medical solid structure. Also nanofibers can be referred to form a firm structure if it is required. For the porous mesh applications, the most adaptable and highly productive process can be electrospinning. Meshes of porous nanofibers produced by electrospinning have been identified for use in numerous applications [1].

In 1902, electrospinning was first patented in the U.S.[2]. However, this process was not generally used until the 1990s and it became popular with recent nanoscience and nanotechnology development. Researchers started the ES nanofiber again with new research concepts of nanofiber production using electrospinning after the 1990s [3].

Electrospinning method is one of the ways to manufacture nanofibers with polymer because it is low manufacturing cost and simple process. This method is in the attention as a useful method for fabricating functional thin porous product like fibers [4]. Electrospinning method can be different in experiment set up such as collector types or kinds of solutions. Several kinds of collectors exist like using mandrel, simple plate and parallel poles (Figure 1). Net arrangement of nanofibers is different according to the kinds of collectors [5-7]. Polycaprolactone(PCL) was used for the material of produced fibers by electrospinning. It is a bioresorbable polymer with possible applications for bone and cartilage repair. PCL has certain advantages relative to other polymers like PLA(poly lactic acid) and it is more stable in its neighboring biomedical environments. It is also significantly low cost and is freely available in large quantities [8].

In this paper, the detail procedure of manufacturing process for the nanofibers is introduced. The nanofibers are produced with a solution of PCL and Chloroform. The PCL was used as bioresorbable polymers because it's hydrophobic, linear and synthetic that indicates high mechanical strength [9]. Nanofiber production using electrospinning method can be influenced by concentration of PCL, polymer solution feed rate, TCD(tip to collector distance) and applied voltage [10]. This paper provides the environment of experiment and introduces a parametric study of nanofiber fabrication applied with electrospinning for the biocompatible polymers.. If you don't wish to use the Word template provided, please use the following page setup measurements.

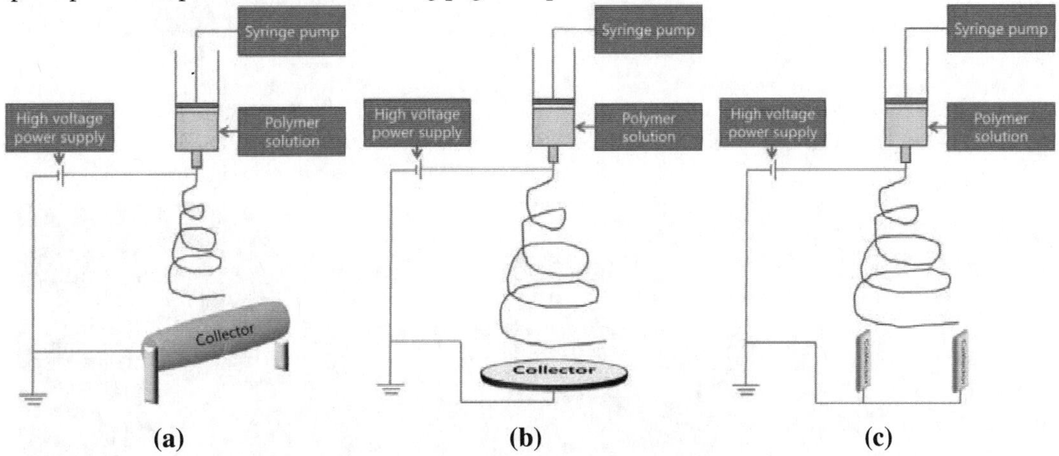

Figure 1. Schematics of collector types in electrospinning: (a) Mandrel type collector, (b) Simple plate type collector, (c) Parallel type collector.

2. Experiment

Please follow these instructions as carefully as possible so all articles within a conference have the same style to the title page. This paragraph follows a section title so it should not be indented.

2.1. Materials and methods

Commercial PCL pellets (Mw=80,000) and Chloroform (Sigma-Aldrich Inc.) were used in the experiments. Figure 2 is a molecular structure of PCL. In this paper, the device of experiment was chosen simple plate type collector in Figure 1 because it can have minimum parameter of electrospinning experiment like Figure 3. Figure 3 shows the concept design and parameters of electrospinning. The Chloroform was used as solution of PCL (Polycaprolactone). The collector was designed as flat aluminum plate with a diameter of 30mm. Fibers were electrospun from polymer solution containing 2, 4, 6 and 8% w/v PCL. The PCL solutions were transferred from 10ml syringes (HSW) and connected to a 32 gauge plastic nozzle. The needle tip was connected to a high voltage generator (NNC-HV30, NanoNC) operating at 5, 10 and 15kV. Polymer solution of PCL was fed into the needle at a rate of 1, 2, 4 and 6ml/hr by a syringe pump (KDS100, kdScientific). TCD (Tip and Collector Distance) was set at 90, 110 and 130mm.

IOP Conf. Series: Materials Science and Engineering **191** (2017) 012012 doi:10.1088/1757-899X/191/1/012012

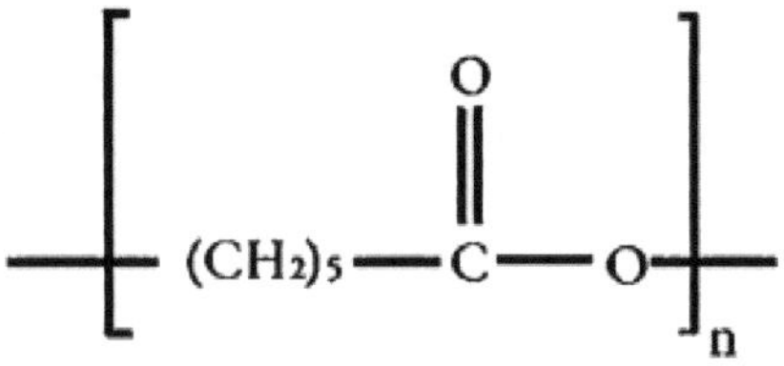

Figure 2. Molecular structure of PCL(polycaprolactone)[11].

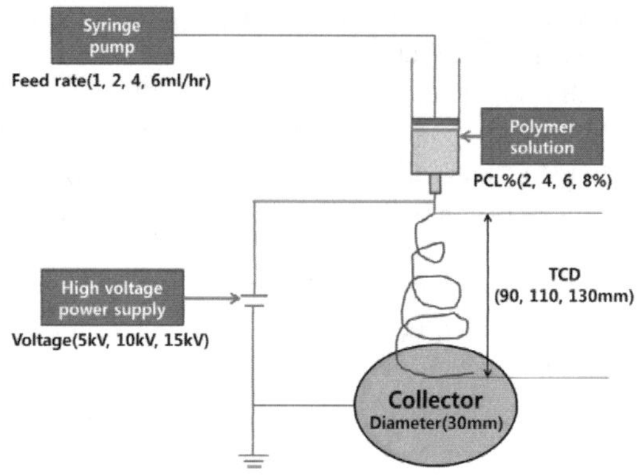

Figure 3. Experimental setup for electrospinning.

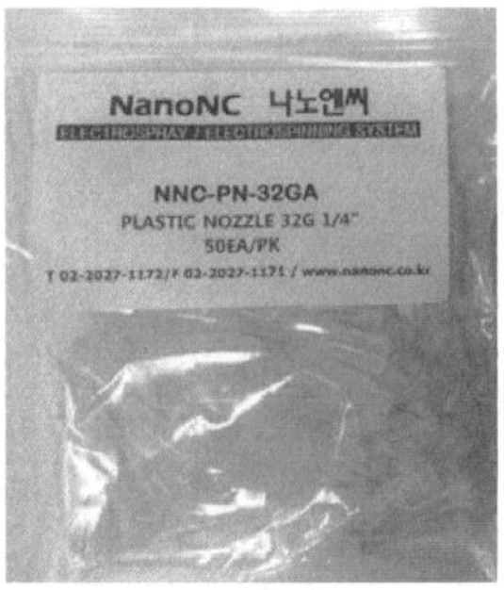

Figure 4. Plastic nozzle of 32 gauge.

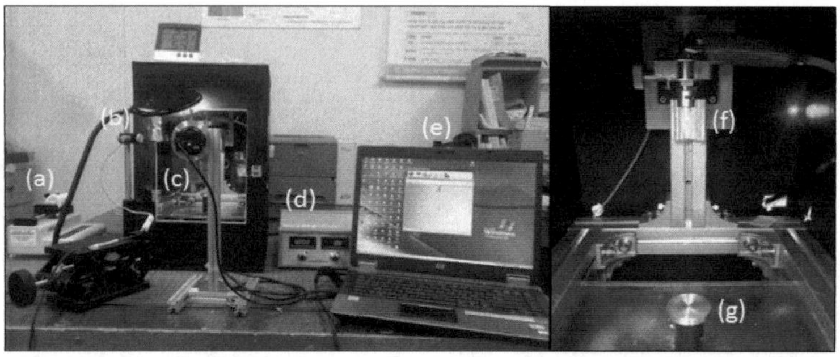

Figure 5. Overall configuration of electrospinning apparatus: (a) Syringe pump, (b) Diode laser(532nm, Green), (c) Camera, (d) High voltage generator, (e) PC, (f) Plastic nozzle, (g) Collector.

2.2. Device of experiment

A syringe pump and a high voltage generator were used for electrospinning with polymer solution of PCL. Capacity of high voltage generator is maximum of 30kV. The installed syringe pump has minimum flow rate of 0.00635cm/hr and maximum of 76.18cm/hr. Syringe has a capacity of 10microliter to 60milliliter. The 532nm wavelength of diode laser was used for the visual monitoring of nanofibers electrospinning. A vision camera (zoom 7000, NAVITAR) was used for the real time monitoring via a computer. In electrospinning process, the material leaves from the nozzle to the collector. The size of nozzle was 32 gauge for the manufacturing of nanofibers as shown in Figure 4. The inner diameter of nozzle was 0.10mm and rack & pinion stage (Edmund Optics Inc.) was used for the movement of nozzle. It can move up to maximum traveling distance of 100mm. Figure 5 explains the setup of electrospinning apparatus.

3. Result

The platform of scaffold with PCL was prepared by electrospinning method. The parameters of ES were PCL concentration, applied voltage, feed rate and TCD, and the details of parametric studies are following.

3.1. PCL concentration

For the experiments, the concentration of PCL was changed from 2w/v% to 8w/v% to fine the optimum solution ratio of electrospun fibers. The samples of electrospun fibers were produced with variable concentration of PCL in solution. The flow rate of 2.0ml/h, TCD of 90mm and applied voltage 15kV were fixed for the experiments. Electrospun fibers were observed like bead in concentration of 2w/v%. When concentration of PCL increased, the size of sample fibers was changed. It formed from bead shape to net shape fibers and the diameter of fibers decreased. In 8w/v% or higher concentration, the size of fibers increased like Figure 6 and Figure 7. Average values of fibers per PCL concentration is summarized with standard deviation values in Table 1. The minimum size of fibers was observed with 6w/v% concentration of PCL.

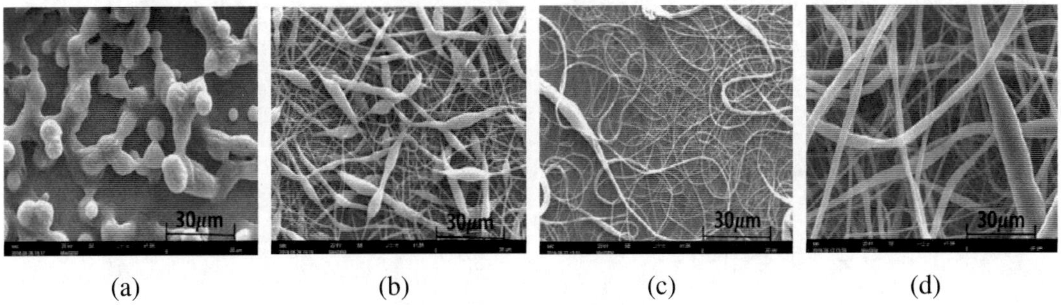

Figure 6. SEM of electrospun PCL composite scaffold with different PCL concentration: (a) 2w/v%, (b) 4w/v%, (c) 6w/v%, (d) 8w/v%.

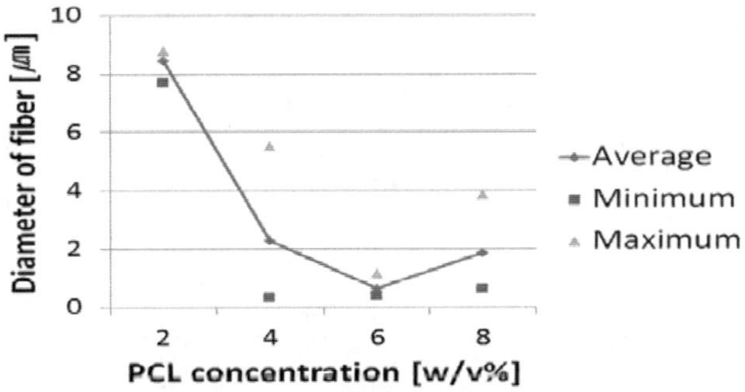

Figure 7. Fibre size distribution per PCL concentration.

Table 1. Average and standard deviation in experiment per PCL concentration.

Concentration of PCL [w/v%]	2	4	6	8
Average of fibres [μm]	8.461	2.293	0.634.	1.846
Standard deviation [μm]	0.447	2.446	0.309	1.206

3.2. Applied voltage

In experiment of variable voltage, the fixed parameters were concentration of 6w/v% PCL, flow rate 8.0ml/h and TCD 90mm. When the applied voltage decreases, the fibres were getting thicker and the fibres were getting thinner in higher voltage. The maximum applied voltage was limited at 15kV for the safety of syringe pump which limits the applied voltage at 15kV. SEM images with variable applied voltage are summarized in Figure 8. Figure 9 is summary of fibre size variation per applied voltages by average, minimum and maximum fibres. The quantities of distribution with standard deviation is summarized in Table 2. The minimum size of fibers was observed with 15kV of applied voltage in this experiment with uniform size of fibers.

IOP Conf. Series: Materials Science and Engineering **191** (2017) 012012 doi:10.1088/1757-899X/191/1/012012

(a) (b) (c)

Figure 8. SEM of electrospun PCL composite scaffold with different applied voltage: (a) 5kV, (b) 10kV, (c) 15kV.

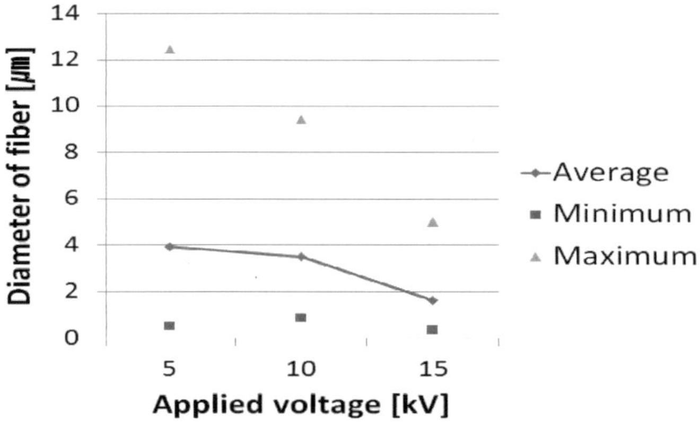

Figure 9. Fibre size distribution per applied voltage.

Table 2. Average and standard deviation in experiment per applied voltage.

Applied voltage [kV]	5	10	15
Average of fibres [μm]	3.932	3.527	1.653.
Standard deviation [μm]	4.999	3.843	1.923

(a) (b) (c) (d)

Figure 10. SEM of electrospun PCL composite scaffold with different flow rate: (a) 1ml/hr, (b) 2ml/hr, (c) 4ml/hr, (d) 6ml/hr.

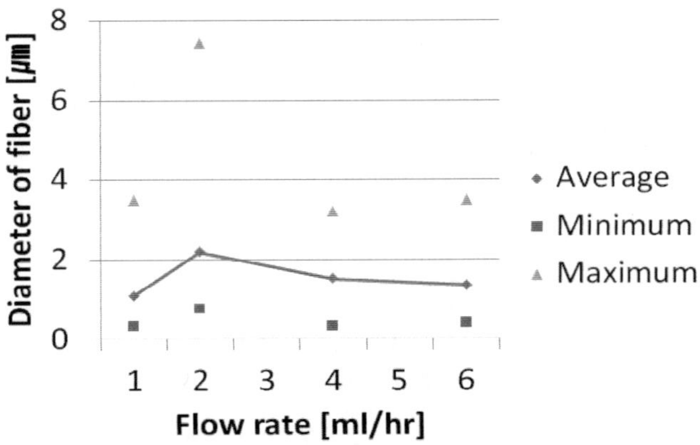

Figure 11. Fibre size distribution per flow rate.

Table 3. Average and standard deviation in experiment per flow rate.

Flow rate [ml/hr]	1	2	4	6
Average of fibres [㎛]	1.107	2.196	1.544.	1.343
Standard deviation [㎛]	1.346	2.928	1.206	1.260

3.3. Flow rate

In this experiment, the flow rate was set as a variable. The concentration of PCL was 6w/v% with applied voltage of 15kV and TCD 90mm in experiment of flow rate. Based on the result of experiment data by the variation of flow rate, the smallest fibre size was observed when flow rate applied 1ml/hr in Figure 10 and Figure 11. The quantities of distribution with standard deviation is summarized in Table 3. The thickest fibre size was observed with 1ml/hr as shown in Figure 11 and Table 3.

3.4. TCD

In this experiment, TCD was set as a variable. The concentration of PCL was 6wt% with flow rate 1ml/h and applied voltage of 15kV. The range of TCD was from 90mm to 130mm. The effect of TCD for fibres size are summarized in Figure 12 and Figure 13. The quantities of distribution with standard deviation is summarized in Table 4. The thickest fibres were observed at 130mm of TCD in this experiment.

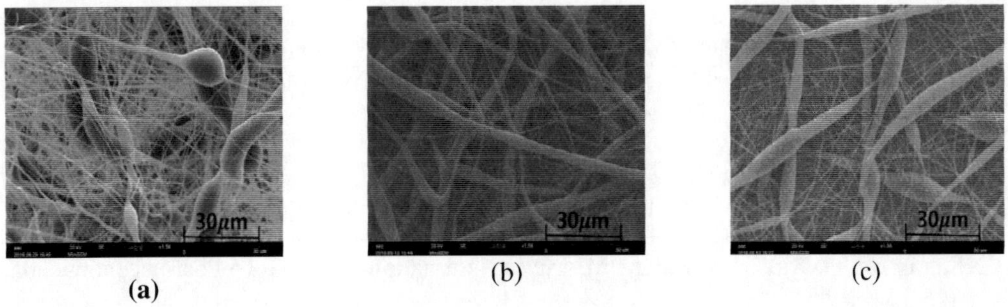

Figure 12. SEM of electrospun PCL composite scaffold with different TCD: (a) 90mm, (b) 110mm, (c) 130mm.

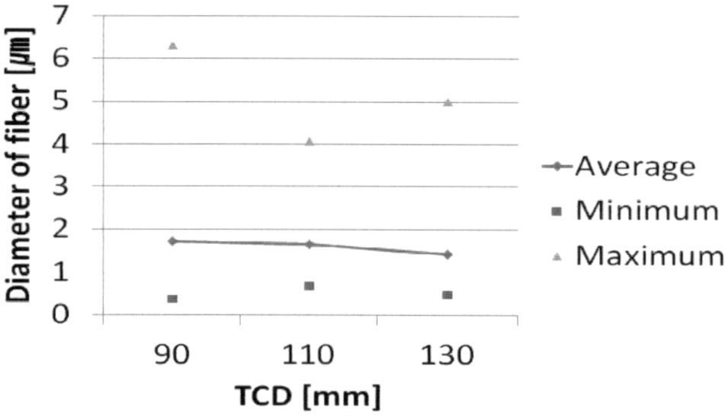

Figure 13. Fibre size distribution per TCD

Table 4. Average and standard deviation in experiment per TCD.

TCD [mm]	90	110	130
Average of fibres [μm]	1.716	1.660	1.439.
Standard deviation [μm]	2.565	1.381	1.986

4. Conclusion

In this study, the general effects of nanofiber size by parameters of electrospinning was introduced. Based on the result analysis, it was found that as TCD and applied voltage increased, the diameter of fibres tended to decrease. In flow rate experiment, the size of fibres decreased when flow rate decreased. The optimized PCL concentration was found at 6w/v% for the minimum of fibres.

Acknowledgments
This research was sponsored by Korea Nation Research Foundation (NRF-2015R1D1A1A01060648).

References

[1] S. Ramakrishna, K. Fujihara, W. Teo, T. Young, Z. Ma, R. Ramaseshan, Materialstoday, 9(3), pp. 40-50, 2006
[2] W. J. Morton, Method of Dispersing Fluids, US patent 705,691, 1902
[3] J. Doshi, D. H. Reneker, Journal of Electrostatics, 35, 151-160, 1995
[4] D. Kang, H. Kang, Applied Surface Science, 387, pp. 82-88, 2016
[5] A. Moheman, M. Alam, A. Mohammad, Advances in Colloid and Interface Science, 229, pp. 1-24, 2016
[6] V. Beachley, X. Wen, Materials Science and Engineering, C29, pp. 663-668, 2009
[7] T. Subbiah, G. Bhat, R. Tock, S. Parameswaran, S. Ramkumar, Journal of Applied Polymer Science, 96(2), pp. 557-569, 2005
[8] J. M. Williams, A. Adewunmi, R. M. Schek, C. L. Flanagan, P. H. Krebsbach, S. E. Feinberg, S. J. Hollister, S. Das, Biomaterials, 26, pp. 4817-4827, 2005
[9] S. Gautam, A. Dinda, N. Mishra, Materials Science and Engineering, C33, pp. 1228-1235, 2013
[10] T. Jiang, E. Carbone, K. Lo, C. Laurencin, Progress in polymer Science, 46, pp. 1-24, 2015
[11] Z. Sultanova, G. Kaleli, G. Kabay, M. Mutlu, International Journal of Pharmaceutics, 505, pp. 133-138, 2016

IOP Conf. Series: Materials Science and Engineering **191** (2017) 012013 doi:10.1088/1757-899X/191/1/012013

Dissolution and characterization of biofunctional keratin particles extracted from chicken feathers

Swati Sharma*, Arun Gupta*, Syed Mohd Saufi Bin Tuan Chik, Chua Yeo Gek Kee and Pradeep Kumar Poddar

Faculty of Chemical and Natural Resource Engineering, Universiti Malaysia Pahang (UMP), Lebuhraya Tun Razak, 26300 Gambang, Kuantan, Pahang, Malaysia

sspandit.89@gmail.com, arungupta10@gmail.com

Abstract. In the present study chicken feathers were hydrolyzed in alkaline environment. The pH value of feather hydrolyzed solution was adjusted according to the principle of isoelectric precipitation. Three kinds of precipitates of keratin polypeptide were collected at pH of 3.5, 5.5 and 7.5 respectively. The keratin solution were freeze dried and denoted as FKP1, FKP2, FKP3 respectively. All keratin particles possessed smooth, uniform and round surface by scanning electron microscope (SEM). FKP1, FKP2 and FKP3 had higher glass transition temperature examined by thermogravimetry (TG). Fourier transform infrared spectroscopy (FTIR) revealed that the extracted keratin retained the most of protein backbone, with the breakage of disulfide cross-links and hydrogen bonds.

1. Introduction

Keratin is present in all mammals having highly conserved amino acid sequence [1]. Keratin is reported for the impact on cell architecture and cell proliferation. Keratin is insoluble, fibrous and structural protein in feathers, which can be applied in many industrial applications. Keratin are three dimensional polymer interlinked by intermolecular bonding of disulfide amino acid and inter and intra molecular bonding of nonpolar and polar amino acids which are the reason for their stability and distinctive physical properties [2].

In food industry, near about $4X10^6$ ton per year are waste of chicken feathers found worldwide [3], some of them are pretreated and used as animal feed [4] and rest disposed in the landfills which cause serious environmental problem. There are mainly three methods of extraction of keratin [5] from the biomass: enzymatic hydrolysis has advantage like less species alteration which is very slow and cannot be used commercially. In the process of acidic hydrolysis the process provides very harsh conditions and can destroy some amino acids. Because of the high hydrolysis mainly alkaline hydrolysis is used for the keratin extraction from chicken feathers. During hydrolysis, chemicals break both types of peptide and disulfide bonds in proteins and as a result the structure of keratin hydrolysate is changed [1]. Reduction hydrolysis can cleave disulfide bonds of protein fibers without any major peptide bond cleavage, thus the microstructure of keratin remain intact [6]. Keratin is one of the most abundant proteins [7, 8, 9, 10]

The main aim of the study is to investigate the transforming of feather waste into bio functional peptide. In the present study three kinds of precipitates of keratin polypeptide were collected at pH of 3.5, 5.5 and 7.5 respectively. The keratin solution was freeze dried and marked as FKP1, FKP2, FKP3 respectively. Chicken feathers were hydrolyzed in sodium sulfide solution and after that by adjusting the pH value of hydrolyzed solution three kind of polypeptide were collected and characterized. The results showed the feasibility of extraction of keratin with different compositions of the polypeptide with applications.

2. Methods and experiments

2.1. Materials

Chicken feathers were collected from chicken processing plant at Jaya Gading, Kuantan, Malaysia. All chemicals were of analytical grade and were purchased from Sigma Aldrich (Selangore, Malaysia). Milli-Q water was used to make solutions and washing. All chemicals were used without any further processing after receipt.

2.2. Hydrolysis and characterization of polypeptide particles

2.2.1. Alkaline hydrolysis of feathers. 25g of feather was immersed into 0.5M of sodium sufide (1L) solution and then digested under 50°C using mechanical stirrer until feather fibers were dissolved thoroughly. The prepared hydrolysis solution was filtered twice for further process.

2.2.2. Isoelectric-point precipitation process of keratin polypeptides. Isoelectric point (pI) is defined as the pH value at which a particular molecule or surface carries zero net electric charge. It is a process in which proteins or amino acids are precipitated at pH value close to their isoelectric points [11,12]. In this study, the pH value of the prepared feather hydrolysis solution was adjusted to pH 7.5 with 2N HCl, a thick layer of precipitates was settled on the bottom after 24 h. The precipitates were collected and marked as Feather keratin polypeptide (FKP3). The rest feather hydrolysis solution was incubated at pH 5.5 and the precipitates after 24h also collected and marked as Feather keratin polypeptide (FKP2). Lastly the remaining solution is precipitated at pH 3.5 and the precipitates were collected after 24h and denoted as Feather keratin polypeptide (FKP1). In order to wash the salts away and any other impurities FKP1, FKP2, FKP3 were dispersed in water and centrifuged 3 times for 10 mins at 10000rpm. Finally the Feather keratin polypeptides (FKP's) were freeze dried to obtain keratin polypeptide powders.

2.2.3. Surface Morphology. The surface morphology of FKP1, FKP2 and FKP3 was analyzed by scanning electron microscope. It is performed by using Hitachi TM3030Plus. In the preparation step, the samples were adhered directly onto an aluminium stub with a thin self adherent carbon film.

2.2.4. TGA analysis. Thermo gravimetric measurements of precipitates were performed on Research Instruments TGA Q 500 under nitrogen atmosphere, in a temperature range between 10°C and 900°C at ramping time of 10°C/min. The samples were vacuum dried at 40°C. Samples with mass 3mg were put in aluminium crucible and thus, the data was analyzed.

2.2.5. FT-IR analysis. Chemical characterization of precipitates of different pH was done by using Fourier transform infrared (FT-IR) spectroscopy. It will help to detect the changes in chemical composition of peptides [13]. Nicolet iS5 from Thermo scientific FT-IR was used for chemical characterization hers in between the $4000cm^{-1}$ and $500cm^{-1}$ wave number range.

3. Results and discussion

3.1. Alkaline hydrolysis of feather and isoelectric precipitation of keratin polypeptide

In the alkaline process of the hydrolysis the disulfide bond (S-S) and partial peptide bonds were broke. Bradford test were used to test the protein content. The hydrolysis was done under the following operating conditions: 0.5M/L sodium sulfide, 50°C and 6 hr. when time was increased to 24 hr and 48 hr, there was not much significant increase in hydrolysis degree of feather keratin. Thus, it can be concluded that prolonged hydrolysis time couldn't increase the degree of hydrolysis at the same temperature and alkali conditions. The purpose of the study was to confirm the possibility of making biofunctional polypeptides from feather keratin using isoelectric precipitation process. The different isoelectric precipitates were collected at different pH range. It has been verify that three types of polypeptides with distinctive compositions were collected successfully. Furthermore, both chemical and physical properties of the polypeptides will be scrutinized through a series of characterization after freeze dry.

3.2. Surface morphology of keratin polypeptide powders

The morphology of the surface layer was studied by SEM analysis. In Figure 1 the regenerated keratin presents small microsphere, this is in accordance with the study on wool and feather keratin in journals [14, 15]. Fig 1 displayed the SEM images of FKP1, FKP2 and FKP3 respectively. The dried keratin is consisted of spherical, tightly packed micro-particles and random arranged porous microstructures.

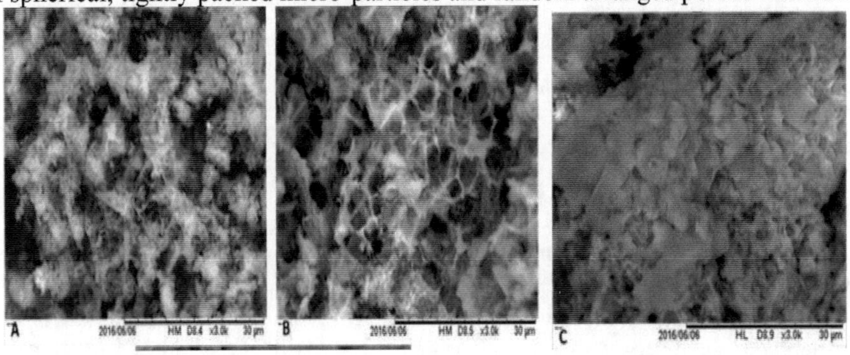

Figure 1. SEM images of (A)Feather keratin polypeptide at pH 3.5 (FKP1) (B) Feather keratin polypeptide at pH 5.5 (FKP2) (C) Feather keratin polypeptide at pH 7.5 (FKP3)

3.3. TGA Measurement

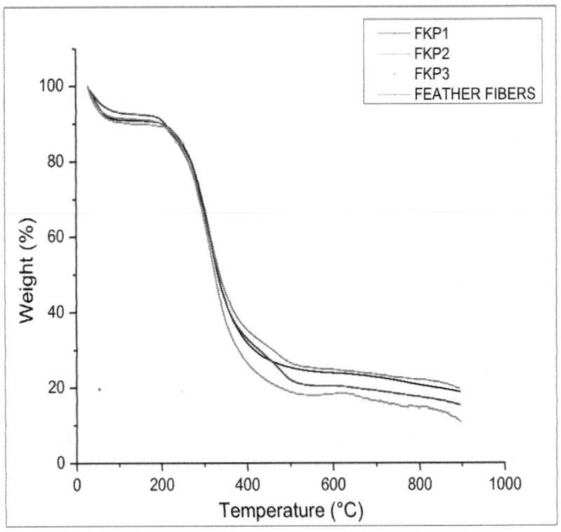

Figure 2. The TG curve of the feather fibers, FKP1, FKP2 and FKP3

The TG curve of the feather keratin polypeptides FKP1, FKP2 and FKP3 as well as feather fibers obtained. Figure 2 displayed the TG curve of feather fiber, FKP1, FKP2 and FKP3. All the samples were quickly decomposed from 250°C to 350°C and produced a significant amount of residue. The total weight loss was about 90%, 82%, 81.4%, 85.7% for feather fiber, FKP1, FKP2 and FKP3 respectively. The thermal decomposition started slowly 150°C with evolution of water and decomposition increased above 220°C showing notable weight loss. The study showed that the feather fiber is more stable as compare of polypeptides.

3.4. FTIR measurement

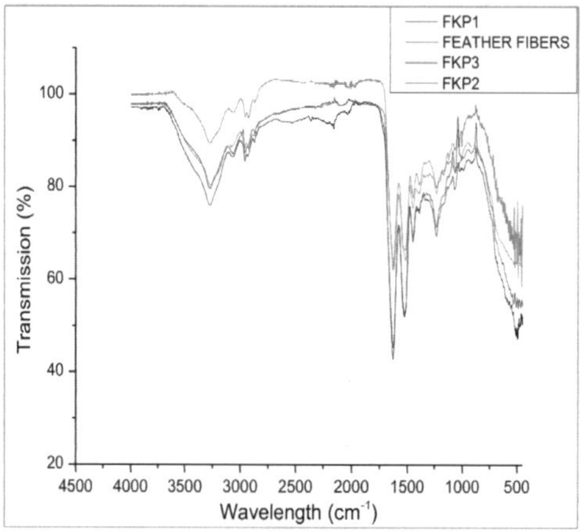

Figure 3. FTIR spectra of the feather fibers, FKP1, FKP2 and FKP3

FTIR measurement was used understand the chemical structure and the results are shown in Figure 3. The obtained results confirm that there are no significant changes in chemical structure of fibers after alkaline treatment. When comparing the absorbance spectra of all samples, their characteristic peaks are similar and are in comparison with other study [16]. On the other hand, the isoelectric precipitation have small effect on the chemical structure of protein, all the treatment showed transmission bands to the peptide bonds (-CONH) and are known as Amide A, Amide I, Amide II, Amide III [17,18]. The broad transmission band region from 3500 cm^{-1} to 3200 cm^{-1} can be attributed to stretching vibration of -O-H and -N-H (Amide A) [19]. The Amide A band which falls at 3282 cm^{-1}, is connected with stretching vibration of N-H bonds. Bands which fall in 3000-2800 cm^{-1} are related to C-H stretching bonds. The strong transmission band is attributed to C=O stretching (Amide I) which occurs in the range of 1700 cm^{-1} to1600 cm^{-1} [13] . The transmission band (Amide II) in the range of 1580 cm^{-1} to 1480 cm^{-1} is for N-H bending and C-H stretching [20]. The weak band between 1300 cm^{-1} to 1220 cm^{-1} is associated with Amide III band which is derived from C-N stretching and N-H bending [21, 22].

4. Conclusion

The proposed procedure for the treatment of feather by using alkali process is cheap and economically viable. Therefore, it could serve as a basis in the development of complex ecologically safe and efficient biotechnology for improved feather wastes utilization applicable in different applications. The aim of the study is the extraction of keratin from chicken feather waste by using reducing agent. Three types of keratin polypeptide precipitations were collected at selected pH value (3.5, 5.5 and 7.5) respectively. The physical and chemical properties of freeze dried FKP1, FKP2 and FKP3 powders were studied through different experiments. SEM image showed spherical and porous microstructures. FTIR spectra showed that the bisulfate bonds were broken and indicted no significant changes in the chemical composition and macromolecular conformation of the hydrolyzed fibers. Though, the feather remainders consisted mostly of β-sheets confirmation. TGA revealed that the polypeptides kept a little lower stable as compare to feather fibers.

In whole, keratin can be extracted by applied method and used in different applications such as coatings, films, packaging, cosmetics and biodegradable composites.

Acknowledgments

The author [SS] greatly acknowledges the financial support from Universiti Malaysia Pahang by providing Doctoral Scholarship Scheme. Authors also want to say thanks to Central Lab and Carif Universiti Malaysia Pahang for their support of there part of the work.

References

[1] P. Hill, H. Brantley, M. Van Dyke: Biomaterials , 31, 585-593(2010).
[2] J.M. Cardamone: J. Mol. Struct., 969(1-3), 97-105 (2010).
[3] A.E. Jiménez-Cervantes, C. Velasco-Santos, A.L. Martínez-Hernández, J.L.Rivera-Armenta, A.M. Mendoza-Martínez, V.M. Casta˜no: J. Compos. Mater., 49, 275-283(2015).
[4] A. Bertsch, N. Coello: Bioresour. Technol., 96, 1703-1708 (2005).
[5] S. Sharma and A. Gupta: Braz. Arch. Biol. Technol., 59, 1-14(2016).
[6] R.D.B. Fraser, T.P. MacRae, G.E. Rogers: Thomus: University of Minnesota, Minneapolis, (1972).
[7] K. Ohgo, C. Zhao, M. Kobayashi and T. Asakura: Polymer. 44(3), 841-846 (2003).
[8] J.A. Matthews, G.E. Wnek, D.G. Simpson and G.L. Bowlin: Biomacromolecules. 3(2), 232-238 (2002).
[9] G.E. Wnek, M.E. Carr, D.G. Simpson and G.L. Bowlin: Nano lett. 3(2), 213-216 (2003).
[10] Z.M. Huang, Y.Z. Zhang, S. Ramakrishma and C.T. Lim: Polymer. 45(15), 5361-5368 (2004).
[11] J.C. Gigliotti, J. Jaczynski, J.C. Tou: Food Chem. 111, 209-214 (2008).
[12] Y.C. Chen, J.C. Tou, J. Jaczynski: J. Food Sci. 74, H31-H39(2009).
[13] A. K. Mohanty, M. Misra and L. T. Drzal: Biopolymers and Biocomposites, CRC Press, (2005).
[14] J. Zhang, Y. Li, J. Li, Z. Zhao, X. Liu, Z. Li, Y. Han, J. Hu and A. Chen: Powder Technol., 246, 356-362 (2013).
[15] Z. Pedram Rad, H. Tavanai, A.R. Moradi: J. Aerosol Sci., 51, 49-56 (2012).
[16] B. Ma, X. Qiao, X. Hou and Y. Yang: Int. J. Biol. Macromolec., 89, 614-621(2016).
[17] H. Xu, S. Cai, L. Xu and Y. Yang: Langmuir., 30, p 8461-8470 (2014).
[18] A. Aluigi, M. Zoccola, C. Vineis, C. Tonin, F. Ferrero and M. Canetti: Int. J. Biol. Macromolec., 41, 266-273 (2007).
[19] D.L. Pavia, G.M. Lampman, G.S. Kriz, and J.A. Vyvyan. Introduction to spectroscopy. Cengage Learning, (2008).
[20] N. Eslahi, F. Dadashian and N. H. Nejad: Prep. Biochem. Biotechnol., 43, 624-648 (2013).
[21] A. Vasconcelos, G. Freddi and A. Cavaco-Paulo: Biomacromolecules, 9 , 1299-1305 (2008).
[22] E. Wojciechowska, A. Wlochowicz: J. Mol. Struct., 511-512, 307-318(1999).

IOP Conf. Series: Materials Science and Engineering **191** (2017) 012014 doi:10.1088/1757-899X/191/1/012014

Green synthesis of magnetite (Fe_3O_4) nanoparticles using *Graptophyllum pictum* leaf aqueous extract

I P Sari[1,a] and Y Yulizar[1,b*]

[1]Department of Chemistry, Faculty of Mathematics and Natural Sciences, Universitas Indonesia, Kampus UI Depok 16424, Indonesia

E-mail: [a]indah.permata51@ui.ac.id, [b*]yokiy@ui.ac.id

Abstract. Magnetite nanoparticles (MNPs) attracted the attention of many researchers due to their unique properties. In this research, nanoscale magnetite particles have been successfully synthesized through an environmentally friendly method using aqueous extract of *Graptophyllum pictum* leaf (GPLE). In MNPs formation, GPLE acted as a base source and capping agent. Alkaloids in GPLE were hydrolyzed in water and hydroxilated Fe^{2+} to form Fe_3O_4 nanoparticles powder through calcination. After the addition of leaf extract, MNPs formation was observed by color change from pale yellow to dark brown. The synthesized nanoparticles were characterized using UV-Vis spectrophotometer, X-Ray diffraction (XRD), and Fourier transform infra red (FTIR) spectroscopy. The results confirmed that MNPs formation indicated the surface plasmon resonance at a maximum wavelength, λ_{max} 291 nm. The average crystallite size is 23.17 nm. The formed MNPs through green synthesis method promise in various medical applications such as drug carrier and targeted therapy.

1. Introduction

Recently, researches on magnetite nanoparticles (MNPs) have been widely developed. MNPs were reported to have some advantages in biocompatibility, biodegradability [1], and easily encapsulated [2]. MNPs are used in various application such as catalyst [3,4], biosensor [5], magnetic resonance imaging (MRI) [6,7] and targeted drug delivery [8,9].

Several conventional methods including co-precipitation and thermal decomposition are widely employed for MNPs synthesis. Co-precipitation method is used due to its simplicity [10] whereas thermal decomposition is a good method in controlling nanoparticles size and morphology [11]. Those methods use chemicals as reagent which are not environmentally friendly. In contrast with the conventional method, green synthesis is more friendly to environment because it utilizes plants extract as the chemicals substitute.

There are several successful studies in synthesizing of MNPs using plant extract, for instance, MNPs synthesis using banana peel ash and *Colocasia esculanta* leaves [12], soya bean sprout [13], and *Datura Inoxia* leaves [14] extract. Unfortunately, some of them still used chemicals such as sodium hydroxide and ammonium hydroxide as a base source. From research, there is limited report of MNPs synthesis using Indonesian medicinal plants.

Graptophyllum pictum is a traditional medicine plants from Papua, Indonesia [15] containing alkaloids [16]. Its secondary metabolite contents act as an alternative base source and capping agent in MNPs synthesis as hazardous chemicals substitute. In previous study, *Graptophyllum pictum* leaf extract (GPLE) was reported contains alkaloids and flavonoids [16]. Alkaloids as a weak base are needed in MNPs preparation, and flavonoids have hydroxil group to prevent MNPs agglomeration

[11]. Both of functional materials are contained in GPLE as base source and capping agents. Therefore, this research is important in MNPs synthesis, which is not reported in previous study yet.

2. Experimental

2.1. Materials

Iron (II) sulphate heptahydrate ($FeSO_4.7H_2O$) was purchased from Merck as precursor of MNPs synthesis. *Graptophyllum pictum* leaves (GPL), were obtained from Cimanggu, Bogor, Indonesia. All the aqueous solutions were prepared by MilliQ water (18.2 Ωcm^{-1}).

2.2. Preparation of GPLE

GPL were cleaned and dried in room temperature for five days. 100 g of leaf powder were soaked in 500 distilled water for 10 days under stirring. The extract was filtered as GPLE using whatman No. 41. The presence of flavonoids, alkaloids, saponin, tannin and terpenoids were tested by general phytochemical method [16].

2.3. MNPs synthesis using GPLE

Synthesis of GPLE-capped MNPs was conducted by mixing 6 mM Fe^{2+} solution with GPLE under stirring at 45°C for three hours until the solution change to dark brown. The mixture was aged for two hours at room temperature, dried in oven at 100°C for one hour and at 70°C for three hours to form MNPs-GPLE powders.

2.4. Characterization

The functional groups presence in active compounds of GPLE and MNPs-GPLE were investigated using FTIR Spectrophotometer (IRPrestige-21 Shimadzu) at range of 400 - 4000 cm^{-1}. The UV-Vis absorption spectrum of MNPs-GPLE was examined using UV-Vis spectrophotometer (Shimadzu 2600) at range 200 - 800 nm. The MNPs band gap was determined using UV-Vis DRS (Shimadzu 2450). The XRD pattern of MNPs-GPLE was observed at 20 mA and 45 kV with Cu Kα radiation (λ = 1.54 Å) at 2θ range of 10 - 80°.

3. Result and discussion

The UV-Vis absorption spectra of Fe^{2+} solution, GPLE and MNPs-GPLE colloids are shown in Figure 1.

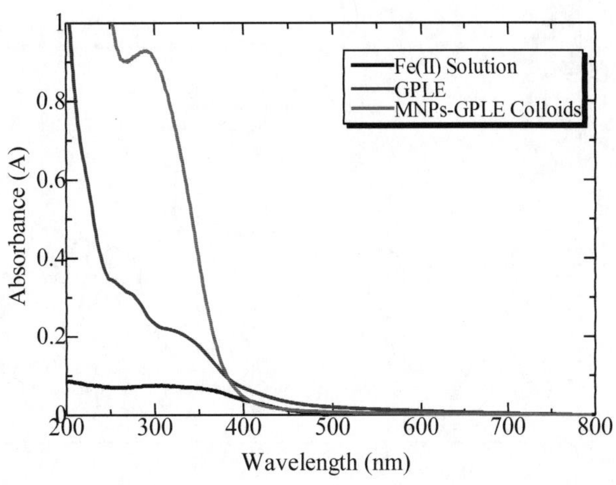

Figure 1. UV-Vis absorption spectra of Fe^{2+} solution, GPLE, and MNPs-GPLE colloids.

The presence of surface plasmon resonance in UV-Vis absorption spectrum at a maximum wavelength, λ_{max} 291 nm indicates the MNPs formation. This λ_{max} is suitable with the previous result of MNPs synthesis using *Datura inoxia* leaf extract [14] and conventional chemical method [17]. Fig. 2 shows linear correlation between F(R) against Eg to determine MNPs-GPLE band gap. Based on graphic equation y = 8.954x - 13.953, the band gap is 1.56 eV.

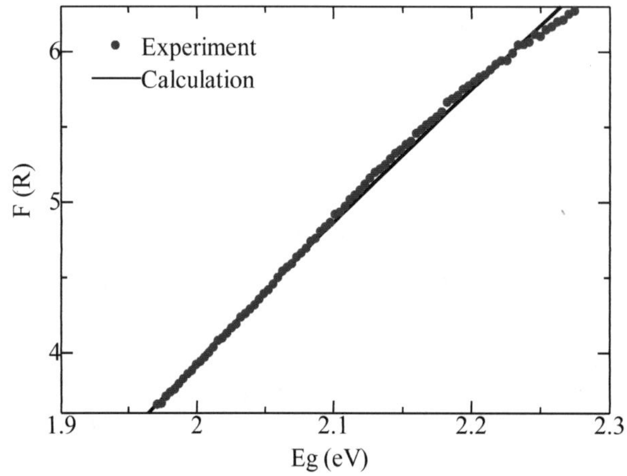

Figure 2. Linear correlation between F(R) against Eg to determine MNPs-GPLE band gap.

FTIR characterization was conducted to study the interaction between functional groups of GPLE and MNPs. Figure 3 showed the wavenumber shifts of GPLE before and after reaction with Fe^{2+}. The band at 3412.0 cm^{-1} of -OH stretch [14] in GPLE shifted to 3051.5 cm^{-1} and occured the decreasing of % transmittance due to MNPs-GPLE formation. The band at 1419.7 cm^{-1} of -NH stretch [14] in GPLE shifted to 1393.6. The band at 1609.7 cm^{-1} of -CO- stretch shifted to 1604.8 cm^{-1} due to interaction Fe in MNPs with GPLE [18]. These results are supported from phytochemical test of flavonoids and alkaloids. The presence of Fe-O was confirmed at 620.0 cm^{-1} [14,17] which was not formed before.

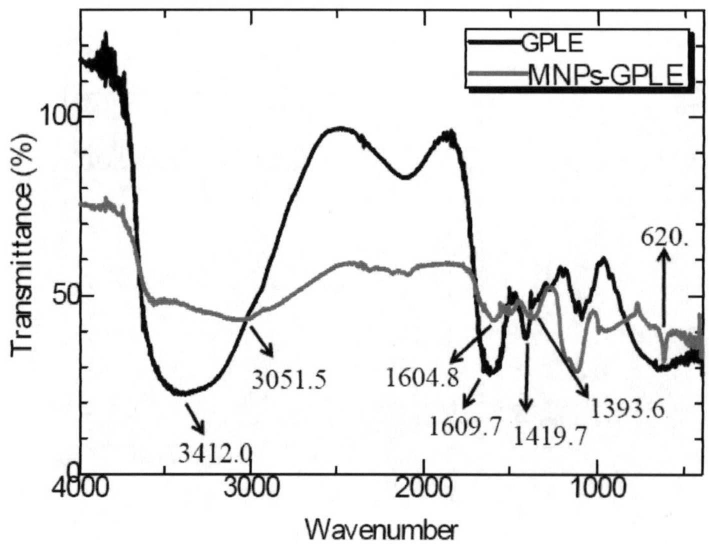

Figure 3. FTIR spectra of GPLE and MNPs-GPLE.

Figure 4 shows XRD pattern of synthesized MNPs-GPLE at 2θ = 17.99°; 30.11°; 35.35°; 43.48°; 53.55°; 55.78°; and 62.18°. These values are adjusted with American Mineralogist Crystal Structure Database (AMCSD) entry number 96-900-5841 (magnetite), 96-901-2693 (maghemite) and 96-101-1170 (wuestite) obtained crystal planes (111), (202), (311), (400), (422), (333), and (404) of magnetite with face-centered cubic (FCC) phase. The crystallite size of MNPs was calculated as 23.17 nm through Debye-Scherrer equation.

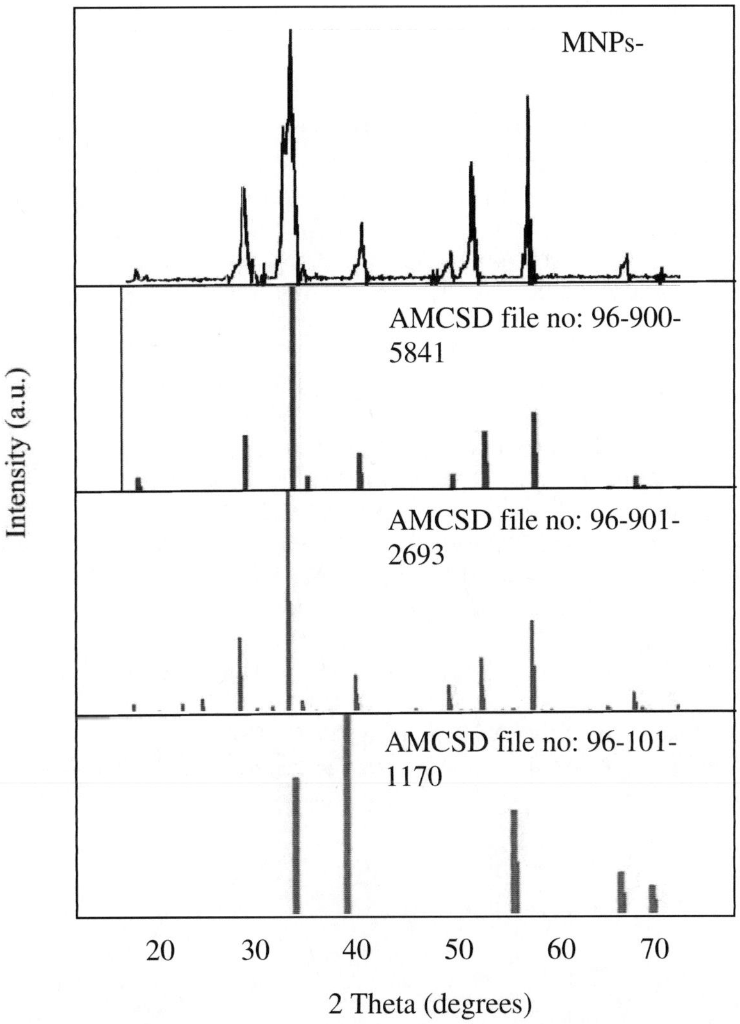

Figure 4. XRD Patterns of MNPS-GPLE, Magnetite, Maghemite, and Wuestite.

The crystalline MNPs were synthesized for the first time using GPLE involved the interactions of -OH and -NH groups in GPLE with MNPs. This result is potential as functional nanoparticles based on environmentally friendly method.

4. Conclusions

Magnetite nanoparticles, Fe_3O_4 (MNPs) were successfully synthesized using *Graptophyllum pictum* leaf extract (GPLE) by a simple and green method. GPLE contains alkaloids as a base source meanwhile flavonoids act as capping agent to prevent the MNPs agglomeration. The XRD analysis confirmed the MNPs crystal size of 23.17 nm. UV-Vis spectroscopy investigated MNPs formation at a maximum wavelength, λ_{max} 291 nm. FTIR spectroscopy showed the presence of biomolecules functional group of GPLE through interaction between -OH and -NH with MNPs. The green synthesized MNPs contribute in medical fields such as drug carrier and targeted therapy. This green synthesis method is environmentally friendly, cheap and potential to produce other metal oxides.

Acknowledgements
This work was funded by Hibah PITTA from Universitas Indonesia through The Directorate of Research and Community Services, Universitas Indonesia (No. 2040/UN2.R12/HKP.05.00/2016) and Lembaga Pengelola Dana Pendidikan (LPDP), Ministry of Finance (No. PRJ-2509/LPDP/2015), Indonesia.

References
[1] Richard A R and Miqin Z 2016 *Mater. Today* 19 157
[2] Evi Y, Sudaryanto, Yoki Y, and Mujamilah 2008 *Indonesian Journal of Materials Science* 228
[3] Manoj G B, Paula S B, and Rajender S V 2013 *Chem. Soc. Rev.* 42 7664
[4] Shelke S N, Bankar S R, Mhaske G S 2014 *ACS Sustainable Chem. Eng.* 2 1699
[5] Kavitha A L, Prabu H G, Babu S A, and Suja S K 2013 *J. Nanosci. Nanotechnol.* 13 98
[6] Haw C Y, Mohamed F, Chia C H, Radiman S, Zakaria S, Huang N M, Lim H N 2010 *Ceram. Int.* 36 1417
[7] Ruirui Q, Chunui Y, and Mingyuan G 2009 *J. Mater. Chem.* 19 6274
[8] Melessa S, Ying X, Alison A, Sohrab R, and Elizabeth R G 2015 *RSC Adv.* 5 37521
[9] Li S, Li H, Liu G, et al. 2012 *Biomaterials* 33 3013
[10] Tang D P, Yuan R, and Chai YQ 2006 *Biotechnol. Lett.* 28 559
[11] Zboril R, Mashlan M, and Petridis D 2002 *Chem. Mater.* 14 969
[12] Thakur S and Karak N 2014 *Mater. Chem. Phys.* 144 425
[13] Cai Y, Shen Y, Xie A Li S, and Wang X 2010 *J. Magn. Magn. Mater.* 322 2938
[14] Das A K, Marwal A, and Verma R 2014 *Nano Hybrids* 7 69
[15] Heyne 1987 *Tumbuhan Berguna Indonesia* (Indonesia: Badan Penelitian dan Pengembangan Departemen Kehutanan)
[16] Olagbende-Dada S O, Ukpo G E, Coker H A B, and Adesina S A 2009 *Afr. J. of Biotechnol.* 8 5979
[17] El-Sigeny S M, Manal, and Taleb A 2015 *Life Sci. J.* 12 161
[18] Yoki Y, Grandprix T M K, Adhitya Y 2016 *AIP Conf. Proc.* 1729 020043-1

IOP Conf. Series: Materials Science and Engineering **191** (2017) 012015 doi:10.1088/1757-899X/191/1/012015

High stability of few layer graphene nanoplatelets in various solvents

X Xu[1,*], J Zhou[1] and G lubineau[1,*]

[1]King Abdullah University of Science and Technology (KAUST), Physical Science and Engineering Division, COHMAS Laboratory, Thuwal 23955-6900, Saudi Arabia

[*]xuezhu.xu@kaust.edu.sa; gilles.lubineau@kaust.edu.sa

Abstract. Dispersion of few-layer graphene nanoplatelets (GNPs) in liquid media is a crucial step for various applications. Here, we highlight a simple, nondestructive method for preparing stable aqueous colloidal solutions with GNP powder quickly dispersed in 5 wt.% sodium–hypochlorite- (NaClO) and sodium-bromide- (NaBr) salted solvent by bath sonication. This method makes it possible to easily prepare a highly concentrated colloidal solution (1 mg·ml^{-1}) of GNPs that can easily be re-dispersed in water (treated GNPs). The aqueous suspension we prepared remained stable for longer than a few weeks. We made similar tests with various solvents and dispersibility appeared to decrease with decreasing polarity. High-concentration suspensions using our facile dispersion method could be of particular interest to the large community using graphene for a diversity of applications.

1. Introduction

The properties of dispersions are highly dependent upon the nature of the material and the dispersing media interface. All graphitic carbon nanostructures without chemical modification are highly hydrophobic. This hydrophobicity prevents them from forming stable dispersions in aqueous media. Efforts aiming to resolve this limitation are increasing because of the growing interest in applications that require water-based dispersion to promote ecologically friendly processes. Two major methods have been developed to disperse few-layer graphene powder in aqueous media: 1) chemical modification by introducing water-solubilizing functional groups and 2) surfactant wrapping of the graphene sheets to create hydrophilic surfaces. Chemical modification, which includes oxidation or covalent functionalization by introducing sp^3-hybridized sites into the graphitic structure, leads to major alterations in the electronic properties of the structure.[1] Surfactants facilitate the quick dispersion of graphene nanoplatelets (GNPs) in water[2] without destroying the graphitic structure; however, an excessively thick layer of surfactant accumulates on the surface of the structure. Thus, the use of surfactants or other methods is not preferred when the purity of GNPs is critical.

2. Materials & methods

Few-layer graphene nanoplatelets (GNPs, N002-PDR), purchased from Angstronmaterials Materials Inc., were used as received. Average lateral dimensions of GNPs were 10 um and had 95 % carbon content. 5 wt.% Aaqueous sodium hypochlorite at 5 wt.% (NaClO) and sodium bromide (NaBr) powders were purchased from RICCA Chemical Company. Methanol, NMP, THF and polymer monomer (1,6-Hexanediol ethoxylate diacrylate) were purchased from Sigma-Aldrich Company. Water was purified by distillation in a Milli-Q (Advantage A10 model) system. UV-vis measurements of colloidal solutions were recorded from 800 nm – 190 nm at 1-nm intervals using a Cary100 UV-Visible spectrophotometer.

3. Results and discussion

Here, we demonstrate a new method to prepare stable GNP colloidal solutions in aqueous media that does not destroy the sp^2-hybridization of graphene and does not involve the formation of a thick layer of surfactant on the surface of the structure. As illustrated in Fig. 1, the method involves bath sonication of the GNP powder in a sodium-hypochlorite- (NaClO) and sodium–bromide- (NaBr) salted aqueous solution (NaClO:NaBr 1:1 molar ratio) for 60 min (Step 1). The dispersed GNP product is subsequently washed with water via centrifugation and dialysis for 15 d (Step 2). Once the washed GNPs (treated) are freeze-dried (Step 3), they can easily be re-dispersed in various solvents by mild bath sonication to form a stable GNP colloidal solution (Step 4).

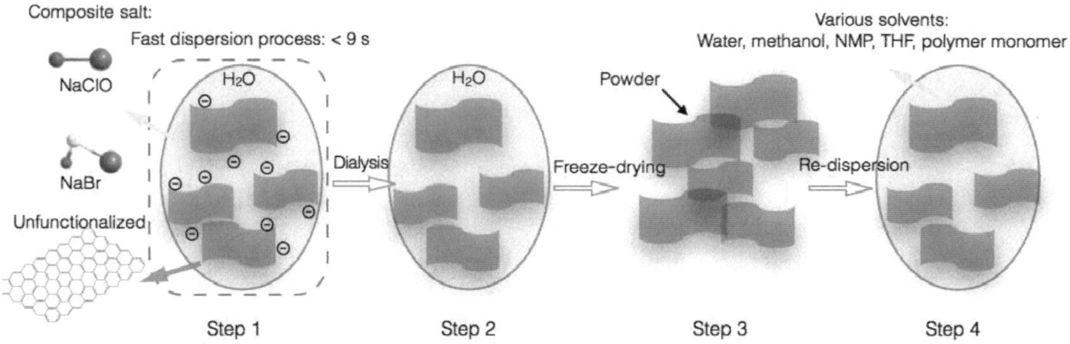

Figure 1. Pristine GNPs are dispersed in aqueous NaClO/NaBr solvent (step 1), purified by dialysis (step 2), freeze-dried (step 3), and re-dispersed in various solvents (step 4).

Figure 2a shows our re-dispersion of treated GNPs in pure water (step 4), which remained as a stable, homogenous colloidal solution (1 mg·ml^{-1}) for several weeks. In comparison, Figure 2b shows a pristine GNP (untreated) aqueous colloidal solution of the same concentration, which was not homogenous and features multiple GNP aggregates on the surface of the bottle and agglomerated GNPs. These photos were taken several weeks after preparation of the solutions. Dispersion of GNPs in water is incompatible due to strong van del walls forces between particles. Although treatment of GNPs left some residual salts, reducing the purity of the solution to some degree, the adsorption of anions (Br^-, ClO^-) by GNPs likely generated electrostatic repulsion between graphitic sheets that eventually stabilized the final aqueous colloidal solution.[5]

During the pre-treatment step (step 1), pristine GNPs have yet to be dispersed correctly in the salted solvent and thus can be efficiently treated. We suspected that the effective dispersion of pristine GNPs in the salted solution could be explained by a close match in surface energies between the GNPs and the NaClO/NaBr-salted solvent. Knowing that graphite has a surface energy of ~70-80 mJ·m^{-2}, we measured the surface energy of our NaClO/NaBr-salted solution at 67.85 mJ·m^{-2}, verifying our prediction. Similarly, N-Methyl-2-pyrrolidone (NMP) was found to be an effective solvent for easily dispersing GNP due to the close match in their surface energies,[3,4] (NMP has a surface energy of 69.50 mJ·m^{-2}).[4]

IOP Conf. Series: Materials Science and Engineering **191** (2017) 012015 doi:10.1088/1757-899X/191/1/012015

Figure 2. (a) 1-mg·ml^{-1} treated GNP/water colloidal solution, (b) 1-mg·ml^{-1} pristine GNP/water colloidal solution. Photos were taken one month after preparation of the solutions.

The treated GNP colloidal solution was freeze-dried to obtain GNP powder and then ~0.8 mg was re-dispersed in ~4.5 ml of water, methanol, NMP, Tetrahydrofuran (THF), or polymer monomer (oil) (0.18 mg·ml^{-1} concentration of GNP powder) solvents to test their dispersibility (Figure 3a). Solvents had the following polarities: water (1.000) > methanol (0.762) > NMP (0.360) > (THF) (0.207) > polymer monomer (0.001). We found that the treated, freeze-dried GNP powder was extremely compatible with water, spontaneously dispersing within about 193 s. After two rounds of 5 s and one round of 120 s sonications, the powder had totally dispersed in the water with no phase separation overnight. Dispersibility appeared to decrease with decreasing polarity, as evidenced by the large aggregates visible with the naked eye in bottles containing THF and polymer monomer solvents (bottles 4 and 5, respectively). The Beer-Lambert law was used to calculate the molar absorption coefficients, α, of each colloidal solution to evaluate their dispersibility:

$$A = \alpha CD \tag{1}$$

where A is the absorbance, C is the concentration of the solute (g·L^{-1}), and D is a 1-cm pathlength. The absorption coefficient α (L·g^{-1}·m^{-1}) can be calculated based on the achieved experimental parameters A, C, and D. Figure 3b shows the absorbance of each colloidal solution as calculated by ultraviolet–visible (UV-vis) spectroscopy at 660 nm for 40 h (Figure 3c). Treated GNP dispersed best in NMP and water, and worst in THF and oil. Figure 3d shows the absorption coefficients of treated GNP powders in each solvent. The absorption coefficient of NMP was highest at 1649 L·g^{-1}·m^{-1}, followed by water and methanol between 1200 – 1100 L·g^{-1}·m^{-1}; the oil and THF had the lowest value at ~ 400 L·g^{-1}·m^{-1}. Thus dispersibility of GNP follows NMP > water > methanol > monomer > THF. Over the 40 h stabilization time, the relationship between the absorbance and the calculated concentration of NMP (mg· ml^{-1}) was linear (Figure 3e).

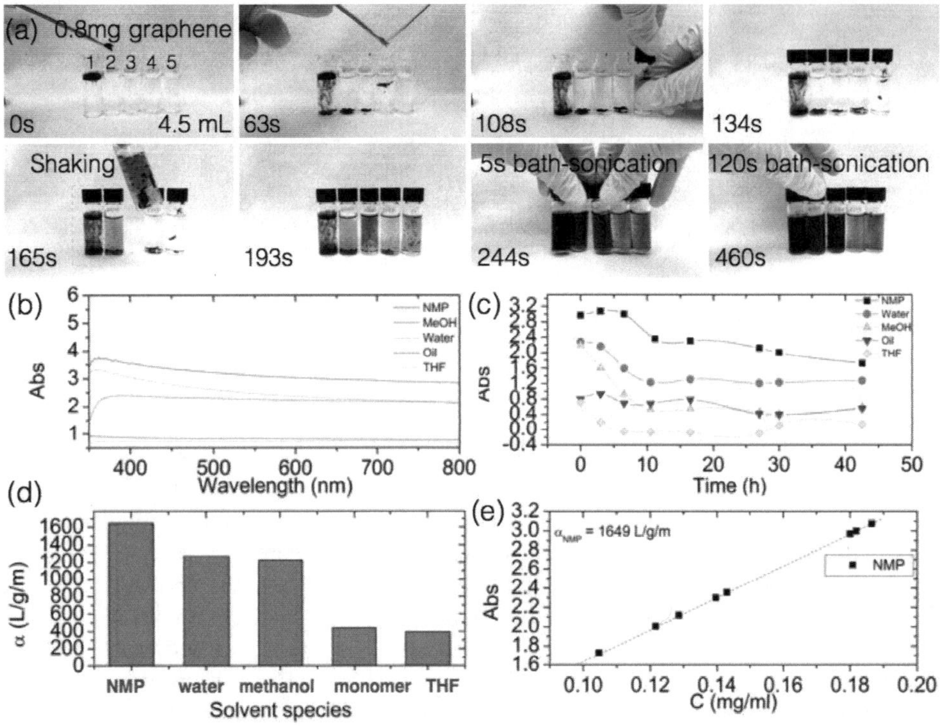

Figure 3. (a) Photos of dispersing treated, freeze-dried GNP in different liquid media. The numbering from 1-5 indicates that the liquid inside the glass bottles is pure water, NMP, methanol, monomer and THF, respectively. (b) UV-vis spectra of as-prepared GNP colloidal solutions as measured instantly. (c) Absorbance changes in UV-vis spectra at 660 nm for treated GNP in NMP, water, methanol, oil and THF suspensions. (d) Calculated absorption coefficients (α) for each solvent. (e) Linear relationship between absorbance and concentration of treated GNP in NMP stabilized for 40 h.

4. Summary

Here, we presented a simple, scalable method of producing a water-soluble few-layer graphene powder by first dispersing the GNPs in a NaClO/NaBr-salted solution. This approach is likely to be of interest to the large body of research/industrial communities.

Acknowledgment

This research was supported by King Abdullah University of Science and Technology.

References

[1] Li D, Müller M B, Gilje S, Kaner R B, Wallace G G 2008 *Nat. Nanotechnol.* 3 101–5
[2] Englert J M, Röhrl J, Schmidt C D, Graupner R, Hundhausen M, Hauke F, Hirsch A 2009 *Adv. Mater.* 21 4265–9
[3] Paton K R, Varrla E, Backes C, Smith R J, Khan U, O'Neill A, Boland C, Lotya M, Istrate O M, King P, Higgins T, Barwich S, May P, Puczkarski P, Ahmed I, Moebius M, Pettersson H, Long E, Coelho J, O'Brien S E, McGuire E K, Sanchez B M, Duesberg G S, McEvoy N, Pennycook T J, Downing C, Crossley A, Nicolosi V, Coleman J N 2009 *Nat. Mater.* 13 624–30
[4] Hernandez Y, Nicolosi V, Lotya M, Blighe F M, Sun Z, De S, McGovern I T, Holland B, Byrne M, Gun'Ko Y K, Boland J J, Niraj P, Duesberg G, Krishnamurthy S, Goodhue R, Hutchison J, Scardaci V, Ferrari A C, Coleman J N 2008 *Nat. Nanotechnol.* 3 563–8
[5] Dimiev A M, Gizzatov A, Wilson L J, Tour J M 2013 *Chem. Commun.* 49 2613–15

IOP Conf. Series: Materials Science and Engineering **191** (2017) 012016 doi:10.1088/1757-899X/191/1/012016

Investigation of plasmonic effects on the metal nanoparticle arrays for biosensor applications

Yuan-Fong Chou Chau[1]

[1]Centre for Advanced Material and Energy Sciences, Universiti Brunei Darussalam, Tungku Link, Gadong BE1410, Negara Brunei Darussalam

E-mail: Chou.fong@ubd.edu.bn

Abstract. Surface plasmon resonance (SPR) on the near field intensity of a two-dimensional (2-D) periodic arrays of bowtie-shape metal nanoparticle (MNP) pairs using the 3-D finite element method (FEM) for biosensor applications are investigated. The peak resonance wavelength can be shifted to red as the filling core media in bowtie-shape MNP pairs have been changed. The tunable and highly sensitive SPR modes are observed due to the hybridization SPR effects. Simulation results show that the peak resonance wavelength (λ_{res}) can be spread in the full wavelength range of 300-1600 nm when the tip edges and different filling media in structure are taken into account. The highly sensitive optical performances give us a qualitative idea of the geometrical and material properties of the periodic array of bowtie-shape MNP pairs on SPRs that can be as a promising candidate for plasmonic biosensor applications.

1. Introduction

The surface plasmon resonances (SPRs) arising from noble metal nanoparticles (MNPs) (e.g., gold or silver) are highly attractive due to their diverse application in nanophotonic field [1-7]. These MNPs mainly can be applied to sense the biological nanoparticles [1]. Tunable peak resonance wavelength (λ_{res}) can be manipulated by dimensions, aspects and the surrounding media of MNPs [2]. Recently, several numerical and experimental works on single or a pair of MNPs has been investigated and the tunable plasmonic effects can be found in the scattering and absorption spectra [3-5]. Dramatic SPR modes are excited at the extremity of the finite chain of MNPs [6]. These results correspond to the surface enhanced Raman scattering by molecules adsorbed on MNPs, which can be applied in plasmonic biosensors [7].

The periodic array of bowtie-shape MNP pairs can be served as plasmonic orientation sensors due to its strong gap enhancement capabilities. Several studies regarding on single or dimer rod-shape MNPs have already demonstrated the rich variety of optical properties [3-5], but rarely deal with the periodic array of bowtie-shape MNP pairs. The optical properties of the periodic array of bowtie-shape MNP pairs with different filling media and structural aspects are still not very clearly understood. This periodic array of bowtie-shape MNP pairs could open up the more challenge to explore the plasmonic effect of the dipolar and cavity plasmon on the near field intensity spectra. This includes a closer study of the influences of the distribution of electric field (E-field) intensities and charge density distributions on the periodic array of bowtie-shape MNP pairs. In this paper, a design strategy to obtain tunable mode field enhancement in the nanoscale focal spots of a 2-D periodic array of bowtie-shape MNP pairs is investigated. The unit cell of the proposed structure is composed of a pair of identical bowtie-shape silver (Ag) with filling dielectric constants (ε) in the structural arms. The influence of its structural and material parameters on the MNP resonance conditions will be examined and compared to their solid case of counterpart. Numerical simulations are performed by using the 3-D finite element method (FEM) of Maxwell's equations. It would be desirable to tune the λ_{res} of these SPR modes to allow the near field intensity spectra to be broadened without changing the outer dimension of MNPs, but it allows for the change of filling media and tip edges in the structures.

2. Simulation method

In the numerical simulations, Ag MNPs described by permittivity data in Ref. [8] was considered. The simulation models are shown in Figure 1(a)-(c), i.e., (a) truncate view of a 2-D periodic array of bowtie-shape MNP pairs, (b) a unit cell of a solid bowtie-shape MNP pair and (c) a unit cell of a dielectric-hole (DH) bowtie-shape MNP pair. The parameters used in the simulations are listed in the inset of Figure 1. Referring to Figure 1, x, P and g are the length of a bowtie-shape MNP pair, period length (P_x along x-axis and P_y along y-axis) and gap distance between two adjacent bowtie-shape MNPs, respectively. For duplicating the periodic array structures, the 2-D arrays are obtained from the unit cell by using periodic boundary conditions (PBC) along the x- and y-axes and the scattering boundary conditions along the electromagnetic (EM) wave propagation direction [7], i.e. the z-axis. The incident EM waves were performed by fixing at $|E_0|$=1V/m and the simulation models are illuminated by normal incidence EM wave with x-polarization (or transverse polarization).

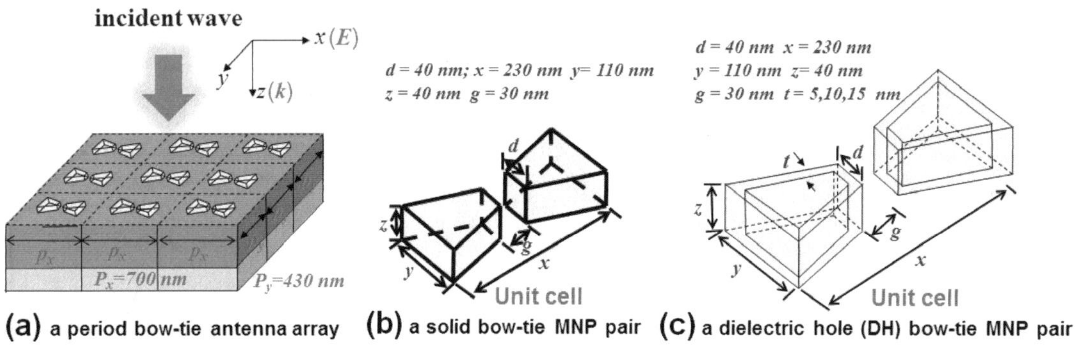

Figure 1. Schematic plot of simulation models.

3. Results and discussion

Figure 2 shows the near field intensity spectra (Figure 2(a)-(c)), E-field), electric field (E-field) intensity distribution (Figure 2 (d)) and surface charge densities (Figure 2 (e)-(f)) of a solid bowtie-shape MNP case and DH bowtie-shape MNP cases with various core media (ε) and shell thickness (t). The λ_{res} of the transverse modes are changed upon varying the core dielectric constant ($1<\varepsilon<5$, in step of 1). The simulated near field intensity spectra obtained at various ε show a red shift of λ_{res} position with the increasing of ε [7], which can be used for the biosensor applications. The sensitivity is related to the spectral position of the MNP resonance and the hybridization of the SPR and core medium effects in cavities [9, 10]. The E-field in MNP gap reaches a peak value at which the Fabry-Perot resonance condition is satisfied [11]. By tailoring the E-field intensity in the MNP gap and core regions, the sensitive operations can be produced from the MNP gaps, making the DH suitable for probing a small number of atoms or molecules localized in the gap and core regions [7, 12, 13]. This finding will be beneficial to the application for single molecule analysis [14].

Next, we use the shell thickness ($5<t<15$ nm), as a tuning parameter and an evident difference on optical properties is observed among the two cases. It can be observed that the resonance peaks in t=5 nm case occur in the wavelength range of 350~550 nm and 675~860 nm at different filling ε in DHs, and the λ_{res} redshift and decrease as the filling ε in DHs is increased. For the case of t=10 nm (Figure 2(b)), the resonance peaks are shrunk in the wavelength range of 375~450 nm and 550~690 nm at different filling ε in DHs, and the λ_{res} redshift and decrease as the filling ε in DHs is increased. It is worthy to note that there are two resonances in bowtie-shape MNP DH cases, i.e., the first one λ_{res} is due to high frequency (short wavelength) oscillation between MNP and incident light [7] and the second one results from the interaction of the two identical MNPs (corresponding to structure resonance) and core medium effect (corresponding to cavity resonances) [15]. In addition, it can be found that when MNP shell thickness is thicker (t=15 nm, Figure 2(c)), the λ_{res} will be shrunk to a shorter wavelength range and their performance is similar to the solid bowtie-shape MNP case, which is due to the shell thickness approaching the bulk Ag mean free path.

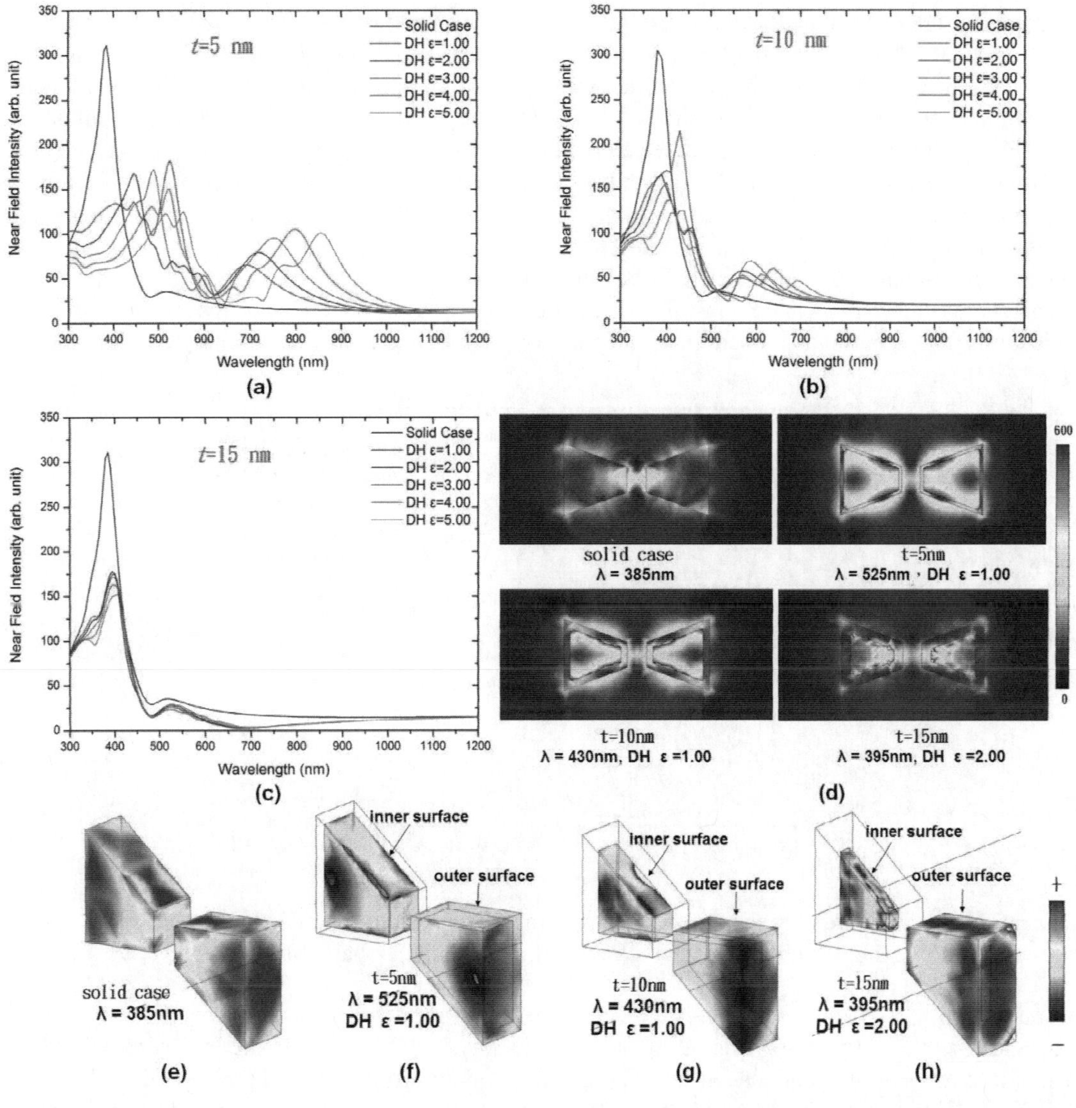

Figure 2. (a)-(c) Near field intensity spectra, (d) E-field intensity distribution and (e)-(f) surface charge densities of a solid bowtie-shape MNP case and the bowtie-shape DH MNP cases with various filling media (ε) and shell thickness (t).

IOP Conf. Series: Materials Science and Engineering **191** (2017) 012016 doi:10.1088/1757-899X/191/1/012016

Likewise, regarding the E-field intensity distribution (Figure 2 (d)), the corresponding charge distributions in two cases are depicted in Figure 2(e)-(h), respectively. The charge distributions on the edge surface of MNPs obtained from the DH cases are larger than those of the solid one. This is because of the hybridization of the void plasmon and liner localized SPR effects [16], namely, both the SPR and cavity modes can be found in the DH cases. It is evident that the local field enhancement can be found on the inner and outer surface of the bowtie-shape MNP DH cases (Figure 2 (d)).

Finally, the E-field enhancement arising from the tip edge will be included in the simulations and the width of tip edge length (w) of 10 nm is considered. The simulation model is depicted in the inset of Figure 3. Figure 3 suggests that we have found that the combination of tip edges and the change of filling ε can significantly tune the λ_{res} in the considered bowtie-shape MNP DH cases while the outer dimension of MNP structure is kept constant. The near field intensity distributions on the inner and outer surface of MNPs show a field enhancement on the surface of tip edges (see the inset of Figure 3). It is evident from Figure 3 that the λ_{res} can be extended in a full wavelength range of 300-1600 nm and this phenomenon cannot be found in the solid case and the bowtie-shape MNP DH cases without tip edges (Figure 2(a)-(c)). It can be attributed to the combination of coupling effects among the incident EM waves, gap regions, core media (ε) and MNP tip edges connecting them is preferable for the positive-negative charge pair excitation on the surface of MNPs [17]. Besides, the coupling effect between two tip apices of MNPs results in edge enhancement in the gap region [18, 19]. Such gap enhancement and environmental sensitivity of the plasmons holds great potential for monitoring local environmental changes during chemical or biological processes. These results may have significant implications for the optimal application of the plasmonic MNP arrays based biosensor configurations.

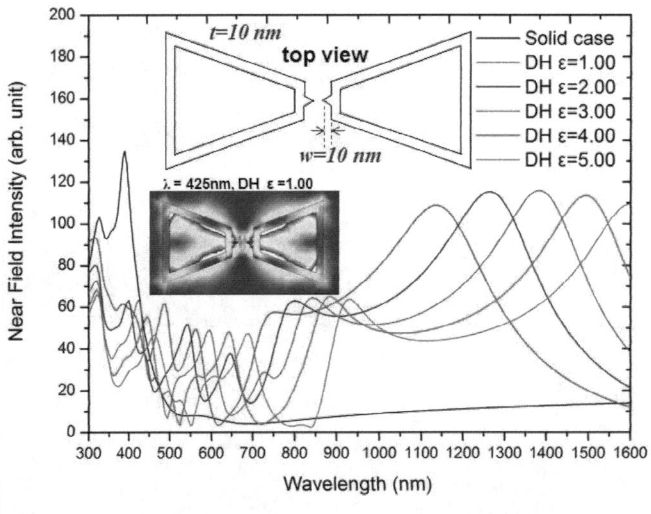

Figure 3. Near field intensity spectra arising from the combination of tip edge and filling core media in the solid bow-tie MNP case and the DH bowtie-shape MNP cases.

4. Conclusion

In conclusion, the purpose of this study is to understand and develop more efficient MNPs for plasmonic biosensor applications. In this paper, the periodic array of bowtie-shape DH MNP pairs which unit cell is composed of two identical nanometals with the filling media and tip edges by using the 3D FEM have been investigated. The tunable and sensitive SPR modes are observed. Simulation results show that the resonant wavelength of the proposed structure can be dramatically tuned over a broad spectral by considering the filling media and introducing the tip edges in the structures simultaneously. This phenomenon cannot be found in the cases without the core-shell and tip edges in structures. The proposed structure promises applications in single molecule surface enhanced Raman spectroscopy (SERS) and plasmonic biosensor owing to its high local field enhancements and large scale manufacturability. In the near future, further experimental works related to the proposed structure will be performed to demonstrate the results in this work more convincing.

Acknowledgments
This work was supported by the University Research Grant of Universiti Brunei Darussalam (Grant No. UBD-ORI-URC-RG331-U01 and UBD/OAVCRI/CRGWG(004)/170101).

References

[1] Chou Chau Y F, Lim C M, Chiang C Y, Voo N Y, Idris N S M, Chai S U 2016 Journal of Nanoparticle Research 18 1-13.
[2] Chau Y F and C Y Jheng 2014 Plasmonics 9 1-9.
[3] Sung M J, Ma Y F, Chau Y F and Huang D W 2010 Applied Optics 49 6295-6301.
[4] Chau Y F, HH Yeh and D P Tsai 2008 Applied Optics 47 5557-5561.
[5] Hsieh L-Z, Chou ChauY-F, Lim C M, Lin M-H, Huang H J, Lin C-T, Idris N S M 2016 Opt. Communications 370 85-90.
[6] Chau Y-F, Yeh H-H, Liao C-C, Ho H-F, Liu C-Y and Tsai D P 2010 Applied Optics 49 1163-1169.
[7] N T R N Kumara, Chou Chau Y-F, Huang J-W, Huang H J, Lin C-T and Chiang H-P 2016 J. Opt. 18 115003.
[8] Johnson P B and R W Christy 1972 Phys. Rev. B 6 4370-437.
[9] Chou Chau Y-F 2009 Plasmonics 4 253-259.
[10] Chau Y-F, Jiang Z-H 2011 Plasmonics 6 581-589.
[11] Chou Chau Y-F, Jiang J-C, Chou Chao C-T, Chiang H-P and Lim C M 2016 J. Phys. D: Appl. Phys. 49 475102.
[12] Chou Chau Y-F, Syu J-Y, Chou Chao C-T, Chiang H-P and Lim C M 2017 J. Phys. D: Appl. Phys. 50 045105.
[13] Chau Y-F, Chen M W and Tsai D P 2009 Applied Optics 48 617-622.
[14] Ashok A D, Samrat M and Edward A L 2008 J. R. Soc. Interface 5 15-45.
[15] Chau Y-F, Yeh H-H, Liao C-C, Ho H-F, Liu C-Y and Tsai D P 2010 Applied Optics 49 1163-1169.
[16] Chou Chau Y-F, Lim C M, Lee C, Huang H J, Lin C-T, Kumara N T R N., Voo N Yg and Chiang H-P 2016 J. Appl. Phys. 120, 093110.
[17] Chau Y-F, Lin Y-J and Tsai D P 2010 Opt. Express 18 3511-3518.
[18] Chau Y-F, Yang T-J and Tsai D P 2004 J. Appl. Phys. 95 3378.
[19] Chau Y-F, Yang T-J and Tsai D P 2004 Jpn. J. Appl. Phys. 43 8115-8125.

IOP Conf. Series: Materials Science and Engineering **191** (2017) 012017 doi:10.1088/1757-899X/191/1/012017

Preparation of arc broadband piezoelectric composite vibrator and its transducer array

Miaojie Lv, Likun Wang, Chao Zhong and Lei Qin

Beijing Key Laboratory for Sensor, Beijing Information Science & Technology University, Beijing 100101, China

E-mail: lvmiaojie1991@163.com, wlikun1962@163.com, zclovelxm@163.com, qinlei@bistu.edu.cn

Abstract. This paper presents an arc broadband piezoelectric composite vibrator and its transducer array. The arc broadband piezoelectric composite vibrator was fabricated by 1-3 piezoelectric composite with matching layer and was prepared by the curved forming process. The vibration characteristics of a new vibrator were simulated via finite element analysis (FEA) and investigated by experiment. First, the acoustic impedance and thickness of the optimized matching layer were determined by theoretical analysis. Second, the arc piezoelectric vibrator was modeled and simulated in air conditions by the FEA method, and the variation law of resonant frequency was analyzed. Finally, the fabrication process of the arc broadband piezoelectric composite vibrator and its transducer array was studied, and the experimental sample was fabricated and measured. The bandwidth of the arc broadband transducer array was found to be up to 56.5 kHz. Results show that the experimental results were in accordance with the simulation results and have good directivity to realize the purpose of bandwidth and beam expansion.

1. Introduction

The multi-beam image sonar can detect, image, and recognize underwater targets and can be applied in underwater engineering, shallow landform mapping, and frogman operations [1, 2]. In practical applications, to obtain a large scope of observation, the transmitting transducer needs to have a wide beam and a high sound-source level and is generally used as an arc transducer. In addition, owing to the wide application of the transducer array, its signal processing system requires the transducer to have not only consistency but also a wideband characteristic. Thus, research on broadband technology has become the frontier field of underwater transducers, which can be summed up into two methods to expand the bandwidth. The first method reduces the mechanical quality factor of sensitive material. The most commonly used method adds flexible polymer into conventional piezoelectric ceramic, thereby increasing material consumption and reducing the Q value, such as the piezoelectric composite [3, 4]. The second method expands the bandwidth using modal coupling [5, 6]. A specific implementation method adds the matching layer to the surface of sensitive material. In response to these urgent needs, vigorous studies have been used to broaden the bandwidth of transducers. Some researchers have achieved remarkable results.

Liu Zhenjun et al. [7] proposed high-frequency underwater transducer curved arrays to improve the transmission voltage response and eliminate the coupling of the vibration mode of the length direction. The transmitting transducer adopted 1-3 piezoelectric composite. Receiver curved array consists of

IOP Conf. Series: Materials Science and Engineering **191** (2017) 012017 doi:10.1088/1757-899X/191/1/012017

181 elements. The transmitting voltage response reaches 161.4 dB, which is –3 dB bandwidth greater than 60 kHz. The sound-source level is up to 206 dB, and the receiver array sensitivity is approximately –197 dB. Benjamin [8] used 1-3 piezoelectric composite to fabricate doubly curved 2D conformal array and placed it in an acoustic test tank. The receiving voltage response is –174 dB for a selected element within the larger side-facing array.

Arc transducer array on the curved forming process requirements is relatively high and requires research focus. To expand the bandwidth and beam angle of the transducer, the multi-element arc piezoelectric vibrator base on 1-3 piezoelectric composite with matching layer is presented in this paper.

2. Design of multi-element arc composite piezoelectric vibrator

2.1. Structure of multi-element arc composite piezoelectric vibrator

According to the aforementioned method, the structure model of this design is proposed, as shown in figure 1. The 1-3 piezoelectric composite is selected as the sensitive material. Then, the matching layer is added to the surface of composites to fabricate the piezoelectric vibrator, and the 5-element arc piezoelectric composite array was prepared by the curved forming process to achieve the wide-band and wide-beam effect of the transducer. In figure 1, the A material is piezoelectric ceramic PZT-4; the B material is epoxy resin; the C material is the matching layer, which is formed by aluminum powder mixed into epoxy resin; the D material is silicone rubber, which is the decoupling material between the elements; *h* is the thickness of the piezoelectric composite; *t* is the thickness of the matching layer; and *w* is the thickness of the silicone rubber between the elements.

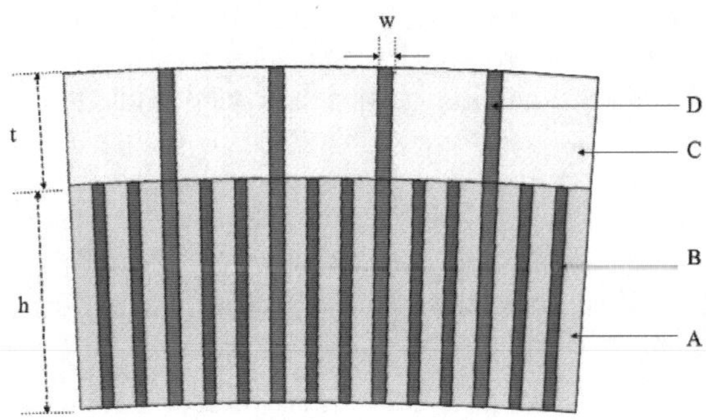

Figure 1. Structure diagram of arc vibrator base on piezoelectric composites

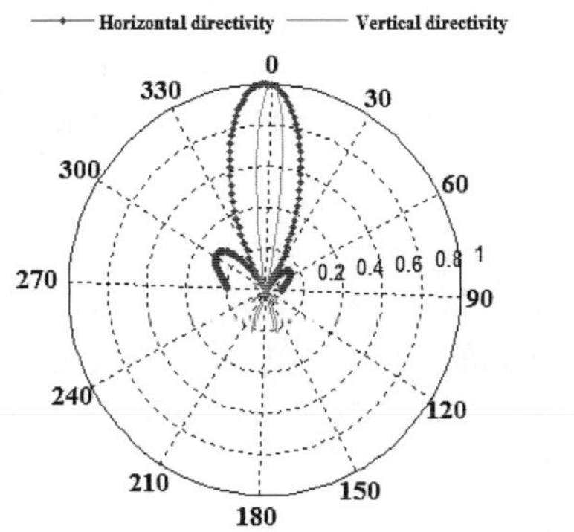

Figure 2. Directivity calculation results of 5-element arc piezoelectric vibrator

2.2. Directivity properties of multi-element arc composite piezoelectric vibrator

The directivity of the piezoelectric vibrator is one of the most important indicators in the design of transducer, and its directivity determines the beam angle of the transducer. The far-field directivity is classified into horizontal and vertical. The vertical directivity [9] of the multi-element arc piezoelectric vibrator array can be calculated according to the linear array directivity in equation (1), and the horizontal directivity [9] is calculated according to the circular array directivity in equation (2).

$$D(\theta) = \frac{\sin(\frac{\pi l}{\lambda}\sin\theta)}{\frac{\pi l}{\lambda}\sin\theta} \tag{1}$$

In equation (1), l represents the vertical height of the radiating surface and λ represents the wavelength of the transducer in water.

$$D(\theta)=\frac{1}{2m+1}\sqrt{\{\sum_{k=-m}^{m}\cos[\frac{2\pi r}{\lambda}\cos(\theta+k\alpha)]\frac{\sin[\frac{\pi d}{\lambda}+\sin(\theta+k\alpha)]}{\frac{\pi d}{\lambda}+\sin(\theta+k\alpha)}\}^2+\{\sum_{k=-m}^{m}\sin[\frac{2\pi r}{\lambda}\cos(\theta+k\alpha)]\frac{\sin[\frac{\pi d}{\lambda}+\sin(\theta+k\alpha)]}{\frac{\pi d}{\lambda}+\sin(\theta+k\alpha)}\}^2} \quad (2)$$

In equation (2), $2m+1$ denotes the number of elements, d denotes the center distance of the element, r denotes the radius of the arc radiating surface, and α denotes the round central angle corresponding to a single element. The horizontal directivity of the arc array is mainly determined by the parameters such as frequency (wavelength), radiating surface radius, and its corresponding round central angle.

In this study, the horizontal and vertical directivities of the transducer are calculated using MATLAB. For the transducer design with a center frequency of 100 kHz, a 5-element arc piezoelectric composite array with radius r = 300 mm, α = 1°, and l = 73.6 mm was selected. The horizontal and vertical directivities at f = 100 kHz are calculated using equations (1) and (2), as indicated in figure 2, which shows a normalized graph in which the vertical directivity –3 dB opening angle is 10° and horizontal directivity –3 dB opening angle is 25°.

3. Finite element simulation

The commercial finite element simulation package ANSYS was used to model mainly the piezoelectric vibrator harmonic response analysis and modal analysis. First, the model of one array element is established and the matching layer is added to the surface of sensitive material. By adjusting the thickness and sound velocity of the matching layer, the dual-mode vibration can be induced between the composite layer and the matching layer so that the transducer bandwidth can be extended. The matching layer parameters can be obtained by theoretical analysis. Second, the ANSYS software was used to establish the model of the 5-element piezoelectric composite vibrator in the air, and the resonant frequency of the multi-element vibrator and admittance variation were studied.

3.1. Finite element simulation of piezoelectric composites

To widen the bandwidth of the transducer, this study uses the method of adding the matching layer on the surface of the 1-3 piezoelectric composite. In figure 3, the piezoelectric vibrator model PZT-4 is chosen as the piezoelectric material, epoxy resin is filled between the piezoelectric pillars, l is the length of the element, a is the width of the element, h is the thickness of the element, and t is the thickness of the matching layer.

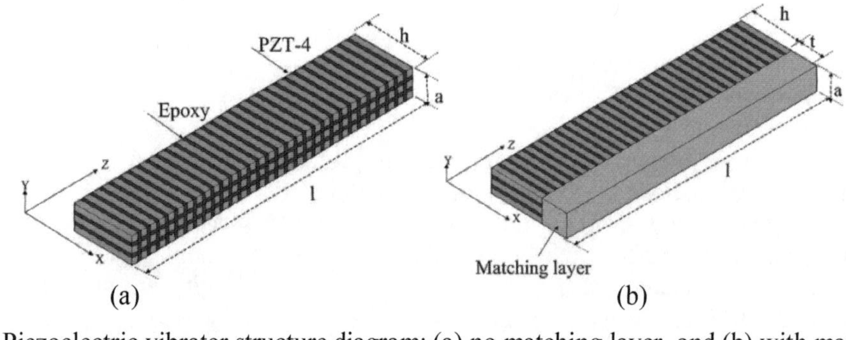

Figure 3. Piezoelectric vibrator structure diagram: (a) no matching layer, and (b) with matching layer

3.1.1. Finite element simulation of piezoelectric composites without matching layer. First, the harmonic response characteristics of the element are simulated in air. The structural parameters of the piezoelectric vibrator in figure 3(a) are as follows: width a = 6.4 mm, thickness h = 15 mm, and length l = 73.6 mm. The ceramic column width is 1.6 mm, the epoxy resin column width is 0.8 mm, and the ceramic volume percentage is 44.4%. The relationship among the conductance amplitude, conductance phase angle, and frequency of the piezoelectric vibrator can be obtained by simulation, as shown in

IOP Conf. Series: Materials Science and Engineering **191** (2017) 012017 doi:10.1088/1757-899X/191/1/012017

figure 4, where point A corresponds to the resonant frequency (f1 = 98 kHz) of the composite element and point B corresponds to the anti-resonant frequency (f2 = 121 kHz) of the composite element.

3.1.2. Design of matching layer. In this paper, the matching layer acoustic impedance characteristic and impedance matching are discussed, and the parameters of the matching layer are designed. The aluminum powder was used as filler, epoxy resin as toughening agent, and curing agent as a matrix to fabricate the matching layer.

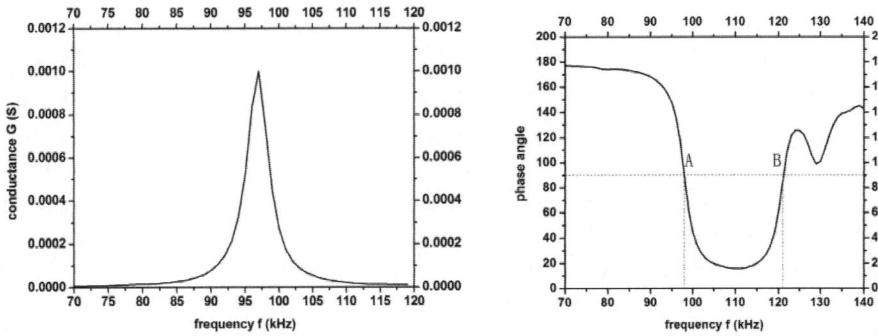

(a) (b)

Figure 4. In the air: (a) conductance amplitude curve, and (b) conductance phase-angle curve

According to the theory of total transmission when the acoustic wave propagates in the medium, the acoustic impedance and thickness of the matching layer are obtained when the acoustic matching is satisfied, where the optimum matching layer thickness in theory is equal to a quarter of wavelength.
(1) Characteristic acoustic impedance of matching layer
The characteristic acoustic impedance of the medium is the main parameter to characterize the acoustic properties of the medium. Here, the formula for the acoustic impedance of the medium is introduced [10]. The impedance of the matching layer is Z, as shown in the following equation:

$$Z = \rho c, \tag{3}$$

where ρ is the matching layer density and c is the propagation velocity of the sound in the matching layer.
(2) Preparation of matching layer
A matching layer was prepared and measured. In this study, the measurement of the matching layer density ρ can be used to determine the measurement result according to the ratio between mass and volume. The acoustic impedance of the matching layer is measured mainly by ultrasonic method. The time between two echoes is measured by digital ultrasonic flaw detector echograph 1095, and then a Vernier caliper is used to measure the thickness of the sample to calculate the speed of sound c of the matching layer. The final matching layer acoustic impedance Z can be calculated by equation (3). The specific parameters of the matching layer are shown in Table 1.

Table 1. Parameters of matching layer

Mass fraction of aluminum powder (%)	Matching layer density (kg/m^3)	Matching layer sound velocity (m/s)	Matching layer acoustic impedance (MRayl)	Matching layer thickness (mm)
10%	1272	2496	3.2	6.4
30%	1383	2826	3.9	7.2
45%	1560	2942	4.6	7.5
60%	1732	3118	5.1	8.0

(3) Acoustic impedance matching of matching layer
According to the traditional theory of acoustic impedance matching requirements, the matching layer of acoustic impedance is between the piezoelectric composite and the water [11]. The acoustic impedance of the piezoelectric composite and water load is Z_3 and Z_0, respectively. The theoretical formula for the acoustic impedance of the matching layer is

$$Z = \sqrt{Z_3 \cdot Z_0} \quad , \tag{4}$$

where the acoustic impedance Z_3 of the piezoelectric composite can be expressed as follows:

$$Z_3 = \rho_3 \cdot c_3 = (\rho_1 v_1 + \rho_2 v_2) \times (2 f_2 \cdot h) \quad , \tag{5}$$

where ρ_1 = 7500 kg/m3 is the density of the piezoelectric ceramic PZT-4; ρ_2 = 1050 kg/m3 is the epoxy resin density; v_1 = 44.4% and v_2 = 55.6% are the ceramic volume percentage and epoxy resin volume percentage, respectively; f_2 = 121 kHz is the anti-resonant frequency of the piezoelectric composite; and h = 15 mm is the thickness of the composite. These values were substituted into equation (5), where the acoustic impedance of the piezoelectric composite is calculated to be 14.2 MRayl and the acoustic impedance of the water load is 1.5 MRayl. Then, by equation (4), the acoustic impedance of the matching layer was calculated as 4.6 MRayl. As shown, the theoretical impedance of the matching layer is exactly the corresponding value when the mass fraction of the matching layer of aluminum powder in Table 1 is 45%. Thus, the optimized matching layer parameters are determined as follows: the mass fraction of aluminum powder is 45%, acoustic impedance is 4.6 MRayl, and matching layer thickness is 7.5 mm.

3.2. Finite element simulation of multi-element arc piezoelectric composite vibrator

The matching layer parameters have been determined by the preceding work. The piezoelectric composite model with matching layer as shown in figure 3(b) is established, and the array elements are gradually added. The modal analysis and harmonic response analysis were conducted by ANSYS software. Moreover, the admittance changes of the arc-shaped piezoelectric composites were observed when the number of elements changed from 1 to 5. The model diagram of the 5-element arc piezoelectric composite vibrator is shown in figure 5(a), where the array element is filled with silicone rubber material. The vibration mode as shown in figure 5(b) shows that the thickness of the vibration mode state is relatively pure.

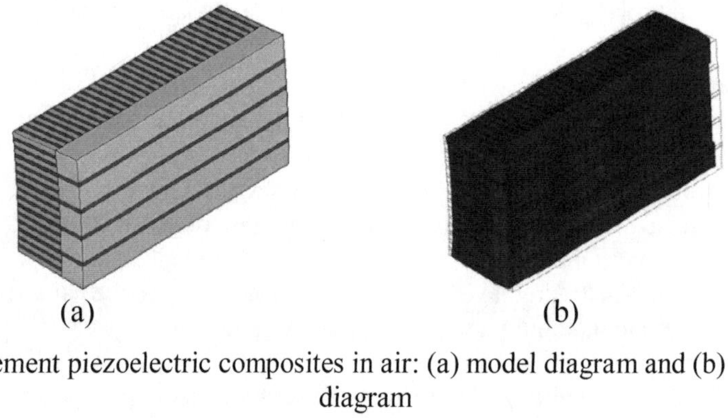

(a) (b)

Figure 5. 5-element piezoelectric composites in air: (a) model diagram and (b) vibration mode diagram

The conductance curve is shown in figure 6. Through the preliminary simulation, two resonant peaks appeared due to the matching layer, which is in agreement with the theory. The frequency spacing and conductance difference of the two resonant peaks directly affect the coupling effect when it is under water.

As shown in figure 6, with the increase in the number of array elements, arc piezoelectric composite vibrator conductance value also increases linearly. The differences of the first and second resonant peak conductance values of the piezoelectric vibrator increasing with a different number of array elements are less than 0.5 and 0.23 mS, respectively, which is due to the parallel relationship between the array elements. When the number of array elements increases, the impedance of the piezoelectric vibrator decreases. As a result, the conductance value increases. Each array element is the same, so the admittance changes are also similar. The specific changes are shown in Table 2.

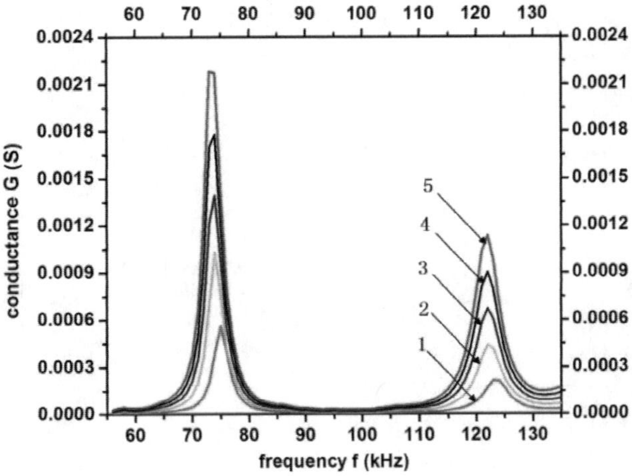

Figure 6. Frequency–conductance curves of different number of array elements

Table 2. Variation of resonant peak of piezoelectric vibrator with different number of array elements

Number of element	Resonant frequency of first resonant peak (kHz)	Resonant frequency of second resonant peak (kHz)	Conductance of first resonant peak (mS)	Difference of conductance between the two elements (mS)	Conductance of second resonant peak (mS)	Difference of conductance between two elements (mS)
1	75	123	0.56	\	0.21	\
2	74	122	1.03	0.47	0.44	0.23
3	74	122	1.39	0.36	0.67	0.23
4	74	122	1.78	0.39	0.90	0.23
5	73	122	2.18	0.4	1.14	0.24

4. Experimental measurement

4.1. Fabrication and measurement of multi-element arc piezoelectric composite vibrator

To verify the results of the finite element analysis, the piezoelectric vibrators with the same structure as the simulation are fabricated in this study. The fabrication process is as follows: The ceramic is cut and then filled with epoxy. The electrode is covered with silver paste. The matching layer is added. Secondary cut is performed. The silicone rubber is filled. Finally, the arc type mold is added.

The flow chart is shown in figure 7. The selected ceramic material is a PZT-4-type piezoelectric ceramic. The thickness of ceramic is 15 mm, the width is 36 mm, and the length is 73.6 mm.

First, the piezoelectric vibrator was measured in air. Their resonant frequencies are measured by an impedance analyzer (Agilent 4294A). The conductance curves are shown in figure 8. The resonant frequencies of the composites with one to five array elements are measured, and the conductance increases with the increase in the number of array elements, which is consistent with the simulation results. The dashed line in the figure shows the simulation curve of the 5-element arc composite vibrator. The simulation and experimental results fit well, and the fluctuation between the simulation conductance and experimental conductance is less than 0.4 mS. The two resonant peaks of simulation conductance curve appeared as 73 and 122 kHz, whereas the two resonant peaks of the experimental conductance curve are 73.8 and 117.6 kHz, respectively; the reason is that the parameters of the piezoelectric ceramics in the simulation have certain inconsistencies with the actual parameters.

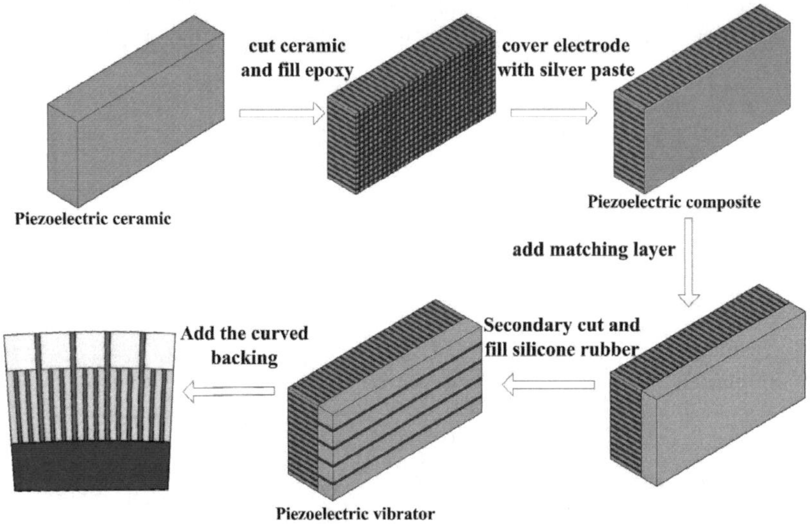

Figure 7. Fabrication flow chart of multi-element arc piezoelectric composite vibrator

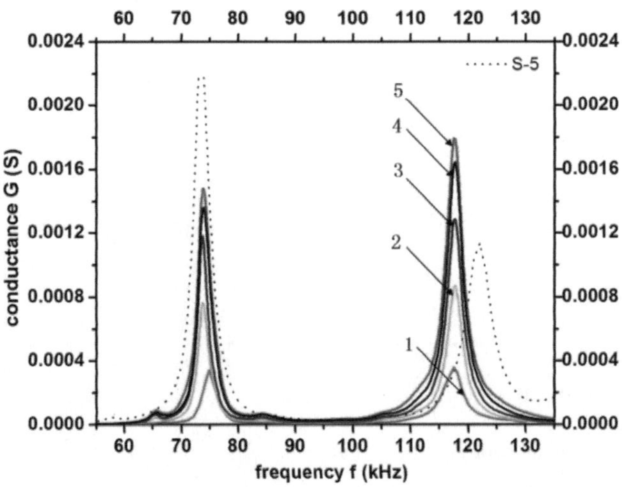

Figure 8. Comparison of simulation and experimental results of conductance curves

4.2. Fabrication and measurement of transducer

The 5-element arc piezoelectric vibrators with matching layer are then made into transducers. The structure is shown in figure 9. One end of the arc piezoelectric vibrator is designed with one sound-absorbing layer, which is made of rigid foam, and the other end is used as the emitting surface to emit sound waves. Then, polyurethane was poured and used as waterproof layer.

The transmitting voltage curves of the transducers can be obtained by the underwater acoustic measurement system. The transducer arrays with 1-, 3-, and 5-array elements were measured. The test results are shown in Table 3 and figure 10.

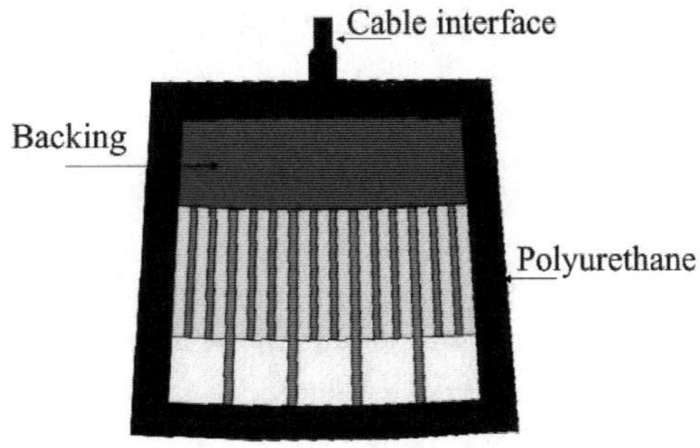

Figure 9. Schematic of transducer structure

Table 3. Performance parameters of transducer

Number of elements	Largest transmitting voltage response (dB)	Transmitting voltage response bandwidth (kHz)
1	148.5	56.5
3	150.5	54
5	151.3	56.5

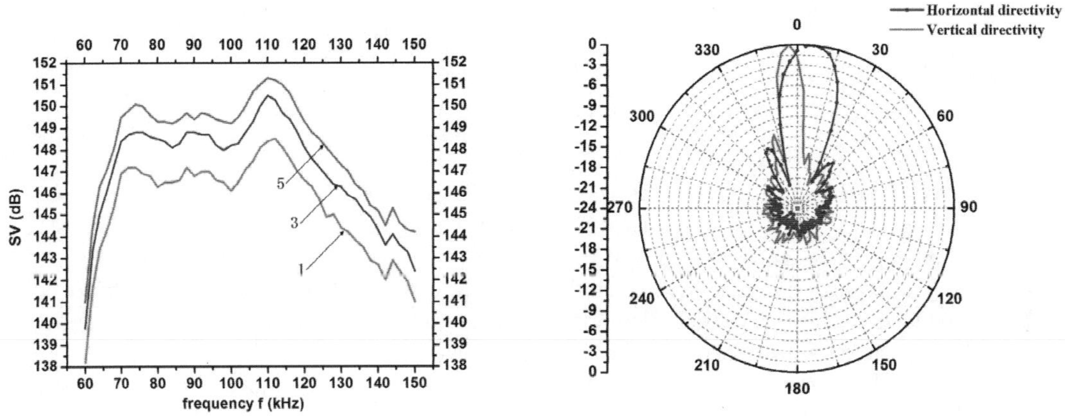

(a) (b)

Figure 10. Multi-element transducer array: (a) transmitting voltage response curves and (b) directivity curves

As shown in figure 10 (a), the value of the transmitting voltage response increases as the number of array elements increases. As the number of array elements increases from 1 to 5, the transmitting voltage response increases by about 3 dB because the sound pressure at each point in the sound field is the superposition of the sound pressure generated by each array element, and the increase in the number of array elements causes the increase of the sound pressure to realize the increase of the transmission voltage response. The matching layer was then added to achieve the expansion bandwidth, where –3 dB bandwidth reached 56.5 kHz.

The directivity of the far field of the piezoelectric vibrator directly determines the wave beam angle of the transducer. Figure 10 (b) presents the horizontal directivity curve and vertical directivity curve of the 5-element transducer array at the center frequency (f = 95 kHz), where the vertical directivity –3 dB opening angle is 9° and the horizontal directivity –3 dB opening angle is 20°. Compared with the theoretical results, the vertical directivity of the opening angle is basically the same. The horizontal directivity opening angle decreased slightly because one end of the arc transducer array is a rigid foam backing and the other end is polyurethane, thereby affecting the propagation of the wave beam and resulting in a decrease in the directivity opening angle.

5. Conclusions

In this study, the 5-element arc broadband piezoelectric composite vibrator and its transducer array were designed and fabricated using the matching principle and curved forming process. First, the optimized parameters of the matching layer and the directivity of the transducer were calculated theoretically. Then, the resonant frequency of the 5-element arc piezoelectric composite vibrator was analyzed using the finite element software ANSYS. Finally, we conducted an experiment for verification. The results show that the experimental results agree well with the theoretical and simulation results. The transmitting voltage response of the 5-element arc transducer array reaches 151.3 dB, the bandwidth of the transducer is up to 56.5 kHz, the vertical directivity of the –3 dB opening angle is 9°, and the horizontal directivity –3 dB opening angle is 20°.

In the future we will expect to study 10-element and 44-element arc broadband piezoelectric composite vibrator and its transducer array, and study the characteristics of broadband and wide beam.

Acknowledgement

This work was supported by the National Natural Science Foundation of China (No.61471047, 61671068, 11604363), and the Beijing College Innovation Capability Promotion Plan of Beijing Municipal Institutions (No. TJSHG201510772015). The project is also supported by The Importation and Development of High-Caliber Talents Project of Beijing Municipal Institutions (No. CIT&TCD201504053).

References

[1] Desilets C *et al* 1997 Composite curved linear array for sonar imaging: Construction, testing, and comparison to FEM simulations *Proc IEEE Ultrason Symp* 2 901-06.

[2] LJ Zhong 2011 Composite spherical transducer with high frequency and wide beam width *Applied Acoustics* 29 330-35.

[3] Abrar A, Zhang D, Su B, Button T W, Kirk K J and Cochran S 2004 1-3 Connectivity piezoelectric ceramic-polymer composite transducers made with viscous polymer processing for high frequency ultrasound *Ultrasonics* 42 479-84.

[4] Rybianets A, Yoram E and Kushkuley L 2006 New low-Q ceramic piezocomposites for ultrasonic transducer applications *Proc. IEEE Ultrason. Symp.* 1 1911-14.

[5] Ramadas S N, O'Leary R L, Gachagan A, Hayward G and Banks R 2009 A wideband annular piezoelectric composite transducer configuration with a graded active layer profile *Proc. IEEE Ultrason. Symp.* 5441479(2009).

[6] CerenElmasli I and Hayrettin Koymen 2007 A Wideband and a Wide-Beamwidth Acoustic Transducer Design for Underwater Acoustic Communications *IEEE Conference Publications* 9810887(2006).

[7] Wu Peirong, ZJ Liu, Y Wang and Y Ya Tao 2012 A high frequency underwater transducer curved array *Technical Acoustics* 31 628-30.

[8] Benjamin Kim C 2001 Recent advances in 1-3 piezoelectric polymer composite transducer technology for AUV/UUV acoustic imaging applications *Oceans Conf Rec IEEE* 1 26-33.

[9] Lu Deming 2001 *Principles of underwater transducer*(Qingdao: Qingdao Ocean University Press)p 394.

[10] Gonghuan Du and Zhemin Zhu 2001 *Foundation of Acoustics*(Nanjing: Nanjing University Press)pp 182-9.

[11] Nuo Feng 1999 *Ultrasound Handbook*(Nanjing: Nanjing University Press)pp 162-6.

Effect on the B-H loop and complex permeability with nickel substitution on Li-Zn ferrite synthesized by the citrate precursor method

S Ibetombi[1] and S Sanatombi[1]

[1]Physics Department, National Institute of Technology Manipur, Langol, Manipur 795004, India

E-mail: ibetombi96@gmail.com

Abstract. $Li_{0.4-0.5x}Zn_{0.2}Ni_xFe_{2.4-0.5x}O_4$ ferrites where $0.02 \leq x \leq 0.1$ in steps of 0.02 have been synthesized by citrate precursor method. The samples were given final sintering at 1040°C after a pre-firing at a temperature of 650°C. Various characterizations using T_C set up, BH loop tracer were carried out on the sintered samples. The X-ray diffraction pattern confirmed the formation of spinel phase. Magnetic properties such as Curie temperature using T_C set up, BH loop parameter and complex permeability using BH loop tracer were studied. Experimental results show that the Curie temperature, saturation magnetization and initial permeability decreases with the increase in the Ni^{2+} concentration. The complex permeability measured as a function of frequency in the range of 20Hz-2MHz show dispersion. Possible mechanism contributing to the above process is being discussed.

1. Introduction

Ferrites had been studied by many workers because of the varied technological application. They are found to be used from microwave to radio frequency ranges, due to their characteristic magnetic properties [1-2]. The magnetic properties of ferrites depend upon various factors like the type of substitutions, synthesis method and the sintering temperature. Substituents play an important role in modifying the properties of basic lithium ferrites and various workers reported that Ni substitution enhances the densification which results in change in the magnetic properties [3-5]. Considering the synthesis method the chemical method known as the citrate precursor method gives samples at low temperature with high purity. However, the densification is less. On the other hand it is learnt that in magnetic applications, a high sintered density, high purity with lesser pores is in demand while fabricating a ceramic material. Therefore a double sintering process is subjected further which is expected to give more densification in addition to the effect of Ni substitution and hence a change in their various properties like Curie temp., BH loop properties complex permeability, dielectric constant, resistivity etc. The study of high curie temperature, high magnetization, permeability gives valuable information regarding the domain response paving way to technical application like microwave device, memory core application etc. Useful frequency range of ferrite is limited by the onset of resonance phenomena for which either the permeability begins to fall at a critical frequency or the losses rise sharply. Hence knowledge of frequency dependence of permeability and permeability loss is also necessary.

In the present work citrate precursor method has been used to prepare LiZnNi ferrite. The sample was further given double sintering and investigation is being carried out on the magnetic properties such as Curie temperature, BH loop parameter, complex permeability etc.

2. Experiment

2.1. *Material synthesis*

Nickel substituted lithium zinc ferrites with compositional formula $Li_{0.4-0.5x}Zn_{0.2}Ni_x Fe_{2.4-0.5x}O_4$ where x=0.02 $\leq$ x $\leq$ 0.1 in steps of 0.02 were prepared by the citrate precursor method. The starting chemicals used were lithium nitrate $LiNO_3$ (Merck Germany), zinc nitrate hexahydrate Zn (NO_3). $6H_2O$ (Merck India), iron (III).$6H_2O$ nitrate nonahydrates $Fe(NO_3).9H_2O$ (Merck India), nickel(II) nitrate hexahydrate $Ni(NO_3)_2$. $6H_2O$(Merck India) and citric acid monohydrates $C_6H_8O_7$. Stoichiometric amount of lithium nitrate, zinc nitrate, iron nitrate, nickel nitrate and citric acid. were mixed to make a solution. The ratio of metal cations to citric acid is 1:1. The solution was mixed homogenously with the help of a magnetic stirrer using a magnetic agitator and the value of P_H was controlled at 7. The P_H controlled solution was refluxed at 40°C with continuous stirring for half an hour. It was then heated at 100°C to form a gel. Finally the dried gel ignite, undergoing a strong auto combustion process giving rise to the ash-synthesized product which is the typical spinel structured LiZnNi ferrite powder. These powder were mixed with polyvinyl alcohol as binder and pressed into torroids with 50 kilo Newton pressure. Final sintering was given to the pressed torropids at 1040^oC after pre sintering at 650°C.

2.2. *Characterizations*

The phase identification of the sintered samples were performed using X-ray diffractometer (Phillips) with CuK_α radiation (λ=1.5405 A°). Curie temperature was measured using Soohoo's method [6]. Saturation magnetization, and coercivity was determined from BH hysteresis loops measured using BH loop tracer (Ferrites India). Complex initial permeability as a function of composition and frequency (20Hz- 2MHz) were carried out. Inductance (L) and the corresponding loss factor tanδ of the sintered toroids were measured in the frequency range of 20Hz – 2MHz using a precision LCR meter E4980A. The measured inductance was used to calculate the initial permeability $\mu_i = L/L_o$ where L is the measured inductance and L_o is the air core inductance calculated using the relation,

$$\mu_i' = \frac{L}{4.6N^2 X \log\frac{OD}{ID} X hX10^{-9}} \quad (1)$$

where N is the number of turns of the coil (N=70), h is the thickness, OD is the outer diameter and ID is the inner diameter of the toroidal specimen.

3. Results and discussion

XRD pattern was used to confirm the spinel phase structure of the $Li_{0.4-0.5x}Zn_{0.2}Ni_xFe_{2.4-0.5x}O_4$ ferrite sample (figure. 1, x=0.08). The diffraction pattern shows peak which corresponds to the lattice plane of (220), (311), (420), (422), (511), (440) spinel structure of LiZnNi ferrite. All the peaks could be indexed to the standard pattern reported by the Joint Committee on Powder Diffraction Standards (JCPDS). There is no extra peak observed showing that there is no impurity present.

Magnetic studies such as Curie temperature, saturation magnetization, retentivity coercivity and complex permeability was carried out. Soohoo's method was used to determine the Curie temperature [6]. The variation of the Curie temperature (T_C) with Ni concentration, show a decreasing trend (table 1).

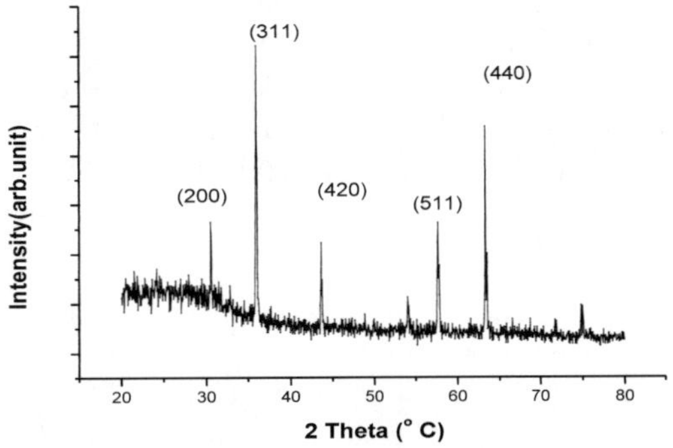

Figure 1. XRD pattern for x=0.08 of $Li_{0.4-0.5x}Zn_{0.2}Ni_xFe_{2.4-0.5x}O_4$.

Table 1. Curie temp. B-H loop parameters and initial permeability of $Li_{0.4-0.5x}Zn_{0.2}Ni_xFe_{2.4-0.5x}O_4$

Conc.(x)	Curie temp. (°C)	Sat. mag. (Gs)	Retentivity (Gs)	Coercivity (Oe)	Initial permeability
0.02(0.78%)	560	2162	1927	6.852	4.1
0.04(1.55%)	551	1160	842	7.420	3.8
0.06(2.32%)	543	1915	1571	7.112	3.7
0.08(3.09%)	530	1751	1466	6.824	2.7
0.1(3.85%)	527	1544	1155	7.917	2.2

This can be understood in terms of the magnetic super exchange interaction proposed by Neel which has a direct relation with the Curie temperature. According to this model the intersublattice AB super exchange interaction dominates over the intrasublattice AA and BB superexchange interactions. The strength of this A-B interaction is determined by the number of linkages $Fe_A^{3+} - O^{2-} - Fe_B^{3+}$ per formula unit and the overall strength of this linkage determines the Curie temperature of the ferrite samples [7]. For the present series of ferrites the cationic expression is expressed as $(Zn_{0.2}Fe_{0.8})[Li_{0.4-0.5x}Ni_xFe_{1.6-0.5x}]O_4$. The magnetic Fe^{3+} ions are being replaced by less magnetic Ni^{2+} ions and hence with progressive substitution of Ni^{2+} ion the number of active $Fe_A^{3+} - O^{2-} - Fe_B^{3+}$ linkage decreases, thereby weakening the AB superexchange interaction. Hence the thermal energy required to offset the spin alignment decreases leading to decrease in the Curie temperature as was observed [7-8].

Figure 2 shows the BH hysteresis loop and table 1 gives the hysteresis loop parameter such as saturation magnetization, retentivity, coercivity of the samples. Experimental result shows a decrease in the saturation magnetization with the increases in Ni^{2+} concentration. These variation can be explained by considering the Neel's molecular field model and the cationic distribution as has been discussed earlier. According to Neels model AB superexchange interaction having the most favourable A-O-B angle of about 125° dominates over intrasublattice AA and BB interaction. Therefore the net magnetization is given by the vector sum of the magnetization of the two sublattices $\mathbf{M_A}$ and$\mathbf{M_B}$. In the present ferrite samples Ni^{2+} ions have strong preference for the B site and Fe^{3+} ions partially occupy the A and B sites. The resultant magnetization, M is the difference of the magnetizations $\boldsymbol{M_B}$ and $\boldsymbol{M_A}$ of the B and A sublattices respectively. The cation distribution can be written as $(Zn_{0.2}Fe_{0.8})_A$ $[Li_{0.4-0.5x}Fe_{1.6-0.5x}Ni_x]_B$ O_4 where parenthesis and square bracket represent tetrahedral (A) and

IOP Conf. Series: Materials Science and Engineering **191** (2017) 012018 doi:10.1088/1757-899X/191/1/012018

octahedral(B) sites respectively. As Li^{1+} and Zn^{2+} is non-magnetic it doesn't contribute so, Ni^{2+} and Fe^{3+} only gives effect on net magnetisation. The magnetic moment of $Ni^{2+}$is $2\mu_B$ and that of Fe^{3+} is $5\mu_B$, therefore, the net magnetisation becomes (8-2.5x+2)-4 =6-2.5x [3,17-18].Consequently, M_s values will decrease and was observed experimentally. However the drastic decrease for Ni=0.04 needs investigation as saturation magnetization depend on several parameters.

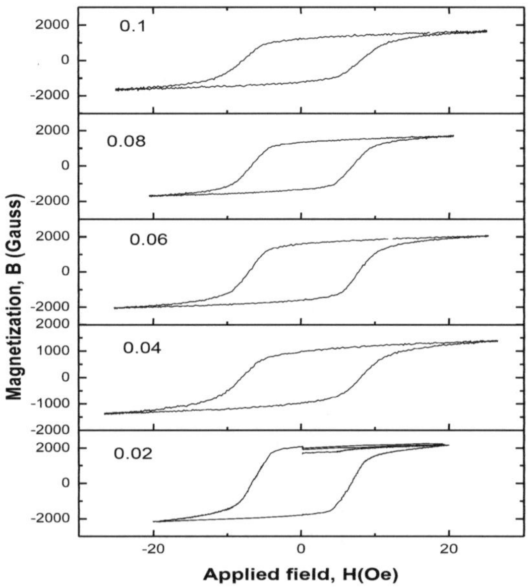

Figure 2. B-H loop for $Li_{0.4-0.5x}Zn_{0.2}Ni_xFe_{2.4-0.5x}O_4$.

The variation of the value of retentivity with increasing Ni concentration also decreases which can be explained in relation to grain size which is known to be inversely proportional to retentively. It was reported in our earlier work for single sintered LiZn Ni ferrite samples that Ni substitution increases the grain size. [10]. However properties of hysteresis parameters are dependent on the various microstructural properties like grain size, porosity, nature of pores etc. which are largely controlled by sintering process. Hence it is difficult to attribute to a single factor and a detail investigation is required.

Investigation on the variation of initial permeability as a function of concentration and frequency is carried out. The complex permeability is given by, $\mu = \mu_i^{'} - j\mu_i^{''}$ where $\mu_i^{'}$ is the real part of permeability known as initial permeability and $\mu_i^{''}$ is the imaginary part of initial permeability known as permeability loss. The initial permeability shows a decrease with increasing Ni^{2+}ion concentration (table 1). The permeability mechanism can be explained in terms of domain wall displacement and rotation of spins in each domain and is found to be dependent on square of the saturation magnetization, ionic distribution, microstructure like grain size, density, such as magneto anisotropy etc. [9-12]. In the present study the saturation magnetization decreases with addition of Ni and the observed overall decreased in the value of initial permeability is expected.

The frequency variation of initial permeability showed dispersion at low frequency, (figure.3a) and becomes nearly constant at higher frequencies. At low frequency, changes of the magnetization direction occur by the motion of domain walls, so that a domain oriented in the direction of the applied field grows at the expense of its neighbours, which are oriented in different direction. As the frequency increases the domain wall is unable to move sufficiently rapidly to follow the alternating field. The overall magnetization vector does not follow the applied field hence accounting for the low value of permeability. The occurrence of resonance at still higher frequency was reported by other workers which was not observed in the present investigation as frequency ranges of the LCR meter being 20Hz

– 2MHz [13]. However the frequency limit up to which the material can be used in a device is given by resonance frequency. This gives us knowledge that the sample can be used for high frequency application.

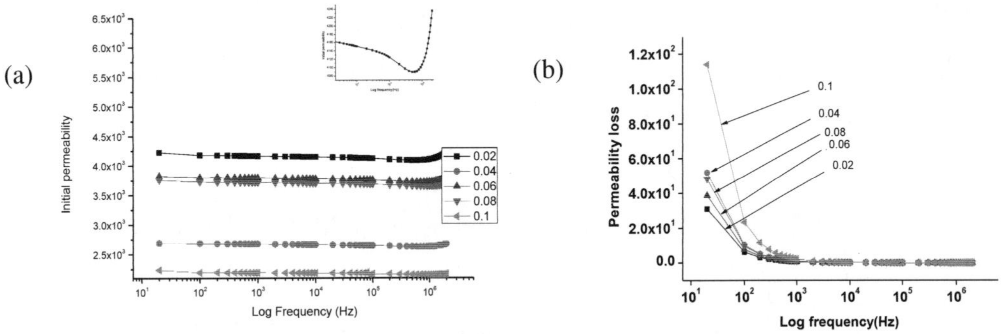

Figure 3. (a) Variation of initial permeability with frequency for $Li_{0.4-0.5x}Zn_{0.2}Ni_xFe_{2.4-0.5x}O_4$ (b) Variation of permeability loss with frequency for $Li_{0.4-0.5x}Zn_{0.2}Ni_xFe_{2.4-0.5x}O_4$.

Figure 3(b) shows the permeability loss as a function of frequency. It show dispersion and at higher frequencies the value becomes small and nearly constant. The loss observed may be due to a lag of the motion of the domain walls which may be attributed to the imperfections in the crystal structure.

4. Conclusion

Ni substituted LiZn ferrite was synthesized by citrate precursor method. XRD confirms the spinel phase structure of the sample. Experimental results show a decrease in Curie temperature, saturation magnetization, retentivity, initial permeability with the increase in Ni^{2+} concentration. Frequency variation of initial permeability and permeability loss show dispersion. It is expected that resonance for real part of complex permeability known as initial permeability might be occurring at higher frequency beyond 2 MHz suggesting the sample may find application in higher frequency.

Acknowledgement
This work was supported by DST GoI (F.No.SR/FTP/PS-095/2012).

References

[1] Ladgaonkar B P, Vasambekar P N and Vaingankar A S 2000 *J. Magn. Mag. Mater.* 210 289.
[2] Sun C and Sun K 2007 *Solid State Commun* 141 258.
[3] Kotnala R K, Dar M M, Verma V, Singh A P and Siddiqui W A 2010 *J. Magn. Mag. Mater.* 322 3714.
[4] Bueno A R, Gregori M L and Nobrega M C S 2007 *Material Chemistry and Physics* 105 229.
[5] Hankare P P, Patil R P, Jadhav A V, Panday R S, Garadkar K M, Sasikala R and Tripathi A K 2011 *J. Alloys. Comp.* 509 2160.
[6] Soohoo R F 1960 *Theory and application of Ferrites* Englewood cliffs Prentice Hall Inc. New Jersey 111.
[7] Neel L 1948 *Ann.Phys.* 3 137.
[8] Smit J, Winkler G, Verweel J, Dillon J F, Weiss R S, Mallinson J C and Middelhoek S 1971 *Magnetic Properties of Materials* McGraw Hill New York.
[9] Smit J and Wijn H P J 1959 Ferrites, Phillips technical library Eindhoven 300.
[10] Ibetombi S, Sumitra P and Chandra P 2009 *J. Magn. Mag. Mater.* 321 2779.
[11] Verma A, Goel T C and Mendiratta R G 2000 *J. Magn. Mag. Mater.* 210 274.
[12] Chauhan B S, Kumar R, Jadhav K M and Singh M J 2004 *J. Magn. Mag. Mater.* 283 71.
[13] Hench L L and West J K 1990 *Principles of Electronics Ceramics* John Wiley and Sons New York 189.

IOP Conf. Series: Materials Science and Engineering **191** (2017) 012019 doi:10.1088/1757-899X/191/1/012019

Theoretical studies on phonon density of states in nonstoichiometric palladium hydrides

D Anzai[1], S Matsuo[1], N Hatakenaka[1*], N Ogita[1], K Matsuda[2] and S Yasuzuka[3]

[1] Graduate School of Integrated Arts and Sciences, Hiroshima University, Higashi-Hiroshima 739-8521, Japan
[2] College of Science and Technology, Nihon University, Tokyo 101-8308, Japan
[3] Research Center for Condensed Matter Physics, Hiroshima Institute of Technology, Hiroshima 731-5193, Japan

*E-mail: noriyuki@hiroshima-u.ac.jp

Abstract. We investigate the phonon density of states in nonstoichiometric palladium hydrides PdH_x, in order to clarify the contribution of lattice vibrations to superconductivity. We employ the Burger-MacLachlan model, which considers a one-dimensional triply-connected diatomic chains with randomly distributed vacancies, focusing on the <111> direction in the rock-salt lattice of PdH_x. The phonon density of states is calculated by the forced oscillator method suitable for random systems. We find that phonon modes at high frequencies change from localized to collective modes with increasing x and the crossover stoichiometry is consistent with that of the appearance of superconductivity. These results support the optical phonon mechanism for superconductivity in PdH_x.

1. Introduction

There has been considerable interest in palladium hydrides (PdH_x) due to their potential applications such as hydrogen storage since Graham discovered that palladium (Pd) absorbs hydrogen (H) [1]. In addition, Skoskiewicz found that the system undergoes a transition to a superconducting state when the stoichiometry $x \equiv \mathrm{H/Pd} \approx 0.81$ [2]. Interestingly, the superconducting transition temperature T_c increases up to 10 K as the stoichiometry x approaches 1 [3]. Thus, the system was considered a promising candidate for a high temperature superconductor before the discovery of copper oxide superconductors.

The mechanism of superconductivity in PdH_x showing such relatively high T_c is believed to be the Bardeen-Cooper-Schrieffer (BCS) theory [4] with *optical phonons* as the glue for the electrons [5] since T_c is proportional to the associated phonon frequency ω_D in the BCS framework given as

$$k_B T_c = 1.13\hbar\omega_D \exp\left(-\frac{1}{N(0)V}\right) \quad (1)$$

where k_B and $\hbar$ are the Boltzmann constant and the reduced Planck constant, respectively. $N(0)$ and V are the electronic density of states at the Fermi surface and the coupling constant, respectively. However, it is not clear whether optical phonon modes formed by the collective motion of a large number of atoms exist in nonstoichiometric PdH_x, which are random or inhomogeneous lattices, because randomness conventionally prevents from emerging collective atomic motion. In this paper, we study the competition between individual excitations and collective excitations in controlling the system's randomness, i.e. stoichiometry and show the development of optical phonons from localized phonon modes as the stoichiometry x increases.

2. The Burger-MacLachlan modek for PdH_x

The crystal structure of palladium is a face-centered cubic (fcc) lattice. Hydrogen atoms adsorbed at surface diffuse into the crystal and end up being randomly distributed in the octahedral interstices, i.e. an interstitial site between six Pd sites in the fcc palladium lattice as shown in Fig.1. Thus, PdH_x is regarded as a disordered lattice. Here we employ the Burger-MacLachlan model [6] to calculate the phonon density of states. In their model, PdH_x is modelled by a one-dimensional diatomic chain consisting of heavy atoms with mass m_H and light atoms with mass m_L, taking only the dominant <111> direction into account [6,7,8]. The schematic diagrams for the model we employ here are shown in Fig. 2. Heavy atoms and light atoms occupy odd sites and even sites, respectively. They are connected to each other in the nearest-neighbor form for heavy-light pairs with force constant C_{HL} and the next-nearest-neighbor form for heavy-heavy and/or light-light pairs with force constants C_H and C_L, respectively. Figure 2(a) shows a simple monatomic chain representing a pure palladium lattice (x = 0.00). Figure 2(b) shows a disordered random chain produced by the absorption of hydrogen atoms randomly located at interstitials at an intermediate stoichiometry. Figure 2(c) represents a triply-connected diatomic chain with fully occupied interstitial sites by hydrogens (x = 1.00).

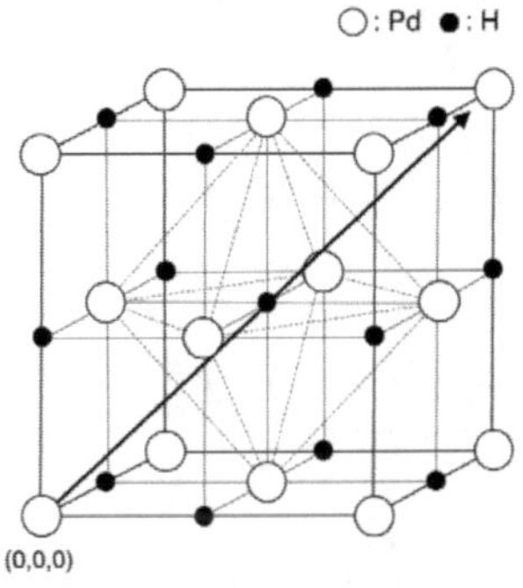

Figure 1. The crystal structure of PdH_x. Open circles and solid circles are palladium and hydrogen, respectively. The arrow from (0,0,0) atom represents the [111] direction. The crystal is modelled by one-dimensional diatomic chain taking only the dominant <111> direction.

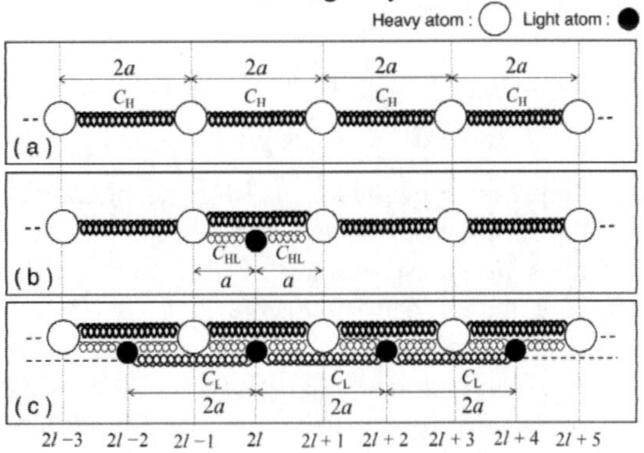

Figure 2. Schematic diagrams of the Burger-MacLachlan model that is triply-connected diatomic chains with randomly distributed vacancies. (a) Heavy atoms (depicted by open circles) occupy odd sites regularly. (b) Light atoms (solid circles) are placed randomly in the even site. (c) The triply-connected diatomic chains with the nearest neighbor and the next nearest neighbor atoms.

Here we summarize the analytical expression of the dispersion relation for x = 1.00, in order to check the reliability of our numerical simulation. The equations of motion for the atoms on $2l$+1th site and $2l$th site in the model shown in Figure 2(c) are given as

$$m_{\mathrm{H}}\frac{d^2u_{2l+1}(t)}{dt^2} = -2(C_{\mathrm{H}}+C_{\mathrm{HL}})u_{2l+1}(t) + C_{\mathrm{H}}[u_{2l+3}(t)+u_{2l-1}(t)] + C_{\mathrm{HL}}[u_{2l+2}(t)+u_{2l}(t)] \quad (2)$$

and

$$m_{\mathrm{L}}\frac{d^2u_{2l}(t)}{dt^2} = -2(C_{\mathrm{L}}+C_{\mathrm{HL}})u_{2l}(t) + C_{\mathrm{L}}[u_{2l+2}(t)+u_{2l-2}(t)] + C_{\mathrm{HL}}[u_{2l+1}(t)+u_{2l-1}(t)] \quad (3)$$

where $u_{2l+1}(t)$ and $u_{2l}(t)$ are the displacement of the heavy atom and the light atom on the $2l$+1th site and $2l$th site, respectively. The dispersion relation for the model shown in Figure 2(c) is given as

$$\omega^2(k) = \omega_0^2 \pm \omega_0^2\left\{1-8\left[2\frac{\omega_{\mathrm{H}}^2}{\omega_0^2}\frac{\omega_{\mathrm{L}}^2}{\omega_0^2}\sin^2(ka)+\frac{\omega_\mu^2}{\omega_0^2}\left(\frac{\omega_M^2}{\omega_0^2}+\frac{\omega_{\mathrm{H}}^2}{\omega_0^2}r_{\mathrm{H}}+\frac{\omega_{\mathrm{L}}^2}{\omega_0^2}r_{\mathrm{L}}\right)\right]\sin^2(ka)\right\}^{\frac{1}{2}}, \quad (4)$$

where we define $\omega_0^2 \equiv 2(\omega_{\mathrm{H}}^2+\omega_{\mathrm{L}}^2)\sin^2(ka)+\omega_\mu^2$, $\omega_{\mathrm{H}}^2 \equiv C_{\mathrm{H}}/m_{\mathrm{H}}$ and $\omega_{\mathrm{L}}^2 \equiv C_{\mathrm{L}}/m_{\mathrm{L}}$. We also define $\omega_\mu^2 \equiv C_{\mathrm{HL}}/\mu$ and $\omega_M^2 \equiv C_{\mathrm{HL}}/M$ with $M \equiv m_{\mathrm{H}}+m_{\mathrm{L}}$ and $1/\mu \equiv 1/m_{\mathrm{H}}+1/m_{\mathrm{L}}$ being the total mass of atoms and the reduced mass, respectively. The ratio of m_{H} to M and m_{L} to M are r_{H} and r_{L}, respectively. The phonon density of states $D_T(\omega)$ can be expressed as

$$D_T(\omega) = \pm\frac{N}{\pi}\frac{\omega}{\omega_\mu^2}\left[\left(\frac{\omega^2}{\omega_\mu^2}-2\gamma\right)(1\pm Z)+\frac{4\sigma^2}{\mathrm{v}^2}\left(1-\frac{\omega^2}{\omega_\mu^2}\right)\right]$$

$$\times\left\{\left(\frac{\omega^2}{\omega_\mu^2}-2\gamma\right)Z\left[\left(\frac{\omega^2}{\omega_\mu^2}-2\gamma\right)(1\pm Z)\left[\frac{8\sigma^2}{\omega_\mu^2}-\left(\frac{\omega^2}{\omega_\mu^2}-2\gamma\right)(1\pm Z)\right]\right]^{\frac{1}{2}}\right\}^{-1}, \quad (5)$$

where

$$\gamma \equiv \frac{\left(\frac{\omega_M^2}{2}+\omega_{\mathrm{H}}^2r_{\mathrm{H}}+\omega_{\mathrm{L}}^2r_{\mathrm{L}}\right)}{\mathrm{v}^2} \quad (6)$$

and

$$Z \equiv \left[1+\frac{4\sigma^2}{\mathrm{v}^2}\frac{\omega^2}{\omega_\mu^2}\frac{\left(2-\frac{\omega^2}{\omega_\mu^2}\right)}{\left(2\gamma-\frac{\omega^2}{\omega_\mu^2}\right)^2}\right]^{\frac{1}{2}}. \quad (7)$$

Here we also use $\mathrm{v}^2 \equiv \omega_{\mathrm{H}}^2+\omega_{\mathrm{L}}^2$ and $1/\sigma^2 \equiv \mathrm{v}^2/(\omega_{\mathrm{H}}^2\omega_{\mathrm{L}}^2)$. Equation (5) can be reduced to well-known expressions for monatomic and diatomic chains when $C_{\mathrm{HL}} = 0$ and $C_{\mathrm{H}} = C_{\mathrm{L}} = 0$, respectively.

3. Forced oscillator method

In the previous section, we obtained analytical formulae for the phonon density of states in specific situations such as $x = 0.00$ and 1.00. However, there is no simple way to obtain similar expressions analytically at intermediate values of the stoichiometry due to randomness. Therefore, we use a numerical method for obtaining the phonon density of states.

In the numerical approach for lattice vibrations, the most essential part of the calculations is to find the eigenvalues and eigenvectors of large matrices [9]. However, it is extremely difficult to find them due to physical limitations of computers, e.g. memory resources, as the size of the matrices increases. Therefore, alternative approaches are highly desirable.

Here, we employ the forced oscillator method (FOM) introduced by Williams and Maris [10] in 1985, and later developed by Yakubo and Nakayama [11,12,13] to calculate the phonon density of states for disordered lattice vibrations. The method is based on resonance phenomena applying a random external force to each atom at rest given by

$$F_l = F_0\cos(\phi_l)\cos(\Omega t), \quad (8)$$

where F_0 is a constant amplitude, the phase ϕ_l and Ω are a random variable with respect to the site number l and a frequency of external force, respectively. These random forces compensate the randomness of the system by averaging over the entire system. In the FOM calculations, the phonon density of states is given by

$$D_F(\Omega) = \frac{8\langle E(\Omega,T)\rangle}{\pi T F_0^2 N}, \tag{9}$$

where $\langle E(\Omega, T)\rangle$ is the averaged total energy of the system after time T and N is the total number of atoms in the system. Thus, we can calculate the phonon density of states only from the time-averaged energy for the forced vibrations without finding all eigenvalues and eigenvectors. Thus, the FOM is suitable for analyzing the disordered system dealing with in the present study.

4. Numerical results

4.1. Parameters

In Fig. 3, we present the phonon density of states at various values of the stoichiometry calculated from the FOM. In the FOM calculations, the number of lattice points was set to N = 100,000. The force constants that we used are $C_{\mathrm{HL}}/C_{\mathrm{H}} = 0.18$ and $C_{\mathrm{L}}/C_{\mathrm{H}} = 0.16$ ($C_{\mathrm{H}} = 373.069 \times 10^{-3}$ kg/s^2, $C_{\mathrm{L}} = 59.691 \times 10^{-3}$ kg/s^{2}and$C_{\mathrm{HL}} = 67.152 \times 10^{-3}$ kg/s^2) estimated from the result of Rowe *et al.* [14] based on neutron diffraction experiments. The mass ratio is $m_{\mathrm{L}}/m_{\mathrm{H}}$ = 1/106 ($m_{\mathrm{H}} = 177.421 \times 10^{-27}$ kg and $m_{\mathrm{L}} = 1.674 \times 10^{-27}$ kg). The FOM solutions are represented as circles with a dashed line, while theoretical solutions are depicted by solid lines shown in both Fig. 3 (a) for a monatomic chain (x = 0.00) and (f) for a diatomic chain (x = 1.00). From those, one can easily identify acoustic phonon modes at lower frequencies and optical phonon modes at higher frequencies. In addition, there is no appreciable difference between the results of the analytical and FOM calculations. This demonstrates the reliability of the FOM calculation.

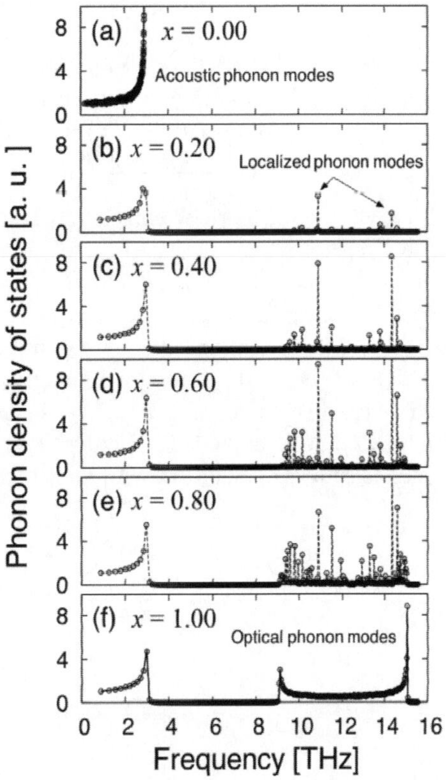

Figure 3. The phonon density of states at various values of the stoichiometry x. The circles and dashed line are FOM solutions, while the solid lines are the analytical solutions given in Eq. (5). The FOM solutions in Fig. (a) and (f) are in good agreement with the analytical solutions.

IOP Conf. Series: Materials Science and Engineering **191** (2017) 012019 doi:10.1088/1757-899X/191/1/012019

4.2. Numerical results

Let us observe the phonon spectra at intermediate stoichiometric ratios. At $x = 0.20$ (Fig. 3 (b)), some peaks appear at high frequencies. These are due to localized lattice vibrations regarded as excitations of individual light atoms. The number of peaks increases with increasing x, with these peaks all lying in the frequency range corresponding to optical phonons (see Fig. 3 (c) to (e)). Figure 4 shows the density of states enlarged the optical phonon part at various x values ranging from 0.60 to 1.00. A large number of localized lattice vibrations are shifted to a collective motion of a large number of atoms called optical phonon modes. These results imply that the optical phonon modes are developed from localized lattice vibrations as x increases.

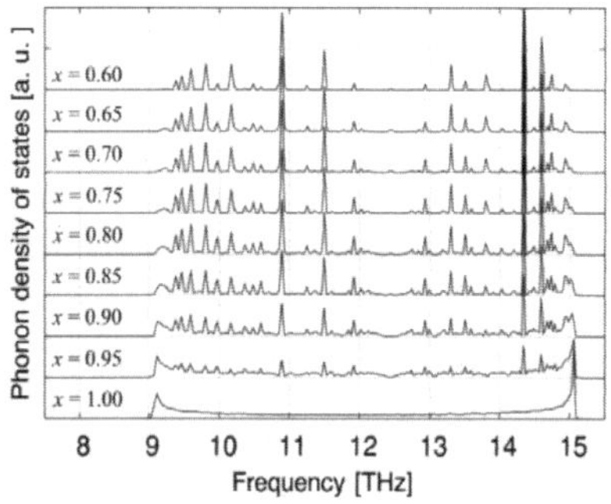

Figure 4. Enlarged drawings of the optical phonon part in the phonon density of states at various values of stoichiometry x ranging from 0.60 to 1.00. A large number of localized lattice vibrations are shifted to a collective motion of a large number of atoms with increasing x.

4.3. Correlations

Now let us consider the crossover from localized lattice vibrations to optical phonon modes as a function of x. To evaluate how similar localized lattice vibrations $D_F(\omega, x)$ to optical phonon modes $D_T(\omega)$ at stoichiometry x are, we introduce the correlation $C(x)$ defined as

$$C(x) = \frac{\sum_\omega D_T(\omega) D_F(\omega,x)}{\sum_\omega D_T^2(\omega)}. \tag{10}$$

There is no correlation if $C(x) = 0$, while perfect correlation if $C(x) = 1$. Figure 5 shows the correlation as a function of x. The circles are the calculated correlations and the solid line is a fitting function $f(x)$ that is a 6th-order polynomial. The correlation first gradually increases at smaller x values. On the other hand, the correlation rapidly increases at large x values. To see this dependence more clearly, we take the derivative of correlations $dC(x)/dx$ in Fig. 6. This figure shows that $dC(x)/dx$ has two different positive slopes in the low stoichiometry range (dot line) and the high stoichiometry range (dash-dotted line). The lower slope indicates the growth of localized phonon modes, while the higher slope implies that of extended optical phonon modes. These distinct slopes crossover at around $x = 0.80$ which is roughly consistent with the stoichiometry for the appearance of superconductivity in PdH_x.

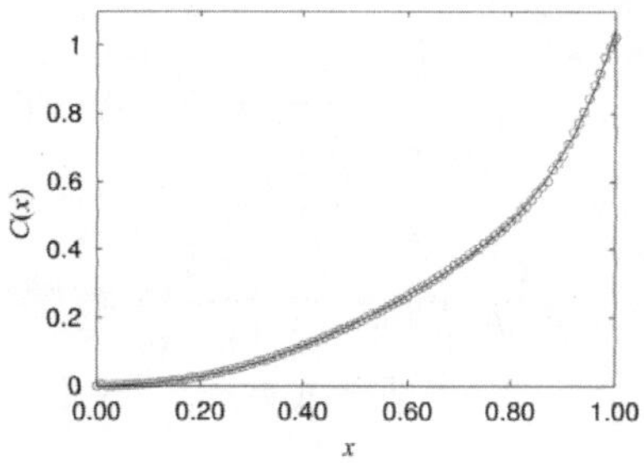

Figure 5. The correlation function defined by Eq. (10) as a function of stoichiometry x. The solid line is the 6th order fitting function $f(x)$.

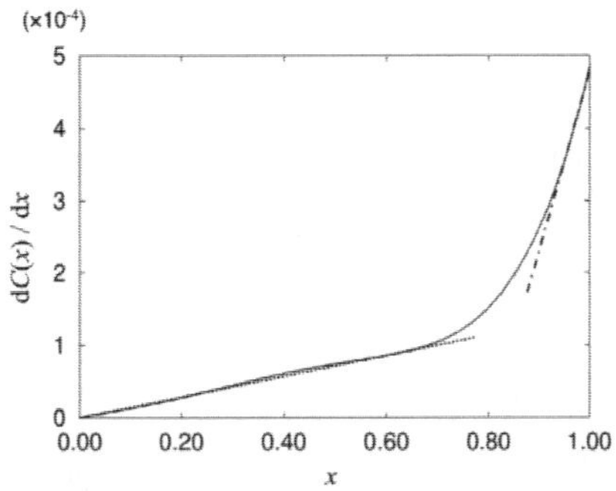

Figure 6. The derivative of correlations as a function of stoichiometry x. There are two different positive slopes, one for the localized phonon range (dotted line) and the other for the optical phonon range (dash-dotted line).

5. Concluding remarks

The mechanism of superconductivity in nonstoichiometric PdH_x has been hypothesized to be *optical phonon*-mediated BCS theory. However, it is not clear how the disordered hydrogen atoms produce the optical phonon modes. Therefore, we have investigated the phonon density of states in the triply-connected diatomic chains with randomly distributed vacancies. The correlation $C(x)$ clearly unveiled two types of lattice dynamics in nonstoichiometric palladium hydrides. This suggests that the localized phonon modes transform into the optical phonon modes. In addition, we found that the crossover stoichiometry is roughly consistent with that of the appearance of superconductivity. This supports the existing optical phonon mechanisms for palladium hydrides. We will extend the one-dimensional Burger-MacLachlan model to three-dimensional cases to confirm these results in more realistic situations.

Acknowledgements

We would like to thank K Yakubo for his FOM programs and technical inputs. We are also grateful to S Tamura, S Ashhab, S Ishizaka and T Inagaki for helpful discussions and suggestions.

References

[1] Graham T 1866 *Philos. Trans. R. Soc. London* 156 399
[2] Skoskiewicz T 1972 *Phys. Stat. Sol. (a)* 11 K123
[3] Skoskiewicz T 1973 *Phys. Stat. Sol. (b)* 59 329
[4] Bardeen J, Cooper L N and Schrieffer J R 1957 *Phys. Rev.* 108 1175
[5] Ganguly B N 1973 *Z. Phys.* 265 433
[6] Burger J P and MacLachlan D S 1976 *J. Phys. (Paris)* 37 1227
[7] Rietschel H 1975 *Z. Phys.* B 22 133
[8] Jacobi N and Caron L G 1978 *J. Low Temp. Phys.* 30 51
[9] Householder A S 1953 *Principles of Numerical Analysis*, (New-York: McGill-Hill)
[10] Williams M L and Maris H 1985 *J, Phys. Rev.* B 31 4508
[11] Yakubo K and Nakayama T 1987 *Phys. Rev.* B 36 8933
[12] Yakubo K and Nakayama T 1989 *Phys. Rev.* B 40 517
[13] Nakayama T and Yakubo K 2001 *Phys. Rep.* 349 239
[14] Rowe J M, et al. 1974 *Phys. Rev. Lett.* 33 1297

IOP Conf. Series: Materials Science and Engineering **191** (2017) 012020 doi:10.1088/1757-899X/191/1/012020

Effect of dividing daylight in symmetric prismatic daylight collector

Shih-Chuan Yeh[1,*], Ju-Lin Lu[2], and Yu-Chin Cheng[1]

[1] De Lin Institute of Technology, Taiwan
[2] Taipei Municipal Wenhua Elementary School, Taiwan

*E-mail: yehshch@gmail.com

Abstract. This paper presented a symmetric prismatic daylight collector to collect daylight for the natural light illumination system. We analyzed the characteristics of the emerging light when the parallel light beam illuminate on the horizontally placed symmetric prismatic daylight collector. The ratio of the relative intensities of collected daylight that emerging from each surface of the daylight collector shown that the ratio is varied with the incident angle during a day. The simulation of the emerging light of the daylight collector shown that the ratio of emerging light is varied with the tilted angle when sunshine illuminated on a symmetric prismatic daylight collector which was not placed horizontally. The integration of normalized intensity is also varied with the tilted angle. The symmetric prismatic daylight collector with the benefits of reducing glare and dividing intensity of incident daylight, it is applicable to using in the natural light illumination system and hybrid system for improving the efficiency of utilizing of solar energy.

1. Introduction

Energy saving and using renewable energy are the most important issues of reducing energy consumption and slow down the pace of global warming. The Energy Performance of Building Directive highlights that the importance of reducing the emerging consumption in building [1]. And the energy consumption by artificial lighting system is a major portion of the total energy consumption in a building, it could account for 30% of the total electrical energy consumption in commercial buildings [2]. Utilization of efficient artificial lighting, such as LED, can reduce the electrical energy consumption [3-5]. An excellent design of interior natural light illumination would contribute to a lower consumption of electrical energy [6].

Solar energy is one of the renewable energies that now providing the lowest cost options for economic and community development in rural regions around the globe. Utilization of solar energy, in addition to the use of solar cell by photovoltaic effect, and solar thermal application, indoor daylight illumination is also an important access to solar energy. Due to the difficulty towards the full implementation of renewable energy, energy saving is crucial for global development. The windows of a building allow the penetration of daylight into a room for illumination; meanwhile, the energy consumption of artificial lighting is reduced. However, the daylight glare probably cause uncomfortable illumination. Kostantzos et al. study daylighting glare probability by simulation and experiment with an HDR camera in office space with dynamic window shades [7]. The perceived level of visual discomfort may vary with time of day, Altomonte et al. study the relationships between visual task difficulty and glare response [8]. The possibility of visual discomfort is likely to arise due

IOP Conf. Series: Materials Science and Engineering **191** (2017) 012020 doi:10.1088/1757-899X/191/1/012020

to the uncontrolled daylighting. Fasi et al. investigated the energy and visual; performance for different types of glazed windows, which including the impact of glare index and daylighting factor [9]. Hee et al. studied the role of window glazing on daylight illumination and energy saving [10]. Huang et al. analyzed the thermal and daylighting performance of glazing and shading design on office building, the results of simulation shown that the performance of windows design on the south orientation getting better as the latitude risen [11]. A natural light illumination system which is compounded by a daylight collector, guiding part and indoor natural light emission part. This system bring daylight into a building as a lighting source not only with the benefits of energy saving and reduced the glare, but also lead to a comfortable working environment and it has the potential to improve human health, mood, performance and productivity [12-16]. The energy conversion efficiency of commercially photovoltaics is about less than 20% at present. The natural light illumination system directly use daylight for indoor illumination which without the loss of conversion in different forms of energy. The possibility of controlling direct daylight by optical designed systems for comfort illumination or energy conservation has been reported by many authors in the past. Daylight is varied with time of day, climate, latitude and the orientation of buildings. One of the challenges of natural light illumination is the difficulty in getting a uniform illuminated condition during a day. For the purpose of getting visual comfort and energy saving, the concept of daylight linked controls which combined with the sensors is also presented [17]. Borisuit et al. proposed a camera like light sensor to monitor luminance and circadian weighted radiance and simulated the potential impact light on non-visual functions on daylight optimization in buildings [18]. Boscarino et al. developed a simulation environment and daylighting control strategy for energy efficient lighting while providing desired lighting level for LED-based [19]. High energy performance building concepts have proposed, combines energy saving and energy recovery from local renewable resources [20], the natural light illumination system can provide part building energy demand. The hybrid systems of natural light illumination coupled with artificial light source or solar cells are also developed for advanced illuminated condition and energy consumption [21-23].

The detailed analysis of transmitted light of daylight collector can improve the performance of natural light illumination systems and aid in the design of hybrid natural light illumination systems. Mohelnikova calculated the light rays transported inside a straight, tubular light guide and their distribution at the output of the light guide. Wittkopf et al. analyzed the luminous intensity distribution curves of anidolic daylighting systems based on the ray tracing model [24, 25]. Prismatic elements, which with the characteristics of re-directing the propagate direction and the potential of dividing the energy of incident light, can be part of a daylight illuminate system located on the roof or facade of buildings and can be used as collectors to collect and guide daylight to reduce glare and save energy. The characteristics of the light emerging from a passive prismatic canopy, which is composed of right-angled prisms have been presented for daylight incident on it at different incident angles over the course of a day based on a mathematical matrix model [26, 27]. It has been shown that these methodologies, which are based on redirecting daylight, can save energy and produce natural light illumination when they are used as daylight guiding systems. The structure of symmetric prisms with the benefits of simple manufacture and dividing intensity of transmitted light more effectively. This paper present a natural light illumination system that is composed of a symmetric prismatic canopy as well as daylight collector, a reflector for re-directing daylight into the room and an indoor reflector as illustrated in figure 1. We analyzed the characteristics of dividing intensity and direction of transmitted light based on the optical theories.

2. Methodology

The symmetric prismatic daylight collectors are made of PMMA, which with refractive index 1.4906. Fig. 2 is the photography of a symmetric prismatic daylight collector with a 60° apex. An illustration of the symmetric prismatic daylight collector is shown in Fig.3. The characteristics of transmitted light when sunshine illuminating onto the symmetric prismatic daylight collector are analyzed in both of geometric optical calculation and optical package program.

IOP Conf. Series: Materials Science and Engineering **191** (2017) 012020 doi:10.1088/1757-899X/191/1/012020

The intensity and propagate direction of transmitted light is calculated based on optical theory when parallel light incident onto the upper surface of the symmetric prismatic daylight collector. The ratio of the area which corresponding to the transmitted light would intersect to the surface left, SL and which would intersect to the surface right, SR of the prismatic daylight collector are obtained by Eq. (1) to Eq. (3), the parameter w is the width of the prismatic unit, h is the height of the prismatic unit, $\phi 2$ is the propagate direction of ST, ϕNT is the normal incident direction of ST, and LSL and LSR are the relative areas that the transmitted light would intersect to SL and SR, respectivelly. The ray trace is obtained by the law of reflection and Snell's law, and the transmitted intensity is determined by Fresnel's equations.

$$\frac{L_{SL}}{L_{SR}} = \frac{W'-L_{\delta}}{W'+L_{\delta}}, \quad (1)$$

$$L_{\delta} = h \cdot \tan\delta \quad (2)$$

$$\delta = \phi_2 - \phi_{NT} \quad (3)$$

Where, $W' = \frac{1}{2}W$

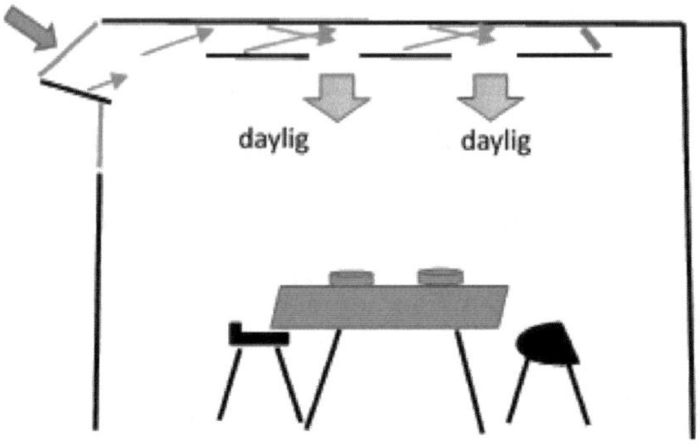

Figure 1. An illustration of a room with a natural light illumination system..

Figure 2. The photography of a symmetric prismatic daylight collector with a 60° apex.

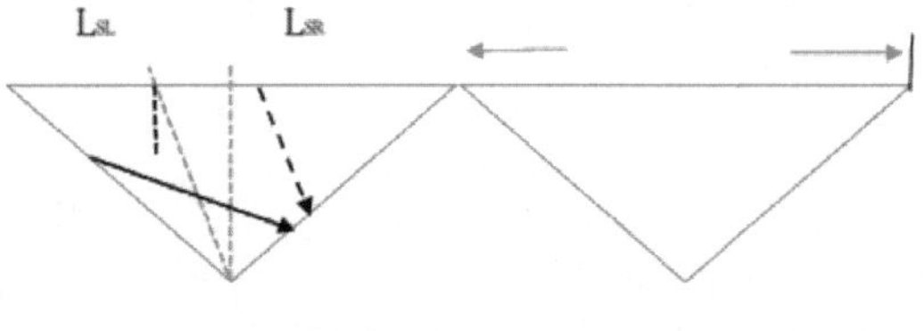

Figure 3. An illustration of the symmetric prismatic daylight collector.

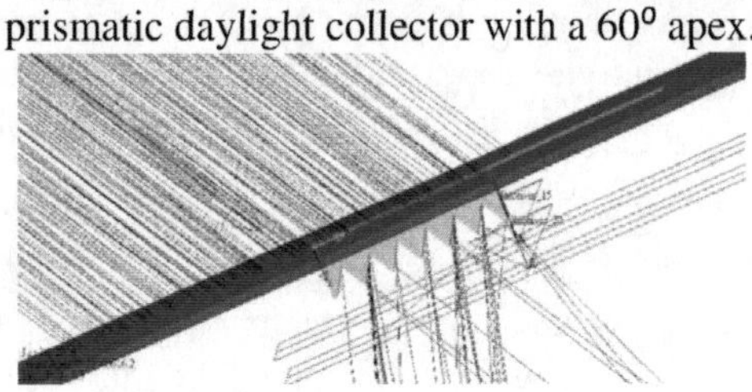

Figure 4. The simulated result of the parallel light incident to a symmetric prismatic daylight collector.

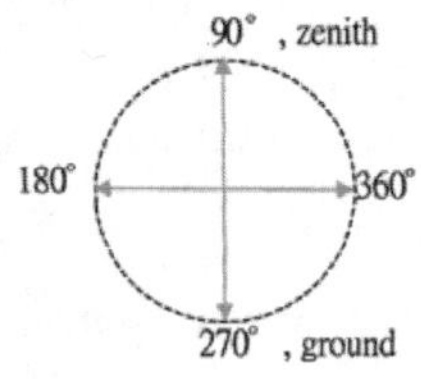

Figure 5. The illustration of the definition of direction of ray.

3. Results

In this paper, we simulated the parallel light incident a symmetric prismatic structure at different angle. Fig. 4 is the simulated result of the parallel light incident to the surface of the symmetric prismatic daylight collector, which with a 60° apex at 30° incident angle. It illustrated the property of dividing intensity of transmitted light when the parallel light incident to the surface of the symmetric prismatic daylight collector. Fig.5 is the illustration of the definition of direction of ray.

The transmitted light of the top surface, ST of the symmetric prismatic daylight collector would be incident to the left surface, SL and right surface, SR of the bottom of the prismatic unit when light illuminated onto the top surface of the symmetric prismatic daylight collector, and the ratio of LSL and LSR is varied with the incident angles. Due to the distribution of emerging light would impact to the design of natural light illumination system, we calculated the transmittance and direction of emerging light that from SR and SL when parallel light incident to the prismatic unit at different direction for improving design the natural light illumination system more efficiently. Surface Right 1, SR1 and Surface Right 2, SR2 which corresponding to the light emerging out from SR, the ray trace of SR1 means the transmitted light from ST and then emerged out from SR; the ray trace of SR2 means the transmitted light from ST and then intersect to SL, and emerging out from SR by the totally internal reflective at SL. Surface Left 1, SL1 and Surface Left 2, SL2 which corresponding to the light emerging out from SL, the ray trace of SL1 is transmitted light from ST and then emerged out from SL; the ray trace of SL2 is transmitted light from ST and then intersect to SR, and emerging out from SL by the totally internal reflective at SR. Fig. 6 is the calculated results of the transmittance and direction of the light which is emerging from SR1 and SL2 by the transmitted light intersect to SR when a parallel light beam incident onto the ST of a horizontally placed symmetric prismatic daylight collector with a 60° apex. The upper part of this figure shown that the main emerging light is transmitted from SR when sunshine illuminated the symmetric prismatic daylight collector at the region of 295° $<\phi 1<$360° , which is corresponding to the time interval during afternoon. Due to the direction of main emerging light is smaller than the direction of light which parallel to the surface SR, it can illuminate to space directly, which ϕSR equals to 330° for a horizontally placed symmetric prismatic unit with a 60° apex. And the other part of this figure shown that the transmitted light from SR by totally internal reflected on SR where inside the prism at the region of $\phi 1>$295° .However, the direction of emerging light is smaller than ϕSL, which light would re-incident to SL. Fig. 7 is the calculated results of the transmittance and direction of emerging light that the transmitted light intersect to SL. This figure shown that the main emerging light is transmitted from directly transmitted from SL when sunshine illuminated the symmetric prismatic daylight collector at $\phi 1<$242° .

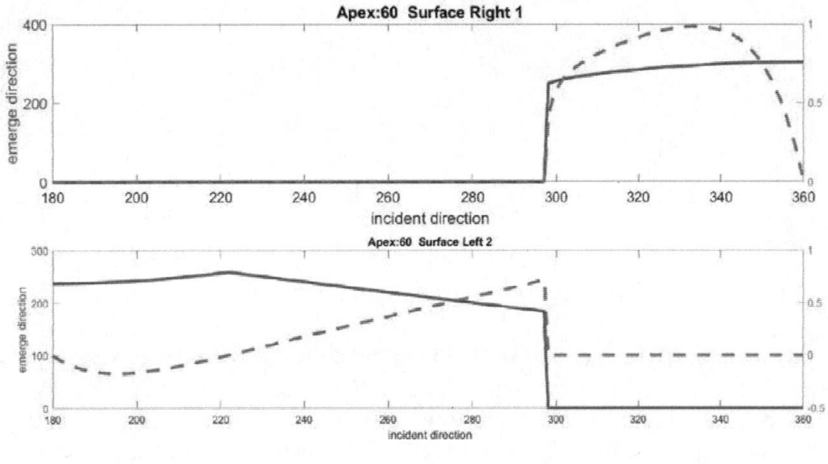

Figure 6. The transmittance and direction of emerging light that the transmitted light intersect to SR.

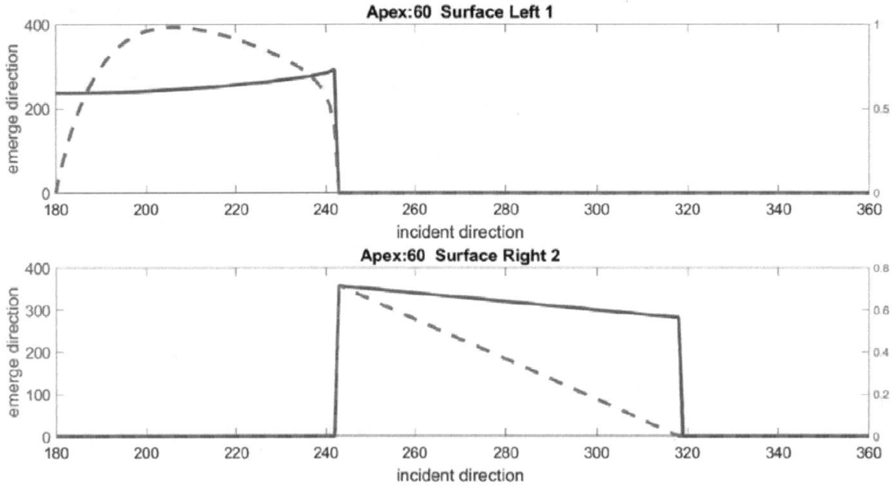

Figure 7. The transmittance and direction of emerging light that the transmitted light intersect to SL.

Fig. 8 is the combination of the emerging light which from SR and SL when the parallel light beam incident to the symmetric prismatic daylight collector. It shown that the transmittance of the emerging light which from SR and SL is varied with the direction of incident light beam, and the intensity is symmetric to $\phi 1=270^{o}$. The result shown that the natural light illumination system would be improving the efficiency of utilizing solar energy by use the symmetric prismatic daylight collector, due to the structure with the potential of dividing emerging light.

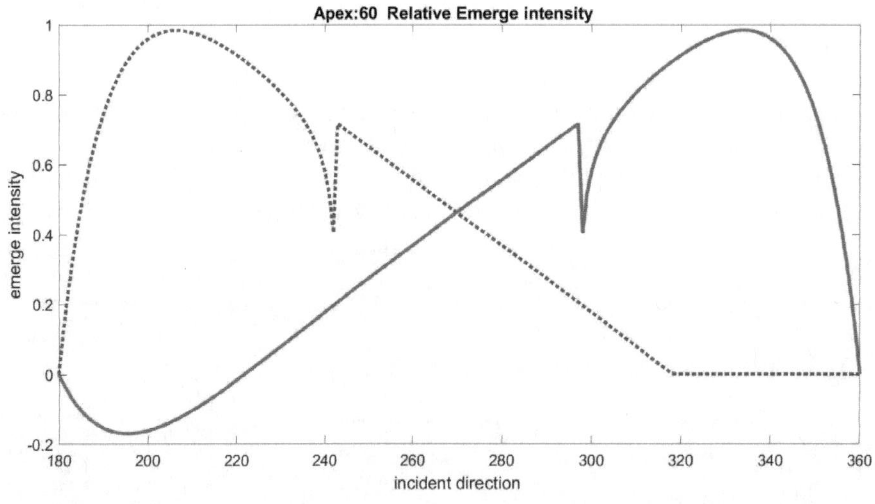

Figure 8. The combination of the emerging light which from SR and SL.

The direction and intensity of daylight beam that incident to the collector is varying with time when sunshine illuminate the collector. A daylight collector that is placed at an applicable tilted angle not only can reduce the glare, but also can increase the efficiency of utilizing of solar energy. Fig. 9 and Fig. 10 are the simulated results of the normalized intensity of emerging light when sunshine illuminate on the symmetric prismatic daylight collector which with the tilted angle 15° and 30°, respectively.

It shown that the ratio of the normalized intensities which from SR and SL are varied with the tilted angle, due to the relative incident angle of incident daylight beam is varied with the tilted angle of the collector. Fig.11 is the calculated result of integrated intensity at the region 270⁰< ϕ1<360⁰, which is refer to afternoon of a day. It shown that the integration of normalized intensity of emerging light from SL is increasing with increasing tilted angle of the symmetric prismatic daylight collector, and the integration of normalized intensity of emerging light from SR is not obviously vary with the tilted angle of the symmetric prismatic daylight collector, which is at a value of 2.2~2.4.

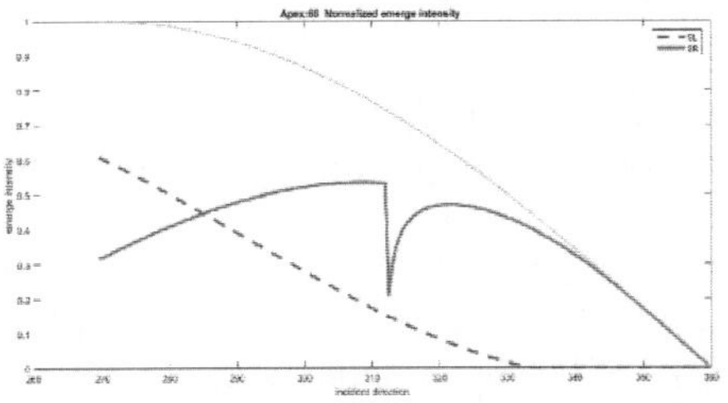

Figure 9. The simulated result of the normalized intensity when sunshine illuminated on the symmetric prismatic daylight collector which with the tilted angle 15° .

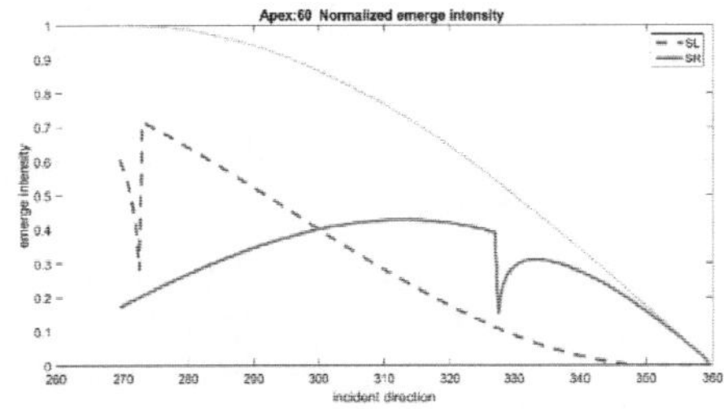

Figure 10. The simulated result of the normalized intensity when sunshine illuminated on the symmetric prismatic daylight collector which with the tilted angle 30° .

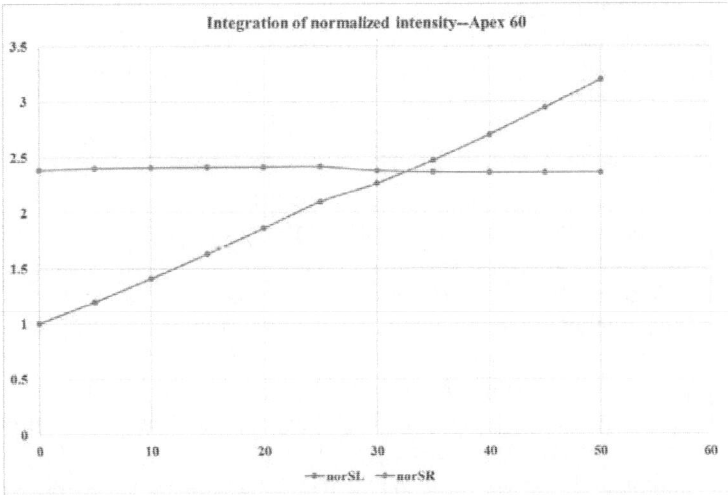

Figure 11. The calculated result of integrated intensity during afternoon.

4. Conclusion

In this paper, we detailed analyzed the characteristics of emerging light when parallel light beam illuminate on the symmetric prismatic daylight collector. The simulation of the parallel light beam incident to a horizontally placed symmetric prismatic daylight collector with a 60° apex shown this structure of daylight collector, symmetric prismatic unit can improve the efficacy of dividing intensity of a hybrid natural light illumination system, due to the ratio of main emerging light which from the SR and SL is varied with the incident angle of the incident beam. The calculated results shown that the transmittance of the emerging light which from SR and SL is varied with the direction of incident light beam, and the intensity is symmetric to ϕ1=270° . The normalized intensity of emerging light shown that the ratio of the integration of normalized intensity of emerging light from SR and SL is varied with tilted angle of the collector when sunshine illuminate on the symmetric prismatic daylight collector which is placed at a tilted angle. A daylight collector that is placed at an applicable tilted angle not only can reduce the glare, but also can increase the efficiency of utilizing of solar energy. The symmetric prismatic daylight collector can improve the efficiently utilizes solar energy for indoor natural light illumination, decrease the discomfort of glare and the application of hybrid natural light illumination system due to the benefits of reducing glare and dividing intesity of incident daylight.

Acknowledgments

The authors gratefully acknowledge the financial support of the Ministry of Science and Technology of Taiwan (MOST 104 - 2221 - E - 237 - 006)

References

[1] European Commission. Directive 2010/31/EU on Energy Performance of Buildings. Brussel: EC, 2010.
[2] Ministerio de Industria, Turismo y Comercio. Plan de Accion 2008-2012 de la Estrategia de Ahorro y Eficiencia Energetica en Espana. Madrid: MITC, 2007.
[3] D. Caicedo, A. Pandharipande, and G. Leus. "Occupancy-based illumination control of LED lighting systems." Lighting Research and Technology 43.2 (2011): 217-234.
[4] L. W. Yeh, C. Y. Lu, C. W. Kou, Y. C. Tseng, & C. W. Yi, (2010). Autonomous light control by wireless sensor and actuator networks. IEEE Sensors Journal, 10(6), 1029-1041.
[5] Y. J. Wen, & Agogino, A. M. (2010). Control of wireless-networked lighting in open-plan offices. Lighting Research and Technology, 1477153510382954.
[6] M. S. Mayhoub, & Papamichael, K. (2016). Cost/benefit analysis for building core sunlighting systems. Energy and Buildings, 118, 37-45.
[7] I. Konstantzos, A. Tzempelikos, & Chan, Y. C. (2015). Experimental and simulation analysis of daylight glare probability in offices with dynamic window shades. Building and Environment, 87, 244-254.
[8] S. Altomonte et al. "Visual task difficulty and temporal influences in glare response." Building and Environment 95 (2016): 209-226.
[9] M. A. Fasi, & I. M. Budaiwi, (2015). Energy performance of windows in office buildings considering daylight integration and visual comfort in hot climates. Energy and Buildings, 108, 307-316.
[10] W. J. Hee, M. A. Alghoul, Bakhtyar, B., Elayeb, O., Shameri, M. A., Alrubaih, M. S., & Sopian, K. (2015). The role of window glazing on daylighting and energy saving in buildings. Renewable and Sustainable Energy Reviews, 42, 323-343.
[11] Y. Huang, Niu, J. L., & Chung, T. M. (2014). Comprehensive analysis on thermal and daylighting performance of glazing and shading designs on office building envelope in cooling-dominant climates. Applied energy, 134, 215-228.
[12] Cheung HD, Chung TM. A study on subjective preference to daylight residential indoor environment using conjoint analysis. Build Environ 2008;43(12):2101–11.
[13] Wittkopf, S., et al., 2010. Ray tracing study for non-imaging daylight collectors. Sol. Energy 84, 986-996.
[14] Rosemann, A., Kaase, H., 2005. Light applications for daylighting systems. Sol. Energy 78, 772-780.
[15] Rosemann, A., Mossman, M., Whitehead, L., 2008. Development of a cost-effective solar illumination system to bring natural light into the building core. Sol. Energy 82, 302-310.
[16] Yeh S. C., Whang A. J., Hsiao H. C., Hu X. D., and Chen Y.Y.,2011. Distribution of Emerged Energy for Daylight Illuminate on Prismatic Elements. J. Sol. Energy Eng. 133, 021007. doi:10.1115/1.4003587.
[17] S. K. Chow, Li, D. H., Lee, E. W., & Lam, J. C. (2013). Analysis and prediction of daylighting and energy performance in atrium spaces using daylight-linked lighting controls. Applied energy, 112, 1016-1024.
[18] A. Borisuit, Kämpf, J., Münch, M., Thanachareonkit, A., & Scartezzini, J. L. (2016).

IOP Conf. Series: Materials Science and Engineering **191** (2017) 012020 doi:10.1088/1757-899X/191/1/012020

Monitoring and rendering of visual and photo-biological properties of daylight-redirecting systems. Solar Energy, 129, 297-309.

[19] G., Boscarino, & M. Moallem, (2016). Daylighting Control and Simulation for LED-Based Energy-Efficient Lighting Systems. IEEE Transactions on Industrial Informatics, 12(1), 301-309.

[20] Thiers S. and Peuportier B. Energy and environmental assessment of two high energy performance residential buildings. Build Environ 2012;51: 276–284.

[21] Tsangrassoulis, A., Doulos, L., Santamouris, M., Fontoynont, M., Maamari, F., Wilson, M., Jacobs, A., Solomon, J., Zimmerman, A., Pohl, W., Mihalakakou, G., 2005. On the energy efficiency of a prototype hybrid daylighting system. Sol. Energy 79, 56-64.

[22] Page, J., Scartezzini, J.L., Kaempf, J., Morel, N., 2007. On-site performance of electrochromic glazing cupled to anidolic daylighting system. Sol. Energy 81, 1166-1179.

[23] Chow, T.T., Qiu, Z., Li, C., 2009. Potential application of see through solar cells in ventilated glazing in Hong Kong. Sol. Enrgy Matrer. Sol. Cells 93, 230-238.

[24] Mohelnikova J., 2009. Tubular light guide evaluation. Building & Environment 44, 965–72.

[25] Wittkopf S., Grobe L. O., Moroder D. G., Compagnon R., Kampf J., Linhart F., Scartezzini J. L., 2010. Ray tracing study for non-imaging daylight collectors. Solar Energy 84, 986–996.

[26] S. C. Yeh, (2014). A natural lighting system using a prismatic daylight collector. Lighting Research and Technology, 46(5), 534-547.

[27] A. J. W.Whang, Lin, C. M., & S. C. Yeh, (2013). Investigation of prismatic daylight collectors with different apexes. Journal of Solar Energy Engineering, 135(1), 011015.

IOP Conf. Series: Materials Science and Engineering **191** (2017) 012021 doi:10.1088/1757-899X/191/1/012021

Numerical model of liquid metal flow in steel making tundish with flow modifiers

Manas Vasantrao More[1], Sandip Kumar Saha[2], Vishal Marje[3], G. Balachandran[3]

[1]Research& Development, Kalyani Center for Technology and Innovation (KCTI), India

[2]Department of Mechanical Engineering, Indian Institute of Technology Bombay, India

[3]Kalyani Carpenter Special Steel Pvt Ltd. (KCSSPL), India

E-mail: manas.more@bharatforge.com

Abstract. The optimum condition for clean steel production in the tundish of a continuous casting process reactor can be obtained using numerical modelling. Five different arrangements of flow modifier in the form of impact pad systems deployed in an eight ton, delta shaped, and two strand bloom caster tundish are analysed and optimum design of the impact pad to improve the inclusion removal efficiency is evolved. Reynolds Averaged Navier-Strokes (RANS) equations with standard k-ε model of turbulence and energy equation are used to study fluid flow and inclusion flotation in the tundish. The inclusion separation efficiency is evaluated by solving the inclusion transport equation. Height variations along with additional notch amongst different impact pads yield best micro inclusion separation efficiency.

1. Introduction

The continuous casting system is comprised of a ladle, a tundish, a mould and flow connectors (sub-merged entry nozzle). Tundish was originally developed as an intermediate vessel placed between the ladle and the mould, to evenly distribute molten steel to different moulds, at constant rate. In addition, the tundish is used to maintain the heat of the molten metal and increase the removal of detrimental non-metallic inclusions [1-2]. Fluid flow equations along with the energy equation were used to predict the temperature distributions under non-isothermal conditions [3-5]. Numerical modelling was used to predict parameters such as residence time distributions (RTD) [5] and distribution of top surface slag layer [6]. Solhed *et al* [6-7] included additional differential equations to describe inclusion trajectories and inclusion number density distribution. Standard *k- ε* model of Launder and Spalding [8] was used to calculate an eddy viscosity. Jha *et al.* [9] studied the effect of different turbulence models on the residence time distribution predictions along with LES model (large eddy simulation) in tundish modelling. Few studies used the low Reynolds number model of Launder and Jones [10-11] to study residence time distributions in a six-strand tundish and found better agreement with the experimental results. The present study concentrates on the performance of the tundish on account of inclusion separation efficiency. Five different design of flow modifier (*i.e.* impact pad) systems are studied to find out the novel design for a given tundish system.

IOP Conf. Series: Materials Science and Engineering **191** (2017) 012021 doi:10.1088/1757-899X/191/1/012021

2. Mathematical modelling

Fig. 1 shows the schematic diagram of the numerical domain of a tundish whose bare capacity was 8 tons, assuming a transverse symmetry plane. The shroud diameter and its immersion height considered were 67 and 400 mm, respectively. The system considers a submerged entry nozzle (SEN) diameter which is 45 mm and with a stopper rod diameter of 120 mm. A dam is present in between the inlet shroud and outlet. The five cases have impact pads having variations in the shape and size.

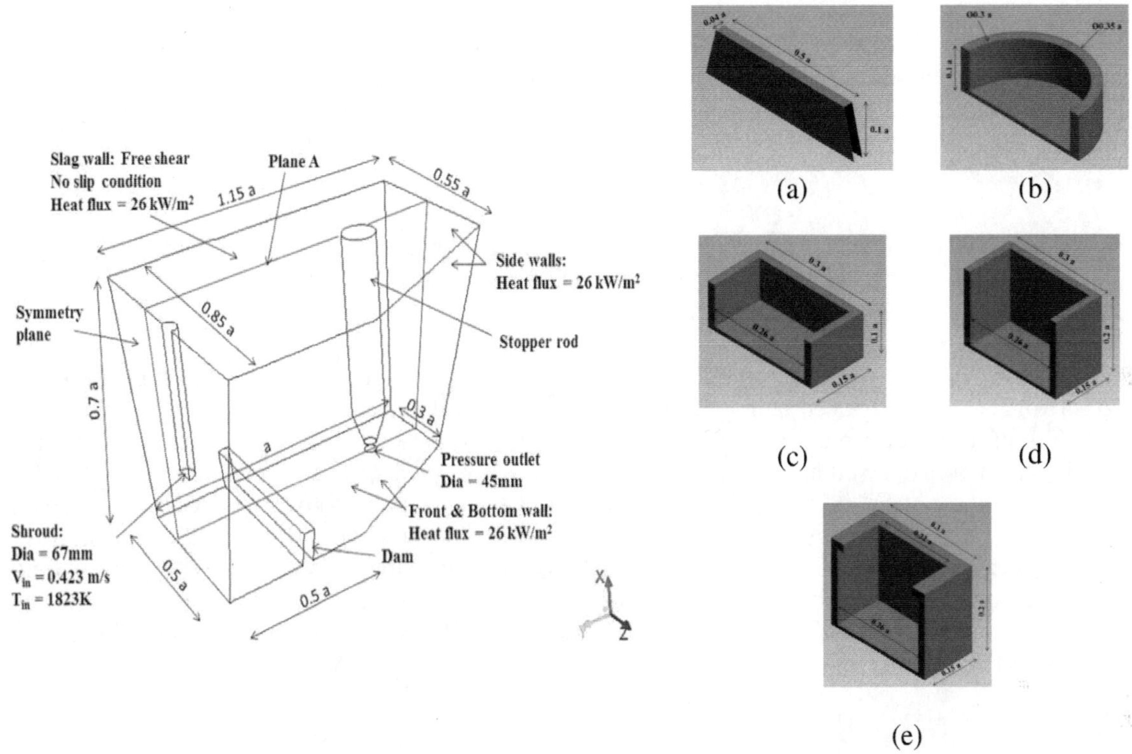

Figure 1. Schematic model and boundary conditions of a two strand tundish with different flow modifiers

2.1. Governing equations

The governing equations considered are the continuity, momentum and energy equations, to evaluate flow behaviour and inclusion trajectories.

Continuity equation:
$$\frac{\partial \rho}{\partial t} + \nabla.(\rho u) = 0 \tag{1}$$

Momentum equation:
$$\frac{\partial(\rho u)}{\partial t} + \nabla(\rho u u) = -\nabla P + \nabla.[(\mu_l + \mu_t)\nabla u] + F_{BY} \tag{2}$$

Energy equation:
$$\frac{\partial(\rho T)}{\partial t} + \nabla(\rho u T) = \nabla.\left[\left(\frac{k_t}{C_p} + \frac{\mu_t}{Pr_t}\right)\nabla \mathrm{T}\right] \tag{3}$$

In equation 2, F_{BY} represents thermally induced buoyancy force term. The acceleration due to gravity (g) acts along the negative direction of x-axis (Fig. 1). Reynolds Averaged Navier-Stokes (RANS) equation and the averaged continuity equation are solved for mean quantities of velocity and pressure. The standard k-ε model of Launder and Spalding [8] is used as the closure equation, which is given as,

Kinetic energy:
$$\frac{\partial(\rho k)}{\partial t} + \nabla.(\rho u k) = \nabla.(\Gamma_k \nabla_k) + G - \rho\varepsilon \tag{4}$$

Dissipation rate:
$$\frac{\partial(\rho\varepsilon)}{\partial t} + \nabla.(\rho u \varepsilon) = \nabla.(\Gamma_\varepsilon \nabla_\varepsilon) + \frac{\varepsilon}{k}(C_1 G - C_2 \rho\varepsilon) \tag{5}$$

Where $\mu_t = C_\mu \rho \frac{k^2}{\varepsilon}$, $\Gamma_k = \frac{\mu_{eff}}{\sigma_k}$ and $\Gamma_\varepsilon = \frac{\mu_{eff}}{\sigma_\varepsilon}$ are turbulent viscosity, diffusion coefficients for turbulent kinetic energy and its diffusion rate, respectively. μ_{eff} is the effective viscosity given by $(\mu_l + \mu_t)$. The turbulent viscosity (μt) & generation term *G* in equation 4 is given,

$$\mu_t = C_\mu \rho \frac{k^2}{\varepsilon} \tag{6}$$

$$G = \mu_t \frac{\partial u_j}{\partial x_i}\left(\frac{\partial u_j}{\partial x_i} + \frac{\partial u_i}{\partial x_j}\right) \tag{7}$$

The recommended values of the constants for the standard k-ε model are *C1= 1.44, C2=1.92, Cμ= 0.09, σk = 1 and σε=1.3* as proposed by Launder and Spalding [8].

2.2. Tundish performance analysis

2.2.1. Residence time distribution (RTD). Residence time distribution (RTD) accounts the fraction of dead volume (*Vd*), dispersed plug volume (*Vp*) and mixed volume (*Vm*) over a total tundish volume (*V*). For efficient performance of any reactor, the dead volume fraction should be the least. In this study, the modified mixed model of Sahai and Emi [13] is adopted. In this technique, a tracer or pulse is added to the system and the response to this pulse is recorded, which gives a typical curve, known as *C* curve for a pulse. *C* curve is represented in the non-dimensional form, in which, the concentration is normalised on the basis of a uniform, well mixed concentration value, while corresponding residence times are normalised with respect to the nominal holding, or theoretical residence time as non-dimensional concentration *C* and non-dimensional time *Θ*, where *V*= volume of tundish liquid and $\forall$= volumetric flow rate of liquid.

$$\Theta = \frac{t}{\bar{t}} = \left(\frac{V_a}{\forall_a}\right)\left(\frac{\forall}{V}\right) \tag{8}$$

$$V_d = 1 - \Theta_a \tag{9}$$

$$V_{dp} = \frac{\Theta_{min} + \Theta_{peak}}{2} \tag{10}$$

$$V_m = 1 - V_a - V_{dp} \tag{11}$$

$$V = V_m + V_p + V_d \tag{12}$$

From the *C* curve, dimensionless mean of RTD $(\bar{\theta})$ & dimensionless variance is calculated as,

$$\bar{\theta} = \frac{\int_0^\infty \Theta C d\Theta}{\int_0^\infty C d\Theta} \tag{13}$$

$$\sigma^2 = \frac{\int_0^\infty (\Theta - \Theta^2) C d\Theta}{\int_0^\infty C d\Theta} \tag{14}$$

2.2.2. Inclusion separation model. Inclusion trajectories are calculated using Lagrangian particle tracking method [12], which solves a transport equation for each inclusion as it travels through the previously-calculated constant molten-steel flow field. The forces acting on a particle of mass (m_p) are the drag force, the buoyancy force and the force due to added mass.

$$\frac{d\bar{u}_p}{dt} = F_D(u - u_p) + \frac{(\rho_c - \rho)}{\rho_c} g \tag{15}$$

$$F_D = \frac{18 \mu_l C_D Re}{24 \rho_c d_c^2} \tag{16}$$

In equation 16, Reynolds number can be obtained as,

$$Re = \frac{\rho d_c |u - u_p|}{\mu_l} \tag{17}$$

$$u'_p = \zeta_i \sqrt{{u'_p}^2} = \zeta_i \sqrt{\frac{2k}{3}} \tag{18}$$

The effect of the chaotic behaviour of non-metallic inclusions in the melt motion model is described by Discrete Random Walk model (DRW) called stochastic model. In this model, a random velocity vector (u'_p) is added to the calculated time-averaged vector $(\bar{u}_p)$ to obtain the inclusion velocity (u_p) at each time step as it travels through the fluid. Each random component of the inclusion velocity is proportional to the local turbulent kinetic energy level (*k*).

The ratio of number of trapped inclusions to the total number of inclusions leaving the inlet shroud gives the separation efficiency in percentage, which can be calculated as,

$$\eta = \frac{N_{in} - N_{out}}{N_{in}} \times 100 \tag{19}$$

2.2.3. Initial and boundary conditions. Liquid steel is considered as a working medium inside the tundish. The inlet velocity and temperature of steel are considered as 0.423 m/s and 1823 K, respectively for the steady and the transient case. Pressure outlet condition is used for the outlet. The detailed boundary conditions are as shown in (Fig. 1). Radiative heat flux $(q^{"})$ is calculated by Stephan-Boltzmann equation considering tundish wall temperature of 1123 K.

At $t = 0$
$$F_l = 0 \tag{20}$$

At inlet:
$$u_{in} = 0.423\,{}^{m}/_{s}, T_{in,l} = 1823\ K, T_{in,g} = 318\ K \tag{21}$$

At outlet:
$$\nabla P = 0 \tag{22}$$

At slag wall:
$$Free\ surface{:}\ \frac{\partial u_{x,l}}{\partial y} = 0\ and\ \frac{\partial u_{x,l}}{\partial z} = 0\ , q^{"} = 26\ kW/m^2 \tag{23}$$

At side walls:
$$No\ slip\ condition{:}\ u_i = 0, q^{"} = 26\ kW/m^2 \tag{24}$$

At symmetry plane:
$$\frac{\partial u}{\partial y} = 0\ and\ \frac{\partial T}{\partial y} = 0 \tag{25}$$

For steady state, the SIMPLE (Semi-Implicit Method for Pressure-Linked Equations) algorithm is used for pressure–velocity coupling [12]. Commercial software ANSYS FLUENT 14 is used to solve the governing equations along with initial and boundary conditions. The coefficients in momentum and *k*-ε equations are calculated using the second order upwind scheme. The turbulence intensity & convergence criteria are taken as 5% & 1×10^{-7}.

Table 1. Thermo physical properties of liquid steel and air used in the numerical analyses

Properties	Liquid steel	Air
ρ (kg/m^3)	7010	1.225
C_p (J/kg·K)	821	1006.43
k_t (W/m.K)	30.5	0.0242
μ (kg/m.s)	0.00648	0.000178

3. Results and discussions

3.1. Residence time distribution (RTD)

Fluid flow in tundish is typically characterized by residence time distribution (RTD) curves. The RTD is frequently plotted as dimensionless concentration versus dimensionless time. This information is graphed as a function of time with respect to tracer concentration that provides the RTD curve, which is also commonly known as a *C* curve. The plug flow fraction is defined by the advent of the first injected tracer volume at the tundish outlet nozzle. The mixed also called dispersed volume is indicated by the tail of the C curve, viz. from the apex to the time defined as 2Θ. Dead flow in RTD curves is generally defined as the portion of the injected tracer persisting beyond 2Θ. It can be noted from the (Fig. 2) & (Table 2) that, in terms of least dead volume fraction, Case 3 is the most efficient model amongst all design. The highest dispersed plug flow volume fraction is obtained for case 5.

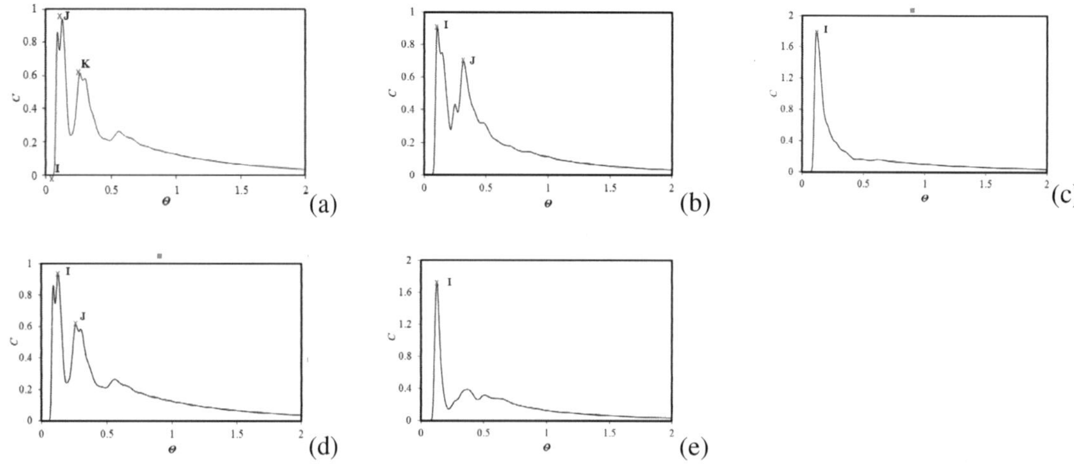

Figure 2. *C* Curve for (a) case 1, (b) case 2, (c) case 3, (d) case 4 & (e) case 5

3.2. Dead volume, dispersed plug flow volume, mixed volume and Inclusion separation efficiency
To study the performance of various flow modifiers based on the RTD curves based on modified mixed model of Sahai and Emi [13] are tabulated in Table 2. The highest dispersed plug flow volume fraction is obtained for case 5.

Table 2. Dead flow, dispersed plug flow and mixed volume fractions for different design cases

Design	Dead volume fraction	Dispersed plug flow volume fraction	Mixed volume fraction
Case 1	0.564	0.096	0.34
Case 2	0.532	0.091	0.378
Case 3	0.485	0.097	0.416
Case 4	0.564	0.096	0.34

Inclusion separation efficiency for different cases is performed using discrete phase model (DPM). The size of the inclusions varies from 1 to 100 µm. High Alumina (Al_2O_3) containing 94-96% aluminium oxide with density of 3960 kg/m^3 is chosen as an inclusion material. For tracking inclusion trajectories, trap boundary condition is used for the slag wall. Stochastic tracking (random walk) model is used to predict the dispersion of inclusion. Table 3 shows the effect of flow modifiers on inclusion removal efficiency of the tundish.

The various sizes of inclusion diameter are analysed by which it can be noted that the highest separation efficiency is achieved for case 5. The important part is the effectiveness of this design in separating micro inclusions (<50 µm), which ultimately determines the steel quality. These effects can reduce the inclusion separation efficiency by 10% from those obtained in real situations.

Table 3 Effect of various flow modifiers on inclusion separation efficiency

Inclusion diameter (μm)	Inclusion separation efficiency (%)				
	Case 1	**Case 2**	**Case 3**	**Case 4**	**Case 5**
1	36.04	40.82	46.52	47.32	62.51
10	36.14	41.67	47.55	48.79	63.29
20	41.67	44.71	49.77	50.52	67.74
30	53.47	49.04	55.81	56	66.52
40	66.04	58.94	62.14	63.97	69.24
50	72.32	68.86	70.91	71.13	73.33
60	77.22	74.81	77.71	75.77	77.43
70	80.79	80.48	81.97	79.67	81.81
80	83.03	84.31	83.67	83.43	86.17
90	87.18	88.74	87.66	87.08	90.29
100	89.16	92.24	90.68	90	92.9

4. Conclusion

A detailed numerical analysis of tundish with different flow modifier is performed to study the effect of different flow modifiers on the fluid flow characteristics under non-isothermal conditions. Different heights of flow modifiers and wide range of inclusion sizes are considered to assess their effects on the inclusion removal efficiency inside the tundish. The use of dam, round and short impact pads is not recommended, as it results in very low values of micro inclusion separation efficiency. Height variations in short box type impact pad from height of 0.1 to 0.2 m yields better results in removal of inclusion. It is found that the long modified box type turbulence inhibitor/ impact pad arrangement (case 5) gives the best solution considering all the performance aspects, namely RTD, inclusion separation and temperature variation inside a tundish. The findings obtained in the present study could be used to formulate the guidelines for designing efficient tundish in steel industries.

Acknowledgements

The authors would like to express sincere gratitude towards Dr. Rajkumar Singh, Kalyani Center for Technology and Innovation, Pune for his continuous support and valuable comments. The authors gratefully acknowledge the extended support provided by Bharat Forge Ltd. and Kalyani Carpenter Special Steels Private Ltd. Management, Pune.

References

[1] Santos A S, Eugenio L F, Carlos R F F, Jaime A S and Amauri G, *Model Simul Mater Sci Eng*, 13 (2005), 1071.
[2] Mazumdar D, *Trans Ind Inst Met*, 66 (2013), 597.
[3] López-Ramírez S, Barreto J De J, Vite-Martínez P, Romero Serrano J A and Duran-Valencia C, *Metall Mater Trans B*, 35 (2004), 957.
[4] Vargas-Zamora A, Morales R D, Díaz-Cruz M, Palafox-Ramos J and Demedices L G, *Int J Heat Mass Tran 46*, (2003), 3029.
[5] Craig K J, De Kock D J, Makgata K W and De Wet G J, *ISIJ Int*, 41 (2001), 1194.
[6] Solhed H, Jonsson L and Jonsson P, *Metall Mater Trans B*, 33 (2002), 173.
[7] Solhed H and Jonsson L, *Scand J Metall*, 32 (2003), 15.
[8] Launder B E and Spalding D B, *Lectures in Mathematical Models of Turbulence*, Academic Press, NY (1972).
[9] Jha P K, Ranjan R, Mondal S S and Dash S K, *Int J Numer Method H*, 13 (2003), 964.
[10] Jha P K and Dash S K, *Int J Numer Method H*, 12 (2002), 560.
[11] Robert A and Mazumdar D, *Steel Res*, 72 (2001), 97.
[12] *ANSYS-FLUENT 14 User Guide*.

IOP Conf. Series: Materials Science and Engineering **191** (2017) 012021 doi:10.1088/1757-899X/191/1/012021

[13] Sahai Y and Emi T, *ISIJ Int*, 36 (1996), 667.

Experimental investigation on erosive wear behaviour of plasma spray coated stainless steel

Girisha K G[1*], K.V.Sreenivas Rao[2], Anil.K.C[3], Sanman.S[4]

[1]Research Scholor, [2]Professor, Department of Mechanical Engineering, Siddaganga Institute of Technology, Tumakuru
[3]Assistant Professor, Department of Industrial Engineering and management, Siddaganga Institute of Technology, Tumakuru.
[4]Assistant Professor, Acharya Institute of technology, Bengaluru.

E-mail: girishamse@gmail.com

Abstract. Slurry erosion is an implicit problem in many engineering industrial components such as ore carrying pipelines, slurry pumps and extruders. Even the water turbine blades are subjected to erosive wear when the water contains considerable amount of silt. In the present study, Al_2O_3-40%TiO_2 powder particles of average particle size of 50 micrometer were deposited on EN56B martenistic stainless steel by atmospheric plasma spray technique. Ni/Cr was pre coated to work as bond coat for good adhesion between coating and the substrate material. A coating thickness of 200 micrometer was achieved. Coated and un-coated substrates were subjected to slurry erosion test as per ASTM G-119 standard. Slurry erosion test rig was used to evaluate the erosion properties at room temperature condition by varying the spindle speed. Scanning electron microphotographs were taken before and after the slurry erosion test. Microstructures reveal uniform distribution of coating materials. Eroded surface shows lip, groove, and crater formation and dense coating resulting in less porosity. Micro hardness test was evaluated and reported. EDX analysis confirms the presence of Al, Ti and O_2 particles. It was observed that, Al_2O_3-40%TiO_2 coated substrates exhibit superior erosion resistance as compared to un-coated substrates due to higher hardness and less coating porosity.

1. Introduction

Martenistic stainless steel are general purpose steels which are used in the fabrication of engineering components that require high mechanical properties and decent resistance to corrosion, such as valve parts, fasteners, cutting tools, hydroelectric turbine blades etc. [1]. Martenistic steels have good shock resistance and high plasticity, but due to less hardness and poor wear resistance they are sometimes not preferred for industrial application [2]. Thermal spraying is frequently considered as alternative prospective to the conventional coating processes for obtaining of wear and corrosion resistant coatings [3].Thermal spraying process is one of the largely used method of coating, which includes many processes and materials. Alumina-Titania composite ceramic coatings were extensively coated by the atmosphere plasma spray process for the components very much needed in wear, corrosion, erosion resistant environment [4]. Thus, plasma spraying is most versatile process used to produce coating of ceramic materials like Al_2O_3, Cr_2O_3, TiO_2 particles. They have properties like high hardness because of their pure ceramic nature. So, they were unconcerned to may erosion and corrosion environments [5]. Sanjeev Bhandari et.al have developed the Al_2O_3-13%TiO_2 and Al_2O_3 coatings on CF8M steels using D-gun spraying, and evaluated slurry erosion behavior with different process

Published under licence by IOP Publishing Ltd

IOP Conf. Series: Materials Science and Engineering **191** (2017) 012022 doi:10.1088/1757-899X/191/1/012022

parameters. They conclude that, Al_2O_3-13%TiO_2 shows superior resistance as compared to Al_2O_3 coated steel [6]. KVS Rao et al have studied the slurry erosion behaviour of plasma sprayed TiO_2 coatings on 410 steel substrate by varying the slurry rotational speed and reported that TiO_2 coated substrates possess more erosion resistance as compared to uncoated substrates [7].

In the light of the above, the objective of the present study is to evaluate the slurry erosion behavior of the plasma sprayed Al_2O_3-40%TiO_2 composite coating on the EN56B martenistic stainless steel by varying the slurry rotational speed at room temperature. These studies could help in increasing the performance of the components which operate in slurry erosive environment.

2. Experimentation details

2.1. Selection of substrate material and feed stock material

Martenistic stainless steels are general purpose steels which are used in the manufacturing of many engineering components, specifically in the hydro electric power generation sectors. During rainy season in India, the Himalayan region power generation units are facing erosion problems. In the present work EN56B martenistic steel substrates were selected as base material to develop the coatings for studying the erosion behavior.The substrate material was cut into standard dimension of 25x25x8mm from the steel sheet and subjected to grit blasting using alumina grit of size 60 micrometer. In the present study, Al_2O_3-40%TiO_2composite powder particles of average particle size 50 micrometer has been selected as feed stock materials. Fusing, blending and crushing operations have been done using ball mill to obtain this composition of powder particles. Ni/Cr powder particles in the ratio 80:20 are used as bond coat between the substrate and coating materials.

2.2. Coating method

Atmospheric plasma spray technique was used to develop the composite coating on the steel substrate. The plasma spray system is well known versatile process for ceramic coatings in which the plasma system consists of a house of power supply, control unit, heat exchangers, water supply unit for supplying the purified water to cool the sprayed gun. The System also has two electrodes(anode and cathode) in which high voltage current is capable of generating the arc between these and allow the primary and secondary(Ar and H2) gases into the arc chamber, which cause the ionization of the gases that leads to high temperature(about 150000K) plasma flame. Al_2O_3-40%TiO_2powders were introduced into the high temperature plasma flame through feeder. Flame from the tip of the nozzle melts the particles and impact on the target substrate with high velocity. Coating thickness of 200 micrometer was achieved successfully. The process parameters were kept constant throughout the operation. Ni/Cr particles of ratio 80:20 were made to deposit primarily on grit blasted substrates of average thickness 50 micrometer which acts as bond coat. Above this bond coat the surface coating of Al_2O_3-40%$TiO_2$was made using the same process.

2.3. Microstructure, micro hardness, porosity

One of the drawbacks of the plasma spray coating is formation of micro cracks and small pores on the surface of the substrates. In the present study scanning electron microscope was used to examine surface morphology of the coated and un coated surfaces before and after test. Microphotographs reveal the consistent scattering of the powder particles. EDX analysis was carried out for coated substrates and confirms the presence of Al and Ti particles. Porosity of the coated substrates was examined using optical microscope with image analysis software. In the present study micro hardness of the coated and uncoated substrates were examined using Vickers micro hardness tester.The micro hardness of the uncoated samples shows hardness of 259 HV0.1 and Al_2O_3-40%TiO_2coating shows hardness of 606 HV0.1.

2.4. Slurry erosion test

Slurry erosion test for the coated and un coated steel substrates were conducted using high speed slurry erosion test rig as per ASTM G119 standard. Test rig consists of six rotating spindles(500-1500 rpm) and stainless steel slurry cups of capacity 2 litres. These cups are kept inside the test rig containing flat surface which is driven by electric motor for moving upward and downward. Test samples of standard size 25x25x8mm were fixed to the rotating spindles by means of screws and dipped into the slurry cup which contains 3.5% of NaCl powder and a required concentration of silica sand particles in one litre of distilled water. Before subjecting the samples to slurry test, one face of the sample was covered with masking tape to avoid erosion of the other face. In the present work slurry erosion study was carried out by varying the speed of the spindle in rpm(500,1000,1500) keeping silica sand particle size 312 micrometer and concentration 200gms/litre as constant throughout the test.

3. Result and discussion

3.1. Micro structure

Figure 1 and Figure 2 shows the scanning electron microphotographs of the uncoated and plasma coated Al_2O_3-TiO_2 steel substrates. They show the uniform distribution and dense deposition of coating material with good bonding with the substrate. Some deformed (partially melted) particles and lamellar structure was also observed.The optical microscopy analysis with image analysis software reveals a porosity of about 3.5% which describes that the coating obtained is very dense. Porosity is a function of powder particle size and effectiveness of the coating technique. Coarse powder particles are not desirable for obtaining dense coating. Fine particle size also yields good surface finish. Proper fusing of the powder coating also leads to less porosity and dense coating. Denser coatings give rise to higher hardness and obviously greater wear resistance. Composite coatings with a combination of various grades of powder particles were also found to yield good compaction and better hardness values. Figure 3 shows the Al_2O_3-40%TiO_2coating's EDX pattern that reveals the presence of Al, Ti, O_2 elements.

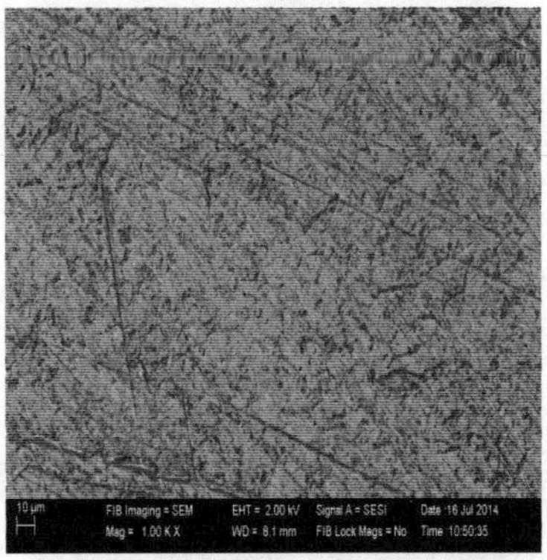

Figure 1. SEM image of base material before test

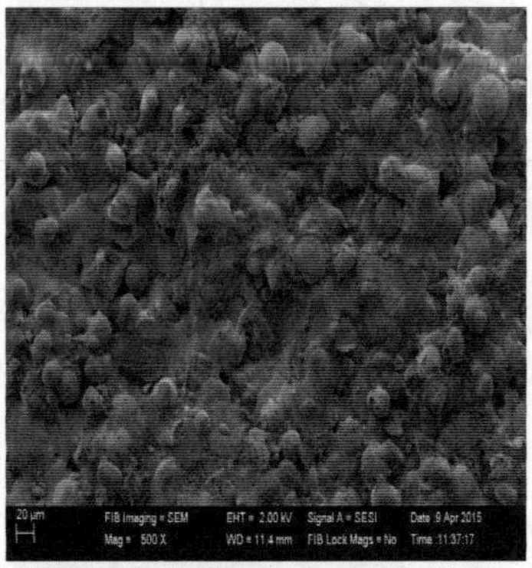

Figure 2. SEM image of Al_2O_3-40%TiO_2 coated substrate

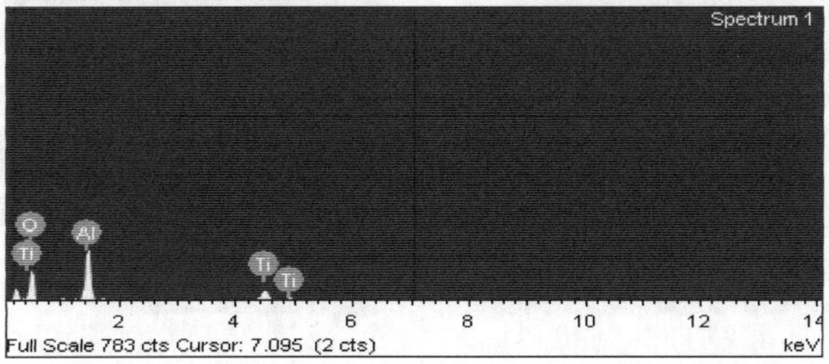

Figure 3. EDX spectrum of Al_2O_3-40%TiO_2coating

3.2. Slurry erosion test

The coated samples were subjected to slurry erosion test using slurry erosion test rig. The test rig consists of a double walled water cooled stainless steel container with an internal diameter of 120 mm. The container houses six stainless steel spindles for mounting the test specimens. The spindle speed can be varied from 100 to 1500 rpm. The silica sand slurry was prepared by using distilled water with 3.5% NaCl and 20% silica sand. The Al_2O_3-40%TiO_2 coated test specimens were mounted on the stainless spindle. The erosion test was conducted for three different spindle speeds of 500rpm, 1000rpm and 1500rpm by keeping time as 5 hour. Slurry erosion rate was measured using weight loss technique. SEM images of the worn surface of uncoated and Al_2O_3-40%TiO_2coated steel are shown in Figure 4-7. Figure 8 shows the effect of spindle rotational speed on the erosion rate of the test surface. It is observed that, the mass loss for both uncoated and coated substrates increases with increase in the slurry rotational speed. Anyhow, coated substrates shows improved resistance to erosion as compared to uncoated surface. Al_2O_3-40%TiO_2 coated substrates shows better erosion resistance due to higher crack propagation resistance and toughness [3, 4]. From the experiment it was confirmed that, maximum mass loss occurs at higher speed (1500rpm). Because, at this speed impinging speed of the silica sand particles on to the substrates will increase that leads to irregular impingement causing more weight loss as compared at lower speeds. More plastic deformation was found to occur on both the uncoated and coated substrates causing the formation of grooves, lips and crater at higher rotational speeds as evidenced by the surface morphology of the test surface. Comparatively the plastic deformation on the Al_2O_3-40%TiO_2coated surface is much less than on the uncoated base material. This is attributed to better intersplat adhesion of Al_2O_3 and TiO_2 particles on the substrate which increases the hardness of the surface that resists wear loss.

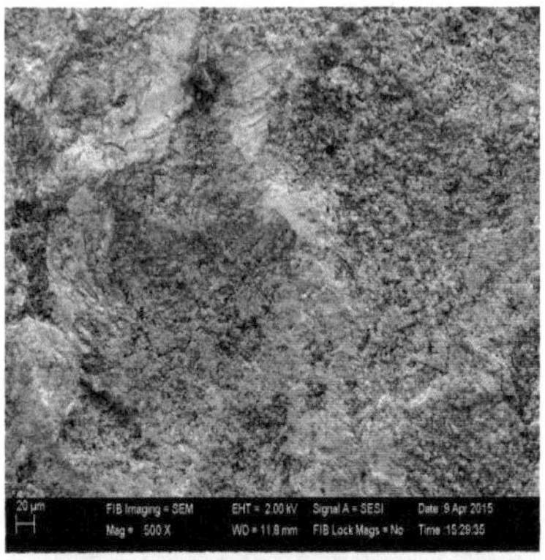

Figure 4. Un coated EN56B Steel at lower speed

Figure 5. Un coated EN56B Steel at Higher speed

Figure 6. Al_2O_3-40%TiO_2coated steel at lower speeds

Figure 7. Al_2O_3-40%TiO_2coated steel at higher speeds

IOP Conf. Series: Materials Science and Engineering **191** (2017) 012022 doi:10.1088/1757-899X/191/1/012022

4. Conclusion

The conclusions composed from the present experiment are as follows:

- Al_2O_3-40%TiO_2powder particles were successfully deposited by plasma spray process on target material with dense coating and less porosity. EDX analysis confirms the presence of feedstock materials.
- Slurry Erosion Rate can be measured by varying different erosion parameters like Time, slurry concentration and particle size of the silica sand.
- Plasma spray coated steel substrates shows superior erosion resistance as compared to uncoated samples due high hardness of the surface.
- With increasing the speed of the slurry rotation, material loss of the exposed surface also increases for both uncoated and coated surface.

References

[1] E. Rolinski, Effect of plasma nitriding temperature on surface properties of austenitic stainless steel, Surface Engineering 3 (1987) 35-40.

[2] I. Alphonsa, A. Chainani, P.M. Raole, B. Ganguli, P.I. John, Surf. Coat.Technol. 150 (2002) 263.

[3] Handbook of Thermal Spray Technology, ASM International, Materials Park, OH, USA, p. 171.

[4] D.A. Gerdman, N.L. Hecht,ArcPlasma Technology in Materials ScienceSpringerVienna,Austria 1972.

[5] V.V. Sobolev, J.M. Guilemany, J, High Velocity Oxy-Fuel Spraying.Theory, Structure-Properties RelationshipsandApplicatns.Maney for theInstitute of Materials,Minerals and Mining,London,2004,p. 27.

[6] Sanjeev Bhandari, Harpreet Singh, “Slurry Erosion Behaviour of Detonation Gun Spray Al_2O_3 and Al_2O_3–13TiO_2-Coated CF8M Steel Under Hydro Accelerated Conditions” Tribol Lett (2012) 45:319-331.

[7] K V Sreenivas Rao, K G Girisha and Y D Rakesh “Evaluation of slurry erosion wear characteristic of plasm sprayed TiO_2 coated 410 steel” IOP Conf.Series: Materials Science&Engineering 149 (2016) 012067

Effect of Spindle Parameters of Woodworking Band Saw on the AE Value of Crack Band Saw Blade in Compound Material Processing (1)

Jin-gui GAO[1], Zhao-fang JIANG[1*] and Lai-peng LUO[1]

[1]Beihua University, Jilin city of jilin province, 132013, P.R.China

E-mail: [*]675375003@qq.com

Abstract. Taking the MJ3210A motion band saw as the research object, the AE value of the band saw blade vibration was obtained by analyzing the VIBSYS vibration signal acquisition and analysis software system in Beijing, and the change of the AE value of the band saw and the crack was found out. The experimental results show that in the MJ3210A sports car sawing machine, the band saw blade with width of 130 mm is used, and the AE value of the cracked band saw blade is well in the high band saw blade AE value. Under the best working condition of the band saw, the band saw blade AE If the value exceeds 104.7 dB (A) above, it means that the band saw blade has at least one crack length greater than 1.38 mm for the crack defect and the need to replace the band saw blade in time. Different species with saw blade of the AE value is different, white pine wood minimum, the largest oak wood; according to a variety of wood processing AE instrument value to determine the band saw blade crack to the situation; so as to fully rational use of band saw blade, The failure and the degree of development to find a new method.

1. Introduction

Due to the saw blade used for cutting, it is the closed cutting in the cutting, the saw blade cutting edge can not observe the change of the state at any time with the eyes, and can not determine the correct replacement time of the saw blade [1-5]. According to the actual survey, wood processing enterprises online inspection usually use two kinds of band saw cutting edge and band saw the overall situation: one is the use of band saw blade cutting time, that is, using time constraints. That is, according to the experience of band saw blade, with a period of time with chalk smear blade root position, and then determine the light development to determine the degree of crack generation, and the second is based on the situation when the sound cutting to determine the work of the band saw and the degree of crack The With reference to the first method of light, check the applied lighting value determination method to check whether the band saw blade crack. In this paper, the focus of the band saw the formation of cracks before and after the expansion of the law. In order to improve the quality of wood surface treatment, the correct judgment of the band saw a reasonable time to provide a scientific basis..

2. The experiment main instruments and equipment

2.1 The experiment main instruments

The apparatus and equipment used in this test are shown in Table 1.

Table 1. Test apparatus and equipment.

Name	Model	Manufacturer
woodworking band saw	MJ3210A	Shenyang Shen He woodworking machinery factory
The large capacity data acquisition instrument quickly	WS – 5942-2-5	Spectrum of Beijing century technology development co., LTD.
Precision multi-function digital AE meter meter	GSM1020A	Shenzhen shigeru source technology co., LTD.

2.2 MJ3210A specific parameters of the woodworking band saw machine

The test and application of Tianjin forestry cutting tool factory production of No. 18 B.W.G band saw blade. Show.

2.3 Cutting wood

The parameters of the test wood are shown in Table 2.

Table 2. Test apparatus and equipment.

Tree species	Manchurian Ash
Moisture content	10%
Cutting thickness	100mm

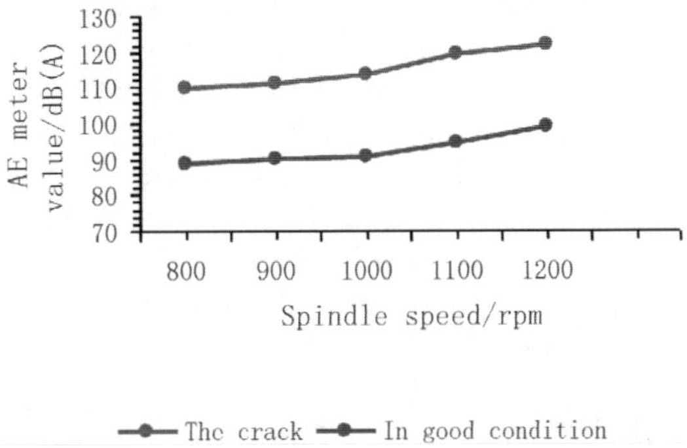

Figure 1. Saw wheel spindle speed under different load crack defects on the intact band saw blade and the band saw blade cutting AE count value

3. Test data processing and result analysis

3.1 Saw wheel spindle speed on the intact and crack defects of band saw blade AE count value

In band sawing machine pressed tuo weight is 93 N, up and down saw card a distance of 600 mm, under the premise of the feed speed of 15 m/min, under the five kinds of saw wheel speed, respectively, the crack is 1.38 mm of defective band saw blade and complete (sound) band saw blade for data collection and analysis, get the band saw blade when cutting AE meter value curve as shown in Fig. 1.

By Fig.1 shows, saw wheel spindle speed impact on the size of the band saw blade transverse vibration displacement is very obvious, saw wheel spindle speed before 1000 RPM with the increase of saw wheel speed saw blade transverse vibration displacement is less, when after saw wheel spindle speed 1000 RPM, with the increase of saw wheel rotation speed of the band saw blade transverse vibration displacement is also growing, and has a crack band saw blade AE meter value Ming is greater than the intact band saw blade AE meter. This is due to the increased with the increase of band sawing machine saw wheel spindle speed band saw blade transverse vibration caused by increased. When saw wheel spindle speed at 1000 RPM sawing machine to achieve the best working state; In

IOP Conf. Series: Materials Science and Engineering **191** (2017) 012023 doi:10.1088/1757-899X/191/1/012023

good band saw blade AE meter value in 88.4 dB (A) 98.9 dB (A), and between the defective band saw blade AE meter value in 109.3 dB (A) ~ 121.4 dB (A).

3.2 The effect of cutting different tree species on the integrity of the wood and the occurrence of cracked defects with band saws

In the determination of saw wheel spindle speed 1000rpm, pressure lump weight 93N, saw blade up and down distance of 600mm, feed rate of 20m / min, under the premise of cutting five kinds of tree birch, color wood, poplar, elm, oak The parameters of the AE scale are shown in Fig. 2 for the data acquisition of the complete saw blade and the defect band saw blade with the crack length of 1.68mm, and the saw blade cutting tree is obtained.

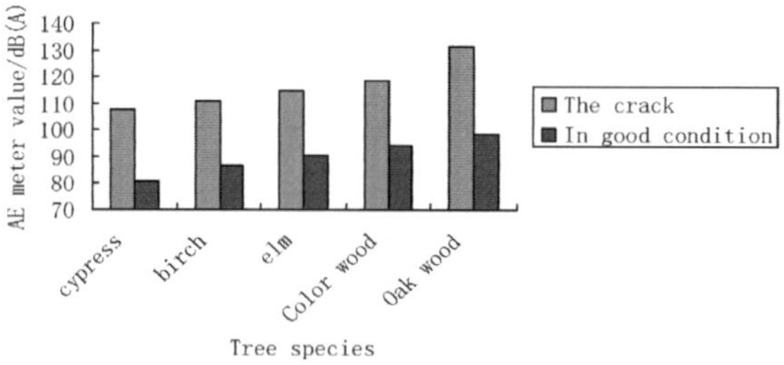

Figure 2. The effect of cutting the AE value when cutting the different saws with the saw blade and the crack

It can be seen from Fig. 2 that the vibration displacement of saw blade with saw blade is smaller than that of other wood, and the largest amount of oak is the smallest. Poplar: complete band saw blade 80.8dB (A), defect band saw blade 107.6dB (A), birch: complete band saw blade 86.8dB (A), defect band saw blade 110..8dB (A), elm: complete band saw blade 90.4 (A), defect band saw blade 114.6dB (A), color wood: complete band saw blade 94.5dB (A), defect band saw blade 118.5dB (A), oak: complete band saw blade 98.6dB (A), defect band saw blade 131.5dB (A) ,; Generate cracks with band saws AE values are generally larger than the full band saw blade AE values. Poplar in these five kinds of trees in the hardness of the smallest, the largest hardness of oak, which shows that with the hardness of the wood cutting is increased, AE value also increases.

3.3 The influence of the upper and lower saw blade distance on the AE value of the good and cracked band saw blade

Under the premise that the speed of the main shaft of the band saw is 1000rpm and the weight is 93N, the band saw blade crack is 1.68mm, and the distance between the saw blade and the saw blade is changed. Now with the saw blade up and down the distance to get the band saw the AE value of the curve shown in Figure 3..

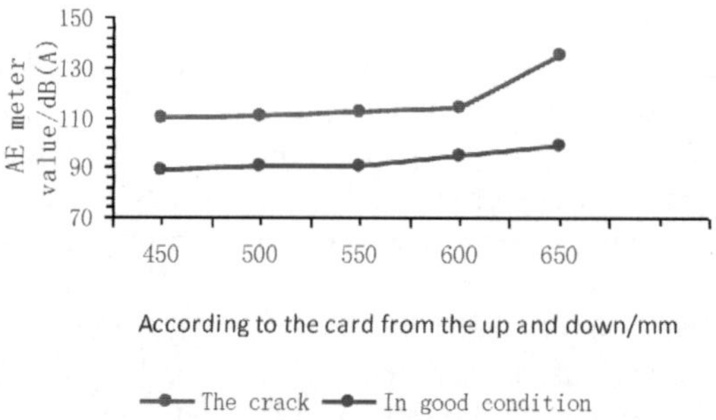

Figure 3. Load under different saw card from up and down on the whole band saw blade and crack defects of band saw blade AE count value

It can be seen from Figure 3, in the upper and lower saw the card before the distance of 600mm, with the increase in the distance from the saw blade saw blade transverse vibration displacement also increased, but the magnitude is smaller, after 600mm with the increase in the distance from the saw blade Saw blades AE increase in the rate of increase; this is due to the increase in the length of the band saw blade. And there is a cracked band saw blade with an AE value greater than the intact band saw blade AE value. (A) -98.3dB (A), the defect band saw blade AE is between 108.6dB (A) and 120.5dB (A).

3.4 Feed speed in good condition and the influence of crack defect band saw blade AE count value

In order to grasp the lateral vibration of the saw blade when cutting, in the case of a good band saw blade and band saw blade crack is 1.68mm, the use of load under the acquisition of data. In the determination of the saw wheel spindle speed 1000rpm, pressing the weight of 93N, up and down the saw blade distance of 600dB (A) under the premise of cutting oak were selected five kinds of feed speed for data acquisition, the analysis results shown in Figure 4.

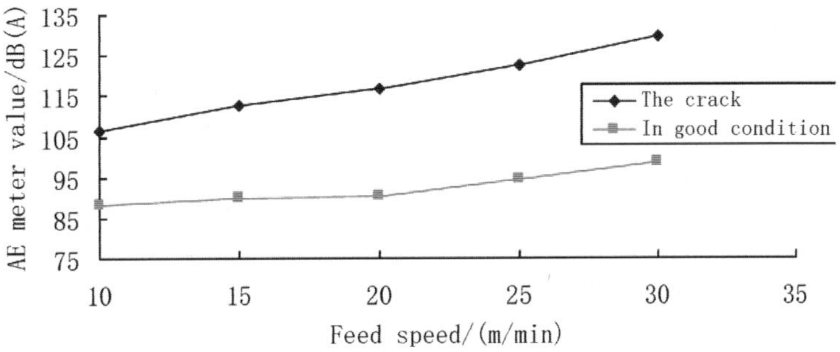

Figure 4. Effect of Different Feeding Speed on AE Value of Cutting Saw Blade and Crack Defect Band Saw Blade

The Fig.4 shows that sports car before the feed speed of 20 m/min, with the increase of car feed speed of transverse vibration displacement distance was generally higher, but the increase amplitude is small, in the car after the feed speed of 20 m/min band saw blade vibration displacement increases gradually, increase rate is higher. And has a crack band saw blade AE meter value Ming is greater than the intact band saw blade AE meter. Good band saw blade AE meter value in 88.3 dB (A)~ 98.6 dB (A) defective blade transverse vibration displacement in 106.6 dB (A)~129.5 dB (A).

3.5 Effect of Different Saw Height on Integrity and Crack Band Saw Blade AE

Select the saw blade tension 8330N, saw the number of revolutions of 1000rpm, feeding speed 20m / min, up and down the saw blade distance of 600mm, cutting oak, in different saw height conditions, the acquisition band saw AE signal and analysis The AE curve of the band saw with different crack length is shown in Fig.5.

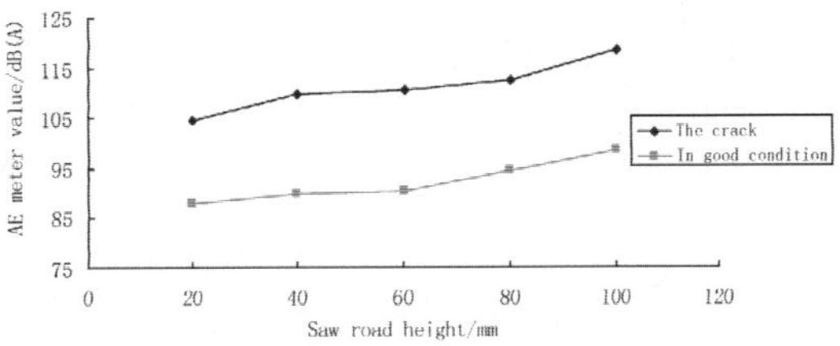

Figure 5. Influence of Different Saw Blade Height of Saw Blade on AE of Saw Blade Cutting

As can be seen from Fig. 5, the height of the saw blade with different saws has a significant effect on the band gauge AE. With the increase in the sawing height of the saw blade, the AE value increases, and the increase in the saw height exceeds 60 mm. And there is a cracked band saw blade with an AE value greater than the intact band saw blade AE value. The full band saw blade AE is between 87.9 dB (A) and 98.4 dB (A). The defective band saw has a value of 104.7 dB (A) to 118.8 dB (A).

3.6 Effect of Different Crack Band Saw Blade on Transverse Vibration Displacement of Band Saw Blade under Load

In the saw blade tension 8330N, the number of revolutions of the saw wheel is 1000rpm, the feeding speed is 20m / min, the distance between the upper and lower saws is 600mm, and the transverse vibration of the band saw blade is taken under the condition of cutting the oak. Value of the signal, the analysis of the crack length and band saw the AE value of the relationship between the curve shown in Figure 6.

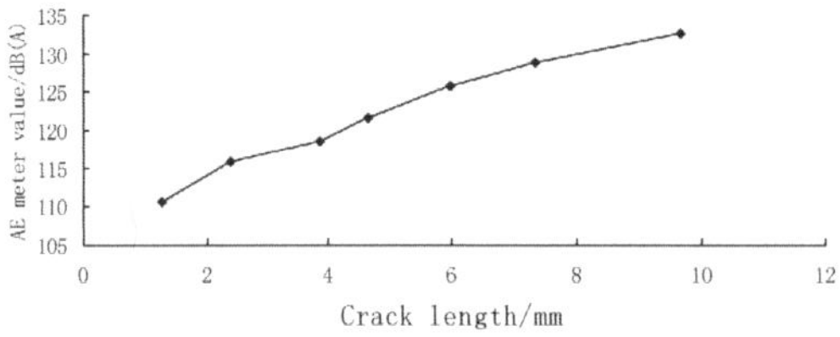

Figure 6. Effect of different crack lengths of band saw blade and AE value of band saw blade cutting

It can be seen from Fig. 6 that the AE value of the saw blade increases with the increase of the crack length of the band saw blade, which is due to the increase of the saw blade. In the best working condition of the saw machine, the lateral vibration displacement of the defective saw blade is between 110.6dB (A) and 132.7dB (A).

3.7 The effect of the weight of the pressure on the AE value of the cracked band saw blade

The speed of the main shaft saw blade is 1000rpm, the distance from the upper and lower sides of the saw blade is 650mm, and the feed speed is 25m / min. Under the premise of changing the weight of the double blade, the complete saw blade and the crack length are 1.68mm Saw blade for data acquisition, will now increase with the weight of the weight of the band saw blade cutting AE value of the curve shown in Fig.7.

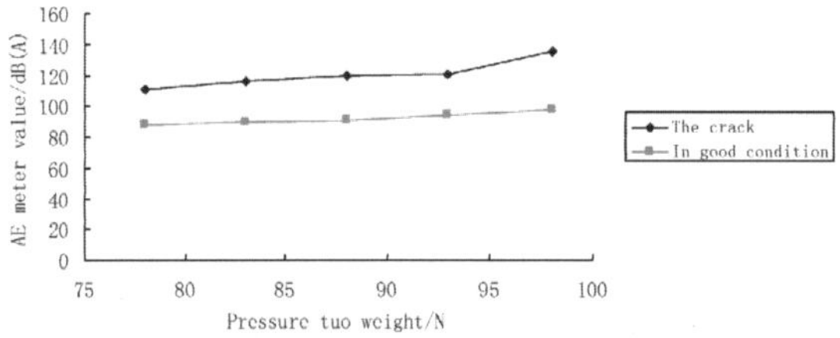

Figure 7. Effect of Different Loads on Load Scales of Saw Blade with Full Band Saw Blade and Crack Defects under Load

As can be seen from Figure 7, before the weight of the pressure mound is 93N, the load increases with the weight of the weight under the load. The band saw is generally raised, but the increase is very small, and the cracked band saw blade has a weight of 93N with the weight of the weight after the increase with the band saw the AE value also increases, the impact is very obvious. And there is a cracked band saw blade with an AE value greater than the intact band saw blade AE value. When the weight of the mound is 93N, the band saw machine works to achieve the best working condition. The AE of the band saw is between 88.3dB (A) and 98.1dB (A). The defect of the saw blade is 110.6dB (A) ~ 135.3dB (A) between.

From the test analysis shows that, for the MJ3210A small sports car wood saws, the use of 120mm width of the saw blade, band saw the crack after its lateral vibration significantly increased, then you can determine the saw blade is the replacement time. When the saw wheel spindle speed, pressure lump weight, feed speed, the upper and lower saw the card distance are adjusted to the most appropriate circumstances, the measured AE value if more than 104.7dB (A) above, then the blade has produced at least one crack The length of more than 1.68mm crack defects, the need for timely replacement of the band saw blade; In addition, the use of band saw blade AE value changes can determine whether the band saw blade crack defects.

4. Summary

After the above MJ3310A sports car woodworking band saw machine testing and analysis, the following conclusions:

(1) Cracked defect band saw blade AE value is significantly higher than the good band saw blade AE value;

(2)The AE value of the birch saw blade is less than the AE value of the other wood, and the AE value of the birch is the smallest, and the AE value of the oak is the largest. Five kinds of tree species in the maximum hardness of oak, with the increase in wood hardness, AE value increases, according to the tree cutting AE value is different to determine the length of the band saw blade crack.

(3) For the MJ3210A sports car saw blade, a 120mmdB width band saw blade is used. The test results show that the AE value of the band saw with crack is significantly higher than that of the complete band saw blade. The change from the AE value of the band saw blade It is necessary to check whether the saw blade is cracked and whether it needs to be replaced in time. If the saw blade spindle speed, the pressure lump weight and the saw blade are adjusted up and down to the most suitable case, the lateral vibration displacement is more than 104.7 dB (A) , It means that the saw blade has produced at least one crack length of more than 1.64mm crack defects, the need for timely replacement band saw blade. The AE scale for the complete band saw blade is between 87.9 dB (A) and 98.9 dB (A). And then determine the range of dB (A) for judging the defect condition of band saw blade, which provides the technical basis for the full and reasonable utilization of band saw blade in actual production.

Funding for the subject project

This topic by the Chinese national natural science fund project funding, its national fund name: based on the analysis of time series of woodworking band saw blade tooth crack and damage

References

[1] Gao Jin-gui, Jian Zhang, Hui. Band saw blade off the horizontal vibration displacement and main frequency variation before and after tooth. Journal of north China university (natural science edition) [J]. 2016, (4) : 549-552.

[2] Gao Jin-gui. Computer simulation and experimental study of mechanical tension system of band saw machine [D]. Northeast Forestry University, 2001.

[3] Tom wang. The relationship between the car interior noise and body sealing study [J]. 2011, Beijing automotive 1999,2 (3) : 30-33.

[4] Gao Jin-gui, ZOU Zhi-ping. fu four thousand. The current situation and development trend of woodworking machinery in China [J]. Fujian Forestry Science and Technology, 2014,41 (3): 115-120.

[5] Zhu Guo-xi, Wang Hua-bin. Modern Chinese lumber production line of research [M]. Harbin: northeast forestry university press, 1989, (9): 2-10.

IOP Conf. Series: Materials Science and Engineering **191** (2017) 012024 doi:10.1088/1757-899X/191/1/012024

Study on the Influence of the Main Parameters of Woodworking Saw Blade on the Illumination Value of Band Saw Blade (1)

Jin-gui GAO[1], Hong-gang ZHAO[1*] and Lai-peng LUO[1]

[1]Beihua University,Jilin city of jilin province,132013, P.R.China

E-mail: [*]1125568071@qq.com

Abstract. In this paper, MJ3310A band saw machine as the research object, through the Beijing VIBSYS vibration signal acquisition and analysis software illumination value analysis, analysis of different circumstances to find good and crack band saw blade illumination value of the law. The results show that the illuminance of the cracked band saw blade is significantly higher than that of the complete band saw blade illumination value. Under the optimum working conditions, if the band saw blade illumination value exceeds 286 Lux, it can be determined that the band saw blade has at least one crack length greater than 1.68 mm Of the defects, the need for timely replacement band saw blade. So as to rational use of band saw blade, band saw blade on-line fault diagnosis provides a technical basis.

1. Introduction

According to the actual production survey, the woodworking band saw blade has its fault state and the eye can not be observed, but can not determine the correct replacement time [1-3]. At present, wood processing enterprises online inspection usually use light to determine the situation of cracks. Whether or not this inspired band saw can be used for online troubleshooting.

According to the new data, on the use of light measurement method (illumination value) for band saw blade crack in-depth study - the occurrence of cracks in the law at home and abroad rarely. From the experimental study of its crack based on the results of less. In order to determine the time of replacement of band saw blades in actual production, it is necessary to study the generation and propagation of cracks before and after cutting under different load conditions. In order to improve the quality of wood surface treatment, to determine the choice of band saw blade reasonable replacement time and online nondestructive fault diagnosis to provide a scientific basis.

2. The experiment main instruments and equipment

2.1. The experiment main instruments

The apparatus and equipment used in this test are shown in Table 1.

2.2. MJ3310A specific parameters of the woodworking band saw machine

The test and application of Tianjin forestry cutting tool factory production of No. 18 B.W.G band saw blade. Show.

IOP Conf. Series: Materials Science and Engineering **191** (2017) 012024 doi:10.1088/1757-899X/191/1/012024

Table 1. Test apparatus and equipment.

Name	Model	Manufacturer
Woodworking band saw	MJ3310A	Shenyang Shen He woodworking machinery factory
The large capacity data acquisition instrument quickly	WS - 5942-2-5	Spectrum of Beijing century technology development co., LTD.
Precision multi-function digital light meter	GSM1020A	Shenzhen shigeru source technology co., LTD.
High brightness adjustable micro digital projector	S6A	Shenzhen shigeru source technology co., LTD.

2.3. Cutting wood

The parameters of the test wood are shown in Table 2.

Table 2. Test apparatus and equipment.

Tree species	Manchurian Ash
Moisture content	10%
Cutting thickness	100mm

3. Test data processing and result analysis

3.1. Cutting wood of different wood species to crack defects of band saw blade illuminance value influence

In the determination of the saw wheel spindle speed 900rpm, pressing the weight of 93N, saw blade up and down distance of 650mm, feed speed of 15m / min, under the premise of cutting five kinds of tree birch, color wood, poplar, elm, oak, respectively The complete saw blade and the cracked length of 1.68mm for the defect band saw blade for data collection, the saw blade cutting tree species are different and the band saw blade illumination value of the collected data drawn as shown in Figure 1

It can be seen from Fig. 1 that the strip saw blade increases with the increase of the feed rate. The band saw blade illumination value changes smoothly, and the cracked band saw blade increases with the feeding speed. The saw blade illumination value also increases; More than the other wood illumination value is small, the largest oak wood, complete band saw blade 32.8LUX, defect band saw blade 454LUX; poplar: complete band saw blade 42.4LUX, defect band saw blade 723LUX; birch: complete band saw blade 43.8LUX, Saw blade 776LUX; elm: complete band saw blade 48.9LUX, defect band saw blade 782LUX; color wood: complete band saw blade 51.7LUX, defect band saw blade 1060LUX; oak: complete band saw blade 58.4LUX, defect band saw blade 1730LUX, cracked band saw blade The illuminance value is generally greater than the illumination value of the complete band saw blade. White pine in the five kinds of tree species in the hardness of the smallest, the largest hardness of oak, which shows that with the increase in the hardness of wood cutting, the illumination value also increases.

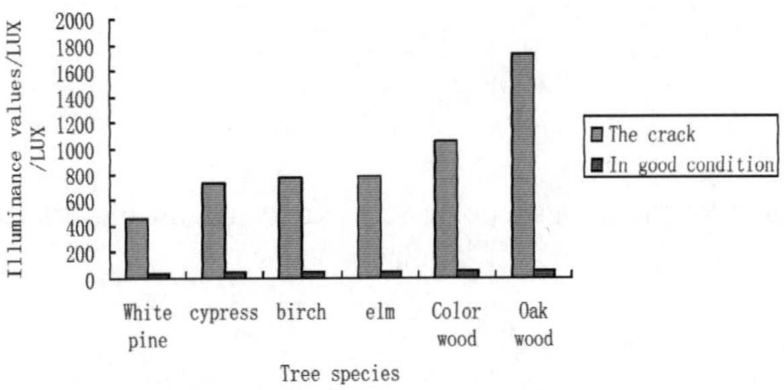

Figure 1. Saw blade cutting different tree species when the full saw blade and crack defects generated by the comparison of the value of the blade

IOP Conf. Series: Materials Science and Engineering **191** (2017) 012024 doi:10.1088/1757-899X/191/1/012024

3.2. Influence of Rotation Speed of Saw Blade on Belt Saw Blade 's Illumination Value

In the band saw machine pressure Tuo weight of 93N, up and down the saw blade distance of 650mm, feed speed of 15m / min, cutting under the premise of wood, in the five kinds of sawing wheel speed, respectively, the crack is 1.68mm defective belt Saw blade and complete (intact) band saw blade for data acquisition and analysis, get band saw blade illumination value curve shown in Figure 2.

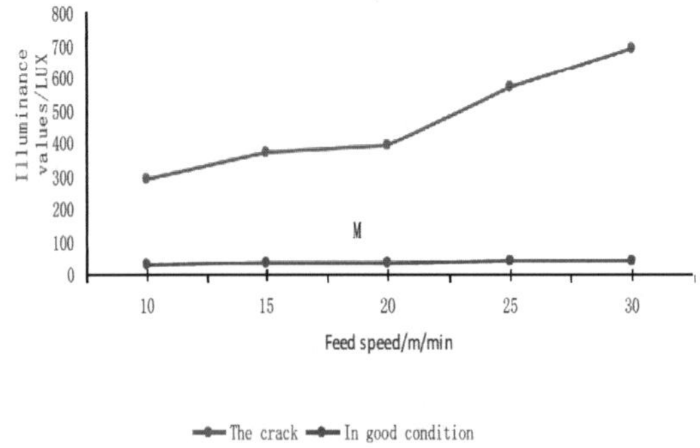

Figure 2. Uunder the load of different saw blade spindle speed on the full saw blade and crack defects in the band saw blade illumination value

It can be seen from Figure 4, the saw wheel spindle speed on the band saw blade illumination value is very obvious, saw wheel spindle speed before 1000rpm with the saw wheel speed increases when the saw blade illumination value increases smaller, saw wheel spindle speed after 1000rpm , With the saw wheel speed increase band saw blade illumination value also increases; intact band saw blade illumination value 26.1LUX-36.8LUX, defective saw blade illumination value between 612LUX ~ 1800LUX.

3.3. Load under different crack band saw blade's influence on the band saw blade illuminance values.

Saw in saw blade tensioning force of 8320 N, rotation number is 1000 RPM, the feed speed of 20 m/min and saw cutting elm card a distance of 600 mm, under different crack length of band saw blade, band saw blade transverse vibration displacement signal sampling got crack length with the band saw blade transverse vibration displacement curve as shown in Fig. 3.

The Fig.3 shows, under the load with the band saw blade saw blade transverse vibration displacement when the increase of crack length is less, the sawing machine to achieve the best work state, defective blade illuminance values between 657 LUX ~ 1325 LUX.

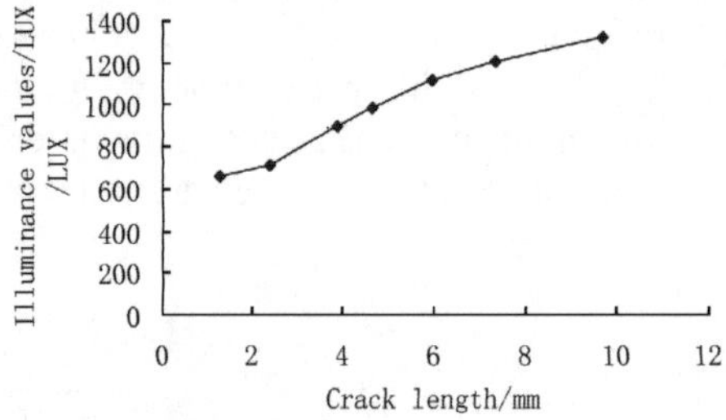

Figure 3. Different crack length band saw blade and the band saw blade the relationship between the illuminance values

3.4. Influence of Weight on the Radius of Saw Blade

The speed of the saw blade is 1000rpm, the distance between the upper and lower sides of the saw blade is 600mm, the feeding speed is 20m / min, under the premise of cutting the color wood, change the weight of the five times, respectively, to the full saw blade and the crack length is 1.68 Mm defect band saw blade for data collection, now with the pressure lump weight increases the band saw blade illumination value of the collected data drawn into a curve shown in Figure4.

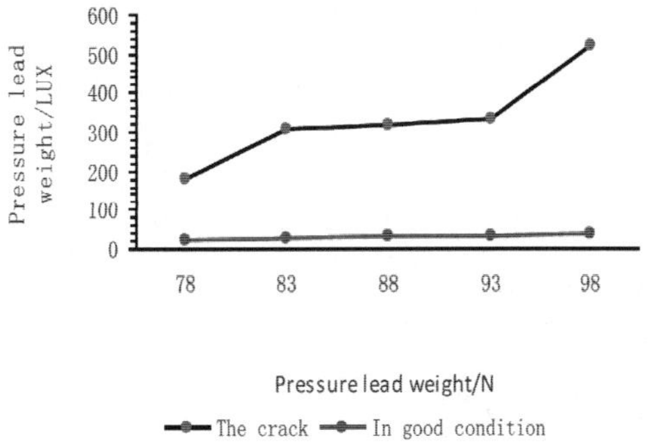

Figure 4 . Under the load of different pressure Tuo weight on the full saw blade and the occurrence of crack defects saw blade illumination value

It can be seen from Fig. 4 that the band saw blade increases with the weight of the pressure mound, and the illuminance value of the band saw blade increases with the increase of the weight of the pressure mound. After the weight of the weight is 93N, the load With the increase of the weight of the pressure mound, the band saw blade illumination value also increases. After the weight of the pressure mound is 93N, the increase of the value of the band saw blade increases with the weight of the weight. The influence is very obvious. Intact band saw blade illumination value 22.4LUX-33.4LUX, defective saw blade illumination value between 711LUX-980LUX.

3.5. Influence of Upper and Lower Saw Card Distance on Illumination Value of Belt Saw Blade

In the band saw machine spindle speed of 900rpm, pressure Tuo weight of 93N under the premise of band saw blade crack 1.68mm, change the distance from the top and bottom of the saw, respectively, the complete saw blade and crack defects of the blade for data collection, Saw card up and down the distance to get the band saw blade illumination value of the collected data drawn into a curve shown in Figure 5.

It can be seen from Fig. 6 that the band saw blade is smooth with the increase of the distance between the upper and lower saws. With the increase of the distance between the upper and lower saws, the value of the band saw blade increases with the increase of the distance between the upper and lower saws. Small, up and down saw the distance from 600mm, with the increase in the distance from the saw blade, the illumination value increases greatly. This is due to the increase in the length of the band saw blade, band saw blade vibration increased the reason. Intact band saw blade illumination value 35.2LUX-44.6LUX defective saw blade illumination value between 721LUX-2690LUX..

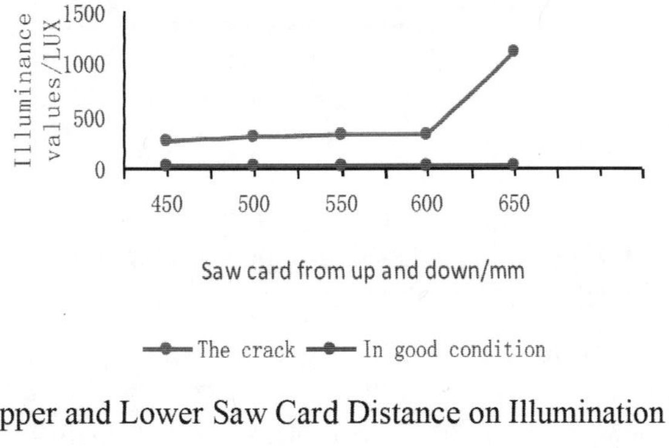

Figure 5. Influence of Upper and Lower Saw Card Distance on Illumination Value of Belt Saw Blade

3.6. Influence of Different Saw Height on the Illumination Value of Crack Band Saw Blade

The saw blade tension force 8330N, the number of revolutions of the saw blade 900rpm, the feeding speed of 15m / min, the distance from the upper and lower saws 650mm, the cutting oak, the transverse vibration of the band saw blade was collected under different height conditions of sawing and the crack length The relationship between the illuminance value and the band saw blade is shown in Fig 6.

It can be seen from Fig. 6 that the band saw blade increases with the height of the saw blade. The change of the band saw blade is smooth. The influence of the height of the saws on the saw blade is very obvious. With the increase of the saw height, The increase. Intact band saw blade illumination value 32.8LUX-58.4LUX, defective saw blade illumination value between 454LUX-1730LUX.

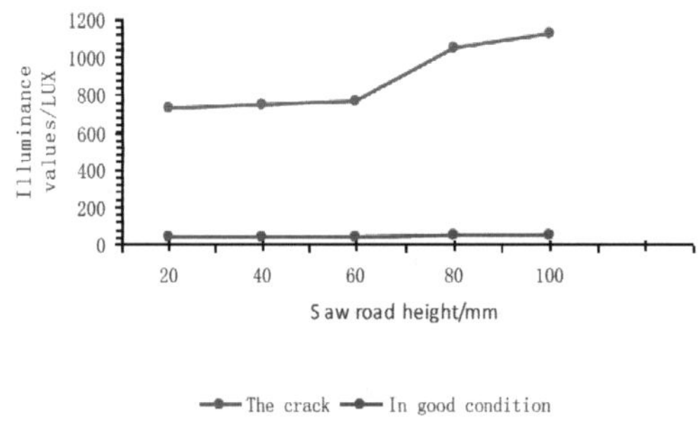

Figure 6. Different band saw blade saw road height on the influence of the band saw blade transverse vibration displacemen.

3.7. Influence of Rotation Speed of Saw Blade on Belt Saw Blade 's Illumination Value

In the band saw machine pressure Tuo weight of 93N, up and down the saw blade distance of 650mm, feed speed of 15m / min, cutting under the premise of wood, in the five kinds of sawing wheel speed, respectively, the crack is 1.68mm defective belt Saw blade and complete (intact) band saw blade for data acquisition and analysis, get band saw blade illumination value of the curve shown in Figure 7.

It can be seen from Figure 7, the saw blade spindle speed on the band saw blade illumination value is very obvious, the saw wheel spindle speed before 1000rpm with the saw wheel speed increases when the saw blade illumination value increases smaller, saw wheel spindle speed after 1000rpm , With the saw wheel speed increase band saw blade illumination value also increases; intact band saw blade illumination value 26.1LUX-36.8LUX, defective saw blade illumination value between 612LUX ~ 1800LUX.

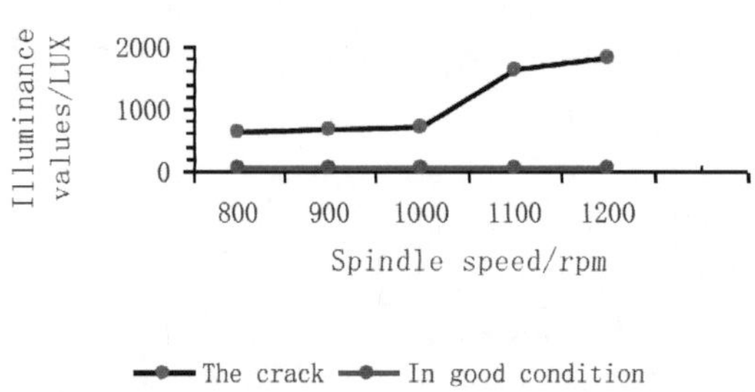

Figure 7. Under the load of different saw blade spindle speed on the full saw blade and the occurrence of crack defects band saw blade illumination value

From the test analysis shows that, for the MJ3310A sports car wood saws, the use of 120mm width of the saw blade, band saw blade after the crack after the lateral vibration significantly increased, this time to determine the saw blade is the replacement time. When the saw wheel spindle speed, pressure lump weight, feed speed, the upper and lower saw the card distance are adjusted to the most appropriate circumstances, the determination of lateral vibration displacement if more than 286LUX above, then the blade has produced at least one crack length of more than 1.68mm Of the crack defects, the need to replace the band saw blade in time; in addition to the use of band saw blade lateral vibration displacement changes can determine whether the band saw blade defects.

4. Summary

Through above MJ3310A type sports car woodworking band saw machine test and analysis, the following conclusions:

(1)The illuminance value of the complete band saw blade is significantly lower than the defective illumination value.

(2) With the increase of the feed rate, the illumination value of the band saw blade is increased and the illumination value increases with the increase of the feed rate.

(3)Sawing white pine band saw blade illumination value than other wood illumination values are small, the largest oak, white pine minimum: crack with band saw blade illumination value is generally greater than the full band saw blade illumination value. White pine in the five kinds of tree species in the hardness of the smallest, the largest hardness of oak, which shows that with the increase in the hardness of wood cutting, the illumination value also increases..

(4) In type MJ3210A sports car woodworking band saw machine, using the 130 mm width of blade, blade judge whether the need to change, can be judged from the change of the saw blade vibration displacement, when saw wheel spindle speed, pressure tuo weight, saw card distance to adjust up and down to the most appropriate cases, the determination of illuminance values if more than 286 lux, then saw blade has had at least one crack length more than 1.68 mm crack defects, need to replace the band saw blade in time.

Funding for the subject project

This topic by the Chinese national natural science fund project funding, its national fund name: based on the analysis of time series of woodworking band saw blade tooth crack and damage formation mechanism and prediction theory (31570556). Key scientific research project of jilin province department (2015204038 NY). Lower level table plate laser cutting spelling solid mu fu joins a floor board, research and development.

References

[1] Jin-gui Gao, Jian Zhang , Hui Li , and so on. Band saw blade off the horizontal vibration displacement and main frequency variation before and after tooth. Journal of north China university (natural science edition) [J]. 2016, (4) : 549-552.

[2] Feng Dong Dong, Wan Yi Shang . Based on laser displacement sensor of band saw blade tooth online detection system research [J]. Journal of tool technology, 2016, 50 (3) : 82-85.

[3] Jin-gui gao, Zhiping Zou. Band saw blade crack before and after vibration analysis [J]. Journal of fujian forestry science and technology, 2014, 9 (3) : 115-120.

IOP Conf. Series: Materials Science and Engineering **191** (2017) 012025 doi:10.1088/1757-899X/191/1/012025

Effect of cutting parameters on surface finish and machinability of graphite reinforced Al-8011 matrix composite

Anil.K.C[1], M.G.Vikas[2], Shanmukha Teja.B[2], K.V.Sreenivas Rao[3]

[1]Assistant Professor, Department of Industrial Engineering and Management,
[2] U.G.Research Scholars, 3Professor, Department of Mechanical Engineering
Siddaganga Institute of Technology, Tumakuru, Karnataka, India

E-mail: anilkc.20@gmail.com

Abstract. Many materials such as alloys, composites find their applications on the basis of machinability, cost and availability. In the present work, graphite (Grp) reinforced Aluminium 8011 is synthesized by convention stir casting process and Surface finish & machinability of prepared composite is examined by using lathe tool dynamometer attached with BANKA Lathe by varying the machining parameters like spindle speed, Depth of cut and Feed rate in 3 levels. Also, Roughness Average (Ra) of machined surfaces is measured by using Surface Roughness Tester (Mitutoyo SJ201). From the studies it is cleared that mechanical properties of a composites increases with addition of Grp and The cutting force were decreased with the reinforcement percentage and thus increases the machinability of composites and also results in increased surface finish.

1. Introduction

Composite materials play an important role in the field of science and engineering as well as advance manufacturing in response to unprecedented demands from technology due to rapidly advancing activities in aircrafts, aerospace, sporting goods, marine and automotive industries[1]. MMCs have higher strength-to-density ratios, better fatigue resistance, better elevated temperature properties (such as high strength and low creep rate), lower coefficients of thermal expansion, high thermal conductivity, good damping characteristics, excellent wear properties and flexibility in design attributes[2]. Some of the typical applications of these MMCs are bearings, automobile pistons, cylinder liners, piston rings, connecting rods, sliding electrical contacts, turbo charger impellers, space structures, etc. [3]. Several researches done experiments on machining of MMCs.[4] have investigated the effect of machining parameters on the surface roughness and tool wear when turning 10% SiCp/Al composites. Results indicated that higher cutting speeds result in relatively better surface finish, but resulting in increased flank wear. [5] Developed the analytical model extending the classical Merchants theory, Slip line theory and Grifith's theory of brittle fracture to the machining of ceramic particle reinforced MMCs. This model predicted the cutting forces and was validated experimentally. They have concluded that the size of reinforcements in the composite material influences roughness of the machined surfaces significantly when its magnitude is comparable to that of the feed rate and tool nose radius employed during machining of the composite. [3] have studied tool wear and surface roughness in machining of particulate aluminum metal matrix composite .it is revealed that the formation of BUE significantly affects the tool wear at low speeds whereas thermal softening plays important role at higher speeds and feed rates. However, no work addresses the machinability of Al-8011–Grp composite produced by stir casting. Hence, the main objective of the present work is to study the effect of reinforcement (2, 4, 6, and 8%) and machining parameters such as cutting speed, feed rate, and depth of cut on surface roughness, and cutting force were analyzed during turning operations.

2. Preparation and characterization composite material

Apart from iron, aluminium / aluminum is currently the next most widely used metal in the world. This is due to the fact that aluminium / aluminum has a unique combination of attractive properties. Properties such as its low weight, corrosion resistance, and easy maintenance of final product, have ensured that this metal and its alloys will be in use for a very long time. In this work aluminum 8011 a wrought alloy is used a matrix alloy Weighed quantity of Al-8011 was melted in electrical resistance furnace to desired superheat temperature of 750 ± 200C in graphite crucible. After attaining the required temperature degassing agent (C_2Cl_6 – solid hexachloro ethane) was added in order to degas the entrapped gases in the melt. Then the molten metal was stirred manually at an average speed of 300 rpm for 200 Sec to form a vortex in the melt. Once the vortex is achieved the preheated (300± 200C) Grp was added by varying weight percentage of 2, 4, 6, 8 respectively. Prepared molten metal then poured in preheated cast iron mould and allowed to solidify. The cast specimens were machined as per the mentioned ASTM standards for measuring tensile and compression test in universal testing machine having 0-20 tons of capacity as per the ASTM-E8M and ASTM-E09 respectively and cylinder samples are polished as per the metallographic procedure to get fine surface before testing their hardness. The hardness of prepared composite is measured by Rockwell hardness test.

Figure 1. Lathe Tool Dynamometer

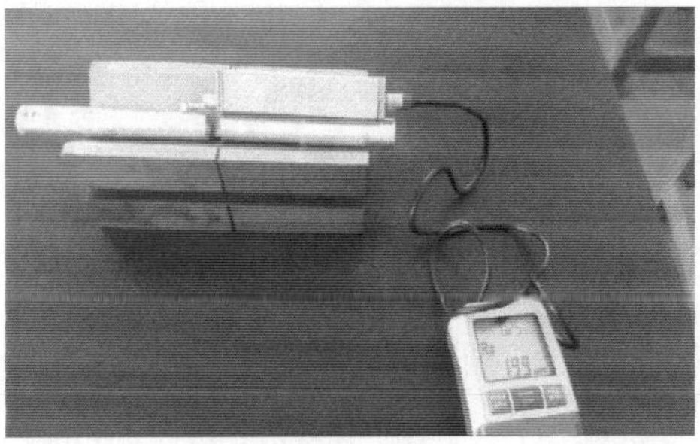

Figure 2. Surface roughness tester

Table 1. Machining parameters

Sl. No	Speed (rpm)	Feed (mm/sec)	Depth of Cut (mm)
1	175	0.067	0.2
2	263	0.111	0.4
3	395	0.167	0.6

The experiments were employed to study the effects of cutting parameters on surface roughness of the materials when machining MMCs work piece. The lathe tool dynamometer which is attached to BANKA-Lathe is used for measuring the cutting forces during the machining are shown in Figure 1. The HSS Cutting tool which is used for machining has following geometry: Rake angle: 0° and clearance angle: 7°, cutting edge angle: 85° and Nose radius: 0.8 mm. The machining parameters which are used for experiments are listed in the table 1. ULI-15 model lathe tool dynamometer having a capacity of 500kg in the X, Y & Z directions with 12mm tool holder and Strain gauge based 350Ω bridge sensor is used for experiment

Cutting forces are measured in each co-ordinate and collectively taken as multi-component forces. The average surface finish (Ra) in the direction of the tool movement was measured in five different places of the machined surface using a surface roughness tester- Mitutoyo SJ201, surface mean roughness (Ra) in microns value of the five locations was considered for the particular trial. A Inductive type Diamond cone Surface roughness tester (Mitutoyo SJ201) having a measuring speed of 0.25mm/s, 0.55 mm/s in forward direction and 0.8 mm/s in returning with a total gauge length of 12.5mm is used for measuring a roughness value of a machined surface.

3. Results and discussions

3.1. Mechanical properties

Machined composite materials are then tested for their mechanical properties as per the ASTM standards. The average results of 5 specimens are tabulated in the table 2. From the results we can say the capacity of withstanding applied tensile load is increased with the addition of weight percentage of reinforcement, this is because of increased bonding between the matrix and the reinforcement. The compression strength of a base alloy Al-8011 is 38.06 KN. The load bearing capacity of matrix material will increased with the addition of reinforcement and the strength of the composite are also implies in hardness test, the indentation resistance of the composite with 8% of Grp is increased to 52.1%.

Table 2. Mechanical properties of composite with different weight percentage of Grp

Composition	Tensile strength (MPa)	Compression Strength (KN)	Hardness (BHN)
Al-8011	91.405	38.06	21.3
Al+2% Grp	105.08	50.87	22.06
Al+4% Grp	122.28	62.96	24.35
Al+6% Grp	145.58	69.16	32.41
Al+8% Grp	160.09	78.50	34.87

3.2. Surface roughness

The surface roughness (Ra) decreases as the cutting speed increases. At low cutting speed, the temperature between the machining interfaces is more enough to create the unstable larger BUE (Built up edge) and also the chips fracture readily producing the rough surface this may cause adhesive wear on tool. As the cutting speed increases, the machining time will reduces and the BUE vanishes, chip fracture decreases, and hence the roughness. Though these phenomena of improved surface finish at higher cutting speeds can be observed in machining conventional materials also, the effect is more pronounced here because removal of hard strengthen particles from the Aluminum matrix becomes easier at higher cutting speeds. Figure 3 shows the effect of cutting speed and reinforcement on surface roughness keeping feed rate and depth of cut constant. The influence of feed rate on the surface roughness is shown in Figure 4. At same level of feed rate the surface roughness value increased when increasing reinforcement wt. percentage. Figure 4 also shows that the increase in feed rate increases the surface roughness (Ra). At the lower feed rates, the BUE forms readily and is accompanied by feed marks resulting in increased roughness. With the increase in feed rate beyond 0.11 mm/rev, the rate of increase in surface roughness (Ra) is less due to the reduced effect of BUE.

The best surface finish was achieved at the lowest feed rate and highest cutting speed combination. Figure 5 shows the effect of depth of cut and reinforcement on the surface roughness (Ra). It shows that when higher depth of cut given its results in poor surface finish. As we known the shear angle and HAZ (Heat affected zone) is depend on the depth of cut, has the depth of cut increases the HAZ and shear angle, thus increase in Cutting force, friction and rise in temperature and causes deposition of removed material on rake face of a tool. Hence, the surface roughness (Ra) increases along with increase in depth of cut.

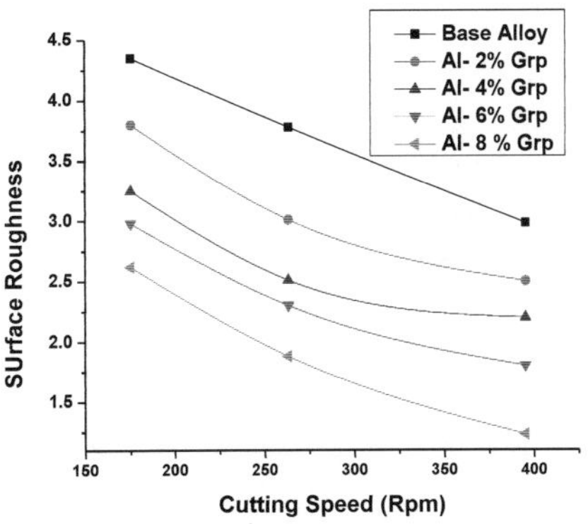

Figure 3. Effect of cutting speed and reinforcement on surface roughness (Feed-0.11mm/Rev and Depth of cut-0.4 mm)

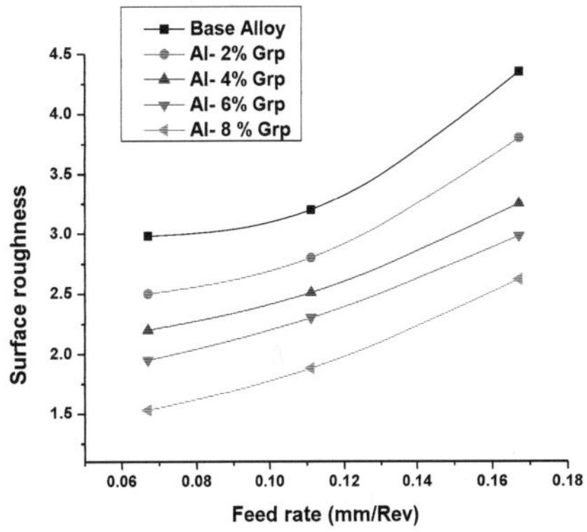

Figure 4. Effect of feed rate and reinforcement on surface roughness (Speed-263 Rpm and Depth of cut-0.4 mm)

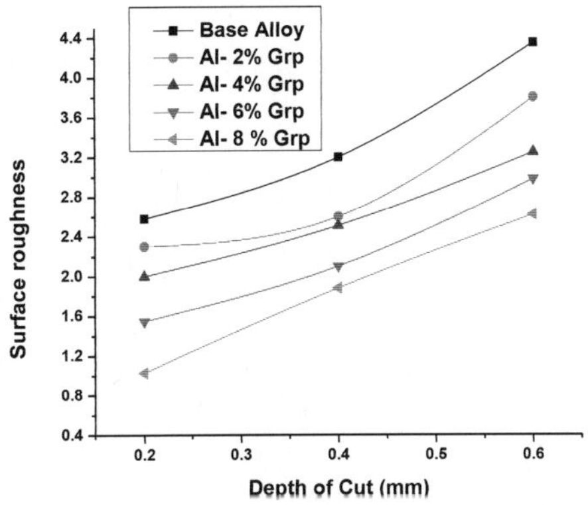

Figure 5. Effect of Depth of cut and reinforcement on surface roughness (Speed-263 Rpm and feed-0.11 mm/Rev)

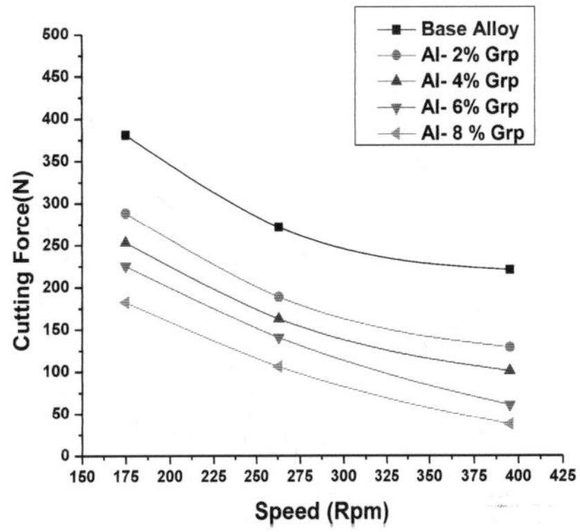

Figure 6. Effect of cutting speed and reinforcement on cutting force (Feed-0.11mm/Rev and Depth of cut-0.4 mm)

3.3. Cutting force

The effect of an increase in reinforcement is evaluated under different speeds, at a constant feed rate of 0.11 mm/rev, a depth of cut of 0.4 mm, and a length of turning of 80 mm. When constant cutting speed was considered, the cutting force is decreased as the reinforcement added. The presence of reinforcement will minimize the build-up edge formation, which reduce the cutting force. It can be observed that the increase in cutting speed will reduce the chip tool contact length therefore cutting force is reduced. Figure 7 shows the trend of cutting force by considering constant Speed-263 Rpm and Depth of cut-0.4 mm for length of 80 mm turning operation. When machining of Al-8011 alloy with higher feed rates the friction win the machining interface will be high this requires high pressure to shear the material thus the cutting force will be more. Figure 7 also reveals that addition of Grp in the matrix material will decreases the cutting force this is because the Grp itself acts a lubricant to avoid the friction in cutting tool and work piece interface. The same trend is followed when machine the alloy and composite material by varying depth of cuts which is showed in figure 8. The contact area of tool – work material interface will directly dependent on the cutting force.

IOP Conf. Series: Materials Science and Engineering **191** (2017) 012025 doi:10.1088/1757-899X/191/1/012025

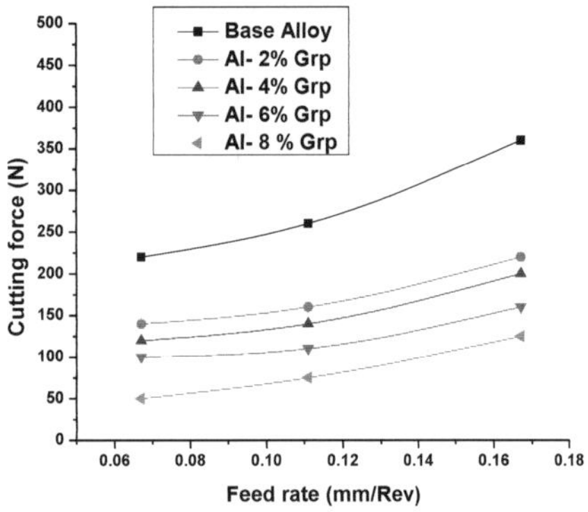

Figure 7. Effect of feed rate and reinforcement on Cutting force (Speed-263 Rpm and Depth of cut-0.4 mm)

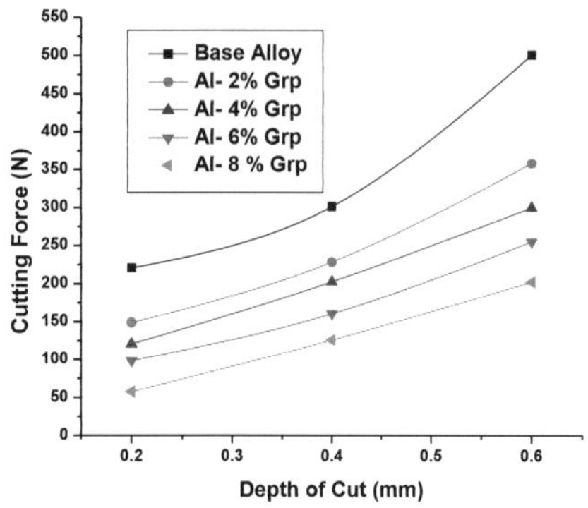

Figure 8. Effect of Depth of cut and reinforcement on Cutting force (Speed-263 Rpm and feed-0.11 mm/Rev)

4. Conclusions

The results of plain turning studies of Al-8011 alloy and Al reinforced Grp composites using HSS-tool by varying machining parameters at three different levels were presented. From the results obtained following conclusions can be drawn:

- Mechanical properties like tensile, compression strengths of Al alloy can be tailored with the addition of Grp by conventional stir casting route
- Higher cutting speed with lower depth of cut and lower feed rates gives good surface finish when compare to lower speed
- Addition of Grp to alloy will gives more surface finish because of its lubricant property
- Increased federates and depth of cuts produces more cutting force and poor surface quality.

References

[1] K. V. Sreenivasrao, Anil. K. C., Girish K. G. and Akash, “Mechanical characterization of red mud reinforced Al-8011 matrix composite” ARPN Journal of Engineering and Applied Sciences, vol. 11, no. 1, January 2016,pp-229-234.

[2] Zhu Y., Kishawy H.A., Influence of alumina particles on the mechanics of machining metal matrix composites, Int J Mach Tools Manuf 45 (2005) 389-398.

[3] Seeman M., Ganesan G., Karthikeyan R. , Velayudham A., Study on tool wear and surface roughness in machining of particulate aluminum metal matrix composite response surface methodology approach Int J Adv Manuf Technol (2010) 48:613-624.

[4] Manna A, Bhattacharayya B Influence of machining parameters on the machinability of particulate reinforced Al/SiC–MMC. Int J Adv Manuf Technol 25 (2005) 850-856.

[5] Pramanik A, Zhang LC, Arsecularatne JA, Prediction of cutting forces in machining of metal matrix composites. Int J Mach Tools Manuf 46 (2006) 1795-1803.

[6] Choudhury IA, El-Baradie MA, Tool-life prediction model by design of experiments for turning high strength steel (290 BHN). J Mater Process Technol 77: (1998)319-326.

IOP Conf. Series: Materials Science and Engineering **191** (2017) 012026 doi:10.1088/1757-899X/191/1/012026

Performance evaluation of NEEM oil and HONGE Oil as cutting fluid in drilling operation of mild steel

Jyothi P. N.[1], Susmitha M.[2], Sharan P.[2]
[1]Professor, Dept. of Mechanical Engineering, K S School of Engineering & Management, Bangalore-109, India
[2]Students, Dept. of Mechanical Engineering, K. S. School of Engineering & Management, Bangalore-10, India

medarametlasusmitha@gmail.com

Abstract. Cutting fluids are used in machining industries for improving tool life, reducing work piece and thermal deformation, improving surface finish and flushing away chips from the cutting zone. Although the application of cutting fluids increases the tool life and Machining efficiency, but it has many major problems related to environmental impacts and health hazards along with recycling & disposal. These problems gave provision for the introduction of mineral, vegetable and animal oils. These oils play an important role in improving various machining properties, including corrosion protection, lubricity, anti-bacterial protection, even emulsibility and chemical stability. Compared to mineral oils, vegetable oils in general possess high viscosity index, high flash point, high lubricity and low evaporative losses. Vegetable oils can be edible or non-edible oils and Various researchers have proved that edible vegetable oils viz., palm oil, coconut oil, canola oil, soya bean oil can be effectively used as eco friendly cutting fluid in machining operations. But in present situations harnessing edible oils for lubricants formation restricts the use due to increased demands of growing population worldwide and availability. In the present work, Non-edible vegetable oil like Neem and Honge are been used as cutting fluid for drilling of Mild steel and its effect on cutting temperature, hardness and surface roughness are been investigated. Results obtained are compared with SAE 20W40 (petroleum based cutting fluid)and dry cutting condition.

1. Introduction

The application of conventional cutting fluids in machining operations is harmful and not eco-friendly. In this context, the present work focuses on environmental friendly cutting fluids such as vegetable oil based cutting fluids (VBCFs). The objective of this work is to determine the influence of vegetable oil based cutting fluids (sesame and coconut oil) with extreme pressure (EP) additive on machining performance [1]. This paper presents the effects of minimum quantity lubrication (MQL) by vegetable oil based cutting fluid on the turning performance of AISI 316L stainless steel as compared to completely dry and wet machining in terms of chip-tool interface temperature, chip formation mode, tool wear and surface roughness [2]. Many researchers have investigated the effect of using different vegetable oils as cutting fluid with different methods of lubrication on different metals for various machining operations. Performances of palm oil and ground nut oil as cutting fluid was evaluated by Sharafadeen Kunle Kolawale [3], and compared with that of mineral oil based cutting fluid during machining operation of mild steel. Palm oil gave the overall highest thickness of 0.27mm probably due to its better lubricating property. Based on these results, ground nut oil and palm oil are being

IOP Conf. Series: Materials Science and Engineering **191** (2017) 012026 doi:10.1088/1757-899X/191/1/012026

recommended as variable alternative lubricants to the mineral oil during machining of mild steel. M. M. A. Khan [4], investigated the effects of minimum quality lubrication (MQL) by vegetable oil based cutting fluid under turning operation of low alloy steel AISI 9310. The results were compared with completely dry and wet machine in terms of tool- chip interface temperature, chip formation mode, tool wear and surface roughness. Results show that MQL provides environmental friendliness and improves the machinability characteristics. Jitendra Kumar Chandrakar [5] showed that lubricants provide smooth operation between movable parts of all machines. It maintains the reliability of machine functions and reduces the risk of failures. Vegetable bio lubricants are non-toxic, degradable, and renewable also possess good lubricating properties. In his work, he reviewed papers on edible oils as cutting fluids. While in few papers non-edible oils such as castor, karanja, mahua were used and proved to have a great potential as lubricant for some of the machining operations. Vaibhav Koushik A V [6] worked on metal working fluids that are widely employed to increase the machining productivity and quality of metal cutting, but the usage poses great threat to the ecology health of workers in industry. In this review paper an effort has been undertaken to provide highlights of vegetable oils over petroleum based oils. The performance of the vegetable oils; palm, sesame, olive and coconut oils were compared under minimum quantity lubrication (MQL) technique by Ahmad Fairuz Mansor [7]. It was concluded that the coconut oil indicates the best machinability in term of highest and uniform chip thickness and least wear on the drill bit under same condition with others. These performances are followed by palm, olive and sesame oil. In additional, the viscosity measurement indicates that coconut oil has the lowest value which can possesses better fluidity and faster cooling capacity than other oils. Overall, coconut oil is recommended as viable alternative lubricants during drilling of stainless steel. Effects of vegetable-based cutting fluids on the tool wear in drilling of AISI 304 austenitic stainless steel with HSSE tool using crude sunflower oil, refined sunflower oil, refined canola oil and commercial semi-synthetic cutting fluid was studied by EmelKuram [8]. Experimental results show that canola based cutting fluid gives the best performance due to its higher lubricant properties with respect to other cutting fluids at the constant cutting conditions. EmelKuram [9], studied the Effects of the three different vegetable-based cutting fluids developed from raw and refined sunflower oil and two commercial types (vegetable and mineral based cutting oils), on thrust force and surface roughness during drilling of AISI 304 austenitic stainless steel with HSSE tool. The results showed Lower thrust force values were obtained with SCF-I and the least thrust force was achieved at spindle speed of 720 rpm. SCF-I generated the highest reduction in thrust force when AISI 304 stainless steel drilled at a feed rate of 0.12 mm/rev and a drilling depth of 21 mm. When AISI 304 stainless steel was drilled, CVCF had low thrust force and the least thrust force was achieved at feed rate of 0.08 mm/rev. The least surface roughness was achieved at spindle speed of 720 rpm using CVCF. SCF-I was the most effective in reducing surface roughness as spindle speed increased. SCF-II had the smallest surface roughness at feed rates lower than 0.12 mm/rev. The least surface roughness was achieved at feed rate of 0.08 mm/rev using SCF-II.

In this work, non-edible oils like Neem and Honge oils are used as cutting fluids during drilling operation of mild steel.

2. Experimental details

In this work Neem and Honge oils are used in their purest forms and their blends with different percentages areused as cutting fluidsThe results obtained for these oils are compared with SAE 20W40 (Petroleum based oils) and at dry conditions. The proportions of these oils used ion this work are 50% Neem & 50% Honge, 33.3% Neem & 66.6% Honge, 66.6% Neem and 33.3% Honge. The oils are choosed based on their physical properties like flash point, fire point, specific heat, adhesiveness and dynamic viscosity. Drilling of mild steel is carried out constant speed of 800 rpm, constant feed rate of 10 mm/rev. CNC machine is used for drilling, and the dimension of the hole is constant with diameterof 13mm to a drilling depth of 30mm.

The temperatures of tool and work piece are measured using IR thermometer. The hardness of the mild steel specimen before machining and after machining is measured using Rockwell hardness tester.

The surface roughness of the specimen is measured for different cutting fluids using surface roughness tester.

(a) (b)

Figure 1. (a) CNC machine is used for drilling (b) continuous supply of cutting fluid during drilling.

Workpiece and cutting tool materials:

The workpiece materials used in with work is AISI 1014 Mild steel which is 75mm length with a diameter of 25mm. The drill bit material is HSS with 10% cobalt with a diameter of 13mm

3. Results and discussions

3.1 Physical properties of cutting fluids

A good cutting fluid should have high flash and fire point so that it can retard the fire at cutting temperature. The flash and fire points of are calculated by using Cleveland's apparatus. The flash point of blend of 50%Neem and 50%Honge is 2560C, which is very high compared to other cutting fluids. Hence blend of 50%Neem and 50%Honge is best in resisting the flame. The blend of 50%Neem and 50%Honge has highest fire point of 290°C is good at resisting the flame, followed by blend of Neem and with 285°c. It should also possess optimum viscosity i.e. if it is high then oils does not carry away heat and chip from machining zone and it will stick the chip in the machining area effecting machining efficiency.

Table 1. Physical properties of cutting fluids.

Sl.No	Type of cutting fluid	Flash point (°C)	Fire point (°C)	Dynamic Viscosity (N-s/m²)	Specific Heat (KJ/Kg. K)	Adhesiveness (g/m2)
1)	Neem	248	285	0.0345	1.6817	687
2)	Honge	220	245	0.0266	1.6761	412
3)	50%Neem 50%Honge	256	290	0.01648	1.6991	359
4)	33.3%Neem 66.6%Honge	228	256	0.0135	1.6703	257
5)	66.6%Neem33.3% Honge	228	264	0.011271	1.6789	367
6)	SAE 20W40	210	215	0.02172	1.97	319

If it is low then it will just pass away from machining zone without lubricating the tool and work piece. Viscosity of selected oils is measured by using saybolt viscometer .viscosity related to motion of oils is important, dynamic viscosity plays an important role, from the table below is it seen that Neem oil has highest value of 0.0345 N-s/m², least is 0.011271 N-s/m² for 66.6%Neem 33.3%Honge and optimum is for 50%Neem 50%Honge with 0.01648 N-s/m². This indicates that the blend of 50%Neem 50%Honge has both cooling and lubricating property. The specific heat is the amount of heat per unit mass required to raise the temperature by one degree Celsius. A good cutting fluid should have high specific heat, so that it can absorb max heat from the machining zone to rise its temperature instead of allowing heat to transfer to tool and work piece. From the table it can be seen that blend of 50%Neem 50%Honge has highest value of 1.6991 KJ/Kg. K followed by pure Neem with value of 1.6817 KJ/Kg. K. This shows blend is capable of absorbing higher heat from the machining zone or acts as good coolant compared to others oils used.

Adhesiveness is the property of cutting fluid referring to ability of fluid to stick to the surface of work and tool during machining and maintain a later separating both elements, so that friction is less. Adhesiveness should not be too high as the fluid will stick to surface along with chips formed. This reduces the life of the tool and machining efficiency. If it is too low then the fluid cannot separate the elements during machining resulting in higher friction and reduction to tool life. A good cutting fluid should have optimum or moderate value of Adhesiveness, so that both the above problems can be overcome. Neem has highest Adhesiveness with value of 687 g/m2, least is 257 g/m2 for blend of 33.3%Neem 66.6%Honge and the optimum is for blend of 50%Neem 50%Honge with the value of 359 g/m2. This shows that 50%Neem 50%Honge is best blend as it separates the machining elements and carries away the heat.

3.2 Temperature of work piece and drill bit during machining

A good cutting fluid should absorb maximum heat from the machining zone, giving very less heat to tool and the work piece, if not tool life will be less and the machining accuracy will be also be affected along with varying the properties of work piece. Temperature measured for tool and work piece with cutting fluid and at dry cutting condition are shown in figure 2. For dry condition temperature of tool and work piece is 56.3°C and 68°C which is very high. This is due to transfer of whatever heat generated to tool and work piece as no coolant is used and also due to high friction between tool and work piece as no lubricant is used. For SAE 20W40 oil the temperature measured for both tool and the work piece is 35.4 °C and 40.2°C, which is less compared dry cutting but higher than non-edible oils used, this due to its lower adhesiveness and lower dynamic viscosity. For 50%Neem, 50%Honge blend the temperature of tool and work piece is 36.2°C and 39.1°C very less value compared to other conditions. This is due high specific heat, high adhesiveness and higher dynamic viscosity. Indicating that, this blend has better cooling and good lubricating property.

3.3 Hardness test

Hardness refers to the resistance of a material to indentations and scratching. This is generally determined by forcing an indenter on to the surface. The resultant deformation in steel is both elastic and plastic. The hardness values of all the specimens before drilling and after drilling operation using different cutting fluids are determined using Rockwell hardness tester and the results are shown in figure 3.

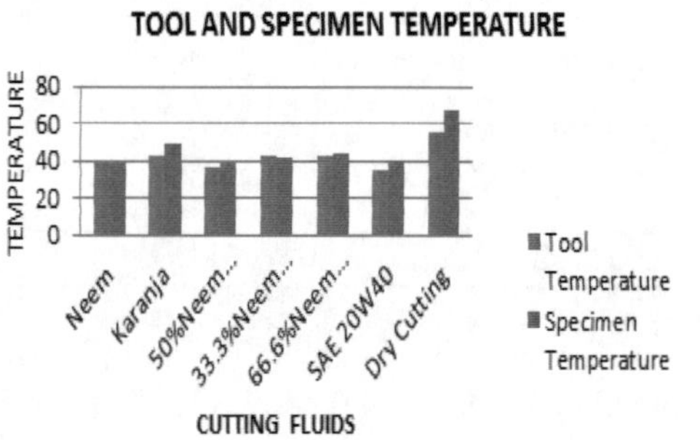

Figure 2. Tool and specimen temperature with different cutting fluids and at dry condition.

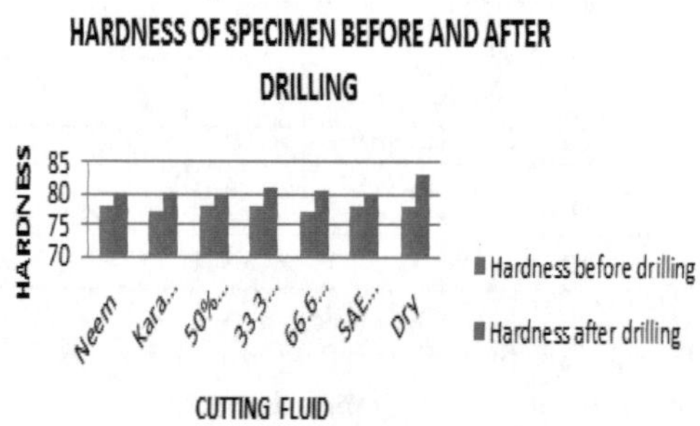

Figure 3. Hardness of specimen before and after drilling with different cutting fluids and at dry condition.

Heat generated during drilling operation due to friction between the work piece and drill bit is given out to tool, chip and work piece if no cutting fluid is used. If the maximum heat is given to work piece then the mechanical properties will be affected and if it is given to tool, then the tool wear is observed leading to reduction in tool life. The function of good cutting fluid is to absorb and carry away the heat generated in the machining zone. In the above table it is observed that for dry condition the hardness value is 78 HRB before drilling and it is 83 HRB after drilling which is very high compared to hardness value before drilling, which indicates that the metal surface has absorbed more heat during machining due to which it has become very brittle increasing its hardness. For petroleum based oil, Neem the hardness value is initially 78HRB which got increased to 80 HRB very less compared to Dry condition, indicating that maximum heat is absorbed by the cutting fluid and only small amount of heat is given to the work piece resulting in lesser hardness value. For blend of 50% Neem and 50% Honge, the hardness value before drilling is 78 HRB and 79.5 after drilling which is very less difference. This may be because of effective cooling and heat absorption capacity of blend used. For other combination of blends, the difference in hardness values are very high indicating that heat absorption is not as effective as blend of 50% Neem 50% Honge.

3.4 Surface roughness of the machined work piece

Surface finish of machined part depends on temperature at machining zone, if the temperature is high the work surface will become brittle and the force required to cut the metal will be high leading very rough surface. Figure 4 surfaces obtained after drilling mild steel work piece with different cutting fluids and at dry condition.

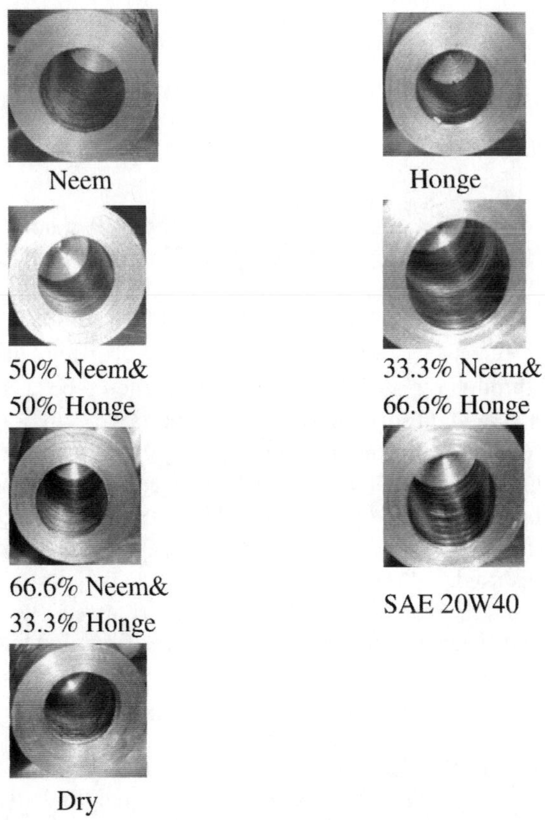

Figure 4. Surface of machined Specimen with different cutting fluids and at dry condition.

Table 2 shows Surface roughness measured with cutting fluid and at dry cutting condition. In dry cutting condition, the chips were discontinuous due to the friction in the tool and work interface. As the heat generated is high, the surface near the drill bit becomes brittle and due high cutting force, machined is very rough with highest value of 4.157 μm .During the use of non-edible oil we found that longer chips can be achieved due to the reduction. For Neem and blend of 50% Neem and 50% Honge, the surface roughness measured 1.279 μm and1.16 μm which is very less compared to other oils. This is due very low temperature at the machining zone and work surface will be ductile in nature leading to lesser cutting force and smooth removal of metal. In case of SAE 20W40 oil as cutting fluid the surface roughness measured is 3.5 μm, which is due to high friction between the drill bit and work piece.

Table 2. Surface Roughness measured on machined surface for different cutting fluids and at dry condition.

Sl. No	Type of cutting fluid	Surface Roughness (μm)
1)	Neem	1.279
2)	Honge	2.689
3)	50%Neem 50%Honge	1.16
4)	33.3%Neem 66.6%Honge	1.332
5)	66.6%Neem 33.3%Honge	1.593
6)	SAE 20W40	3.5
7)	Dry condition	4.157

4. Conclusion

- Blend of 50% Neem- 50% Honge oil is good cutting fluid compared to other oils used as it the highest flash point of 256°C, highest fire point of 290°C, Optimum viscosity of 0.01648 N-s/m^2, high specific heat of 1.6991KJ/Kg. K and optimum adhesiveness of 359 g/m^2.
- The temperature of tool and specimen was very less when the blend of 50% Neem- 50% Honge was used. The temperature of tool was 36.2°C and the temperature of specimen was 39.1°C.
- Due to friction the heat flows through the specimen and this heat should be less if the cutting fluid is effective i.e. cutting fluid should absorb more heat. Properties of the specimen will vary compared to parent work piece and this change in properties depends on the heat given to the specimen during machining. For 50% Neem- 50% Honge, the hardness of machined specimen(79.5 HRB) is very nearer to the value obtained with un-machined specimen(78 HRB)
- Surface roughness measured for Blend of 50% Neem- 50% Honge oil is 1.16 μm, which is less compared to other cutting fluids used.

From the above results it was found that the blend of 50% Neem and 50% Honge is the best cutting fluid compared to other cutting fluids used and it is environmental friendly, bio-degradable, non-hazardous and economical.

Acknowledgement
We would like to express our heart full thanks to our beloved HOD Dr.Shailesh Rao. A, for his continuous support and guidance. Words fall short to express our gratitude towards Dr. S.N.Sridhara, Principal/Director, K.S.School of Engineering and Management for his support and guidance. We would like to thank KSCST for funding our project. This would be incomplete if we wouldn't have been supported by my parents and friends. We are very much grateful people who have directly or indirectly helped us towards the completion of the work.

References

[1] B. Satheesh Kumar, G. Padmanabhan, P. Vamsi Krishna, "Experimental Investigations of Vegetable Oil Based Cutting Fluids with Extreme Pressure Additive in Machining of AISI 1040 Steel", *Manufacturing science and Technology 3(1)*, 1-9, 2015.

[2] Neetu Upadhyay, "Environmentally Friendly Machining: Vegetable Based Cutting Fluid" *S-JPSET: Vol. 7, Issue 2*, ISSN : 2229-7111.

[3] Sharafadeen K. Kolawole, Jamiu K. Odusote, "Performance Evaluation of Vegetable Oil-based Cutting Fluids in Mild Steel Machining", *Chemistry and material research,Vol.3.no.9*,2013.

[4] M. M. A. Khan, M. A. H. Mithu, N. R. Dhar "Effects of minimum quantity lubrication on turning AISI 9310 alloy steel using vegetable oilbased cutting fluid", *Journal of Materials Processing Technology 209 (2009) 5573–5583*, May 2009, page 5573- 5583

[5] Jitendra Kumar Chandrakar and Amit Suhane, The Prospects of Vegetable Based Oil as Metal working Fluid in Manufacturing Application-A Review, *International Journal of Engineering Research and Technology, Vol. 3-Issue 5 (May-2014)* ISSN 2278-0181.

[6] VaibhavKoushik A.V, Narendra Shetty. S & Ram prasad.C, *International Journal on Theoretical and* Applied *Research in Mechanical Engineering (IJTARME) ISSN (Print): 2319-3182, Volume-1, Issue-1,* 2012

[7] M.A. Fairuz, NurulAdlina M.J, A.I. Azmi, M.R.M. Hafieza ,K.W Leong, "Investigation of Chip Formation And Tool Wear in Drilling Process Using Various Types of Vegetable-Oil Based Lubricants", *Applied Mechanics and Materials Vols. 799-800 (2015) pp 247-250*

[8] EmelKuram, M. Huseyin Cetin, Babur Ozcelik ,ErhanDemirbas, Performance Analysis of Developed Vegetable-Based Cutting Fluids by D-Optimal Experimental Design in Turning Process, *International Journal of Computer Integrated Manufacturing, Volume 25, Issue 12*, 2012.

[9] E. Kuram, B. Ozcelik, E. Demirbas, and E. Şık, Effects of the Cutting Fluid Types and Cutting Parameters on Surface Roughness and Thrust Force, *Proceedings of the World Congress on Engineering 2010 Vol II WCE 2010, June 30 - July 2, 2010*, London, U.K.

IOP Conf. Series: Materials Science and Engineering **191** (2017) 012027 doi:10.1088/1757-899X/191/1/012027

Optimizing aerosol jet printing process of silver ink for printed electronics

S Agarwala[1], G L Goh[1,2] and W Y Yeong[1,2]

[1]Singapore Center for 3D Printing, Nanyang Technological University, Singapore
[2]School of Mechanical and Aerospace Engineering, Nanyang Technological University, Singapore

E-mail: SAgarwala@ntu.edu.sg

Abstract. Aerosol jet is an additive manufacturing technique that writes electronic inks directly onto the surface. The process has great potential for printed electronics as it is non-contact method that works well with variety of materials- metals, insulators, semiconductors, epoxy and encapsulation materials for conformal writing. This paper explores the effect of some parameters like ultrasonic power and atomizer flow on overall print quality, thus establishing a process window.

1. Introduction

The growing complexity of devices and emergence of multifunctional structures has made research and industrial community to look for alternate method of fabrication and material processing. Additive manufacturing has emerged as the promising candidate for creating complex 3D structures with easy processing [1]. Additive manufacturing builds an object through layer-by-layer deposition, and can embed functional elements in every single layer [2, 3]. Aerosol jet is an additive manufacturing technique that is based on droplet-deposition method and is widely explored for printed electronics. There are only few reports available where authors exploit aerosol jet for printing polymers and carbon based materials [4-6]. Seifert et al. do talk about differences between inkjet and aerosol jet [7], but none of the reports delve on the parameter influences of aerosol jet. Studying various parameters helps to establish guidelines for materials and contributes towards multi-material printing. Understanding parameter processing also give users an idea on what dimensions can be realistically printed.

This paper explores aerosol jet printing technique for printing silver electronic tracks on polyimide substrate. Both qualitative and quantitative analysis is carried out to determine the process window parameters to have controlled, repeatable and good line quality.

2. Aerosol jet printing

The basic schematic of aerosol jet printer is shown in figure 1. The Aerosol Jet process uses aerodynamic focusing to precisely deposit inks onto the substrates. The ink is placed into an atomizer, which creates a dense mist of material loaded droplets between 2- 5 μm in diameter. The aerosol mist is then delivered to the deposition head, where a sheath gas focuses it. When the sheath gas and aerosol pass though the nozzle, they accelerate and are focused into a tight stream of droplets flowing inside the sheath gas. The sheath gas also serves to insulate the nozzle from material contact preventing clogging. The resulting high velocity particle stream remains focused during its travel from the nozzle to the substrate over a distance of 2 to 5 mm maintaining feature resolution on non-uniform and 3D substrates. The system is driven by standard CAD data.

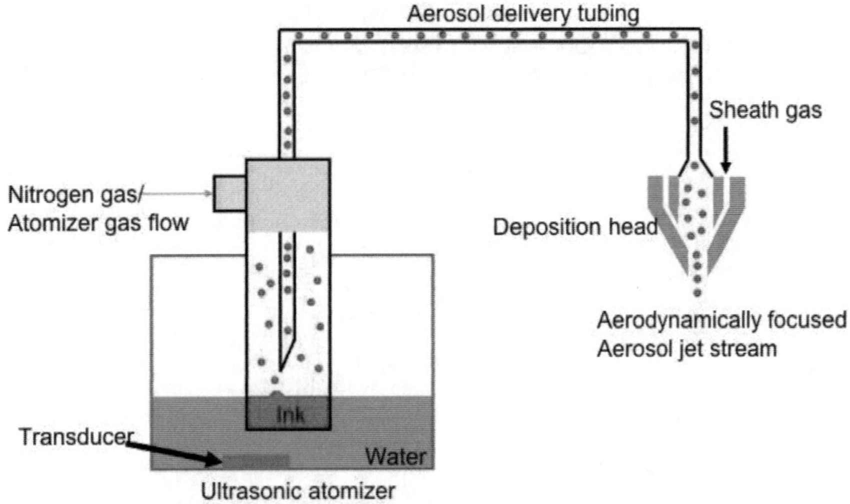

Figure 1. Schematic of Aerosol jet printing process.

Aerosol Jet printing is fast becoming an alternative for screen-printing and direct write methods like inkjet, as it allows smaller print resolution and deposition on non-planer surfaces [8, 9]. Such features give aerosol jet added flexibility, scalability and to print conformal multifunctional features in when compared to traditional technologies. Wide range of ink viscosities (1-1000 cP) and materials like metals, polymers, ceramic powders, semiconductors and insulators can be printed using aerosol jet.

3. Experimental section

Commercial silver nanoparticle ink (PRELECT TPS 50, Clariant) was used to print desired tracks on polyimide film substrate. Substrates were cleaned using ethanol prior to printing. The electronic tracks were designed in CAD and printed using aerosol jet (Optomec). Silver ink was dried at room temperature and later annealed at 200 °C for 60 min. Contact angle measurement was done to determine the wettability of silver ink on polyimide using an Attension Theta optical tensiometer and accompanying Biolin Scientific software. Inverted microscope (Olympus CKX41) was used for imaging the samples. Image J Software was used to quantify and measure the dimensions of printed part. Scanning electron microscope (SEM, Joel JSM-5600LV) was employed to determine surface morphology. All physical measurements were performed at least five times to achieve statistically significant results.

4. Results and discussion

The surface roughness of the polyimide film is measured to be 0.6 µm. The measured contact angle of the silver nanoparticle ink after printing is about 38°, indicating hydrophilic nature of the surface. There are many factors that will influence the quality of the final printed lines. In particular, substrate/ ink combination, number of printed layers and print head speed are relevant to the printing quality. The width control of the aerosol beam and thus the line width of the deposited line are controlled by various parameters, but the most important of them are atomizer flow rate and ultrasonic power. This work investigates the effect of these two parameters on the overall print quality achieved using aerosol jet. The aim is to establish process window to allow printing close to designed values. It is to be noted that sheath gas flow (0), ink temperature (30 °C), platen temperature (70 °C) and process speed (10 mm/s) are kept constant throughout the experiments.

Figure 2a shows how the line width of printed line changes with ultrasonic current at constant atomizer flow of 14 ccm3. The thickness of the printed line increases with increasing ultrasonic current. It is to be noted that 0.65 A is the maximum current that can be achieved with this aerosol jet printer. Also investigated is the change in line width with atomizer flow rate under constant ultrasonic current of 0.5 A (figure 2b). The curve in figure 2b tells us that when flow rate is below 16 ccm3, very less ink is deposited leading to discontinuous and breaking lines. For flow rate above 20 ccm3, excess ink is dispensed leading to bulging features on the substrate.

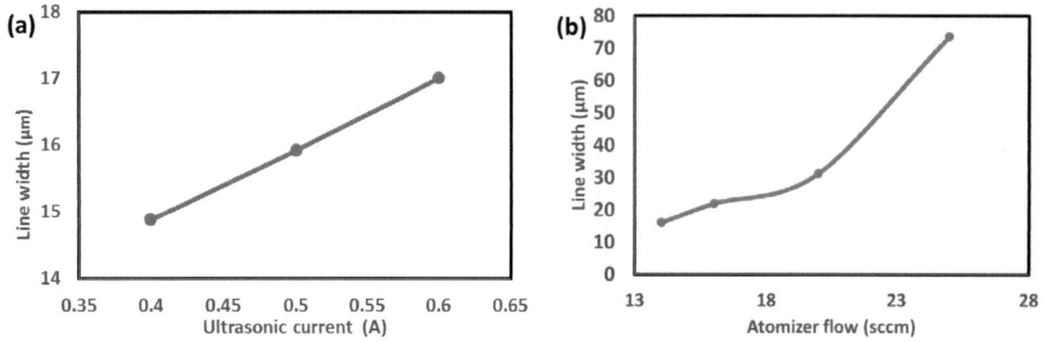

Figure 2. Line width as a function of (a) Ultrasonic current at constant atomizer flow of 14 sccm and (b) Atomizer flow rate under constant ultrasonic current of 0.5 A.

A good quality printed line is homogenous, continuous and adheres to the designed value. Figure 3a shows the process parameter combinations required to achieve a good printed line. Red, yellow and green zones signify discontinuous line, bulging line and good printed line quality respectively. The optical image of various printed lines is shown in figure 3. As discussed earlier, less ink in mist will lead to broken and irregular features. An overspray will form swollen and expanded prints. The overspray phenomenon is usually caused by the smallest droplets in the aerosol. However, changes in atomizer flow rate have much higher impact on the print quality than ultrasonic power. This means a small change in atomizer flow is enough to change the line width as against ultrasonic power. The table gives an indication of parameters that need to be used to have good printability. Atomizer flow rate between 15-20 sccm and ultrasonic current in range 0.4-0.6 A gives the best features for silver nanoparticle ink on polyimide substrate. It is to be noted that these parameters will change if ink, substrate and other parameters (that are kept constant) are varied.

Figure 4a shows a strain gauge pattern printed successfully on polyimide substrate. The aerosol jet has an added capability to write on curved surfaces, as shown in inset of figure 4a. Same strain gauge pattern is printed conformably on the dome shaped object using silver nanoparticle ink. The objective of printing the design is to get good print quality and narrow line-width of around 20 μm. The printed ink is further characterized using SEM to analyze the microstructure. Figure 4b shows sintered silver ink parallel tracks on polyimide. The printed pattern of strain gauge containing parallel lines depicts a good print quality as expected, with line width approaching 20 μm as desired. Such narrow and fine feature size is highly desirable for microelectronic and Nanoelectronic applications. The higher magnification image shows the particle size to be in the range of approximately 20-50 nm (figure 4c). The deposited silver film on polyimide is homogeneous and crack-free. Figure 4d also depicts 2.5 dimensional (2.5D) image of the pattern depicting the consistency of the printed features.

Hence, it was observed that when ultrasonic current was kept in the range of 0.3-0.6 A and atomize flow restricted to 16-20 sccm the print quality for silver is good. The deposited lines are close to design value and homogenous. The established parameters can also be used to write directly on 3D surfaces also, as shown for the dome structure.

(a)	Line width (µm)			
Ultrasonic current (A)	0.3	0.4	0.5	0.6
Atomizer flow (sccm)				
14		14.9	15.2	17
16	20.7	18.8	21.8	22.3
20	37.5	51.2	31.1	42.9
25	90.5	74.5	73.6	55.3

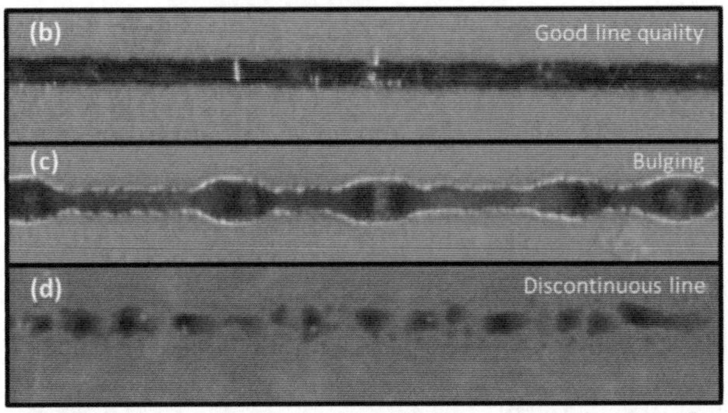

Figure 3. (a) Print quality determination for two parameters in aerosol jet printing. Red, yellow and green zones signify discontinuous line, bulging line and good printed line quality respectively. (b), (c) and (d) Optical image of the good and bad features of printed silver ink.

Figure 4. (a) Image of printed strain gauge on polyimide substrate. Inset shows the same design written on a dome shaped object. SEM micrographs showing printed silver lines after annealing, (a) low- and (b) high magnified image. (d) 2.5D image obtained using optical microscope.

IOP Conf. Series: Materials Science and Engineering **191** (2017) 012027 doi:10.1088/1757-899X/191/1/012027

5. Future works

The future work may involve studying the effect of print passes and thickness variation for fabricating functional devices like antenna, sensors and transistors.

6. Conclusion

Aerosol jet printing is a unique additive manufacturing technique for fabricating miniaturized electronic circuits and components on various substrates for printed electronics. The process works with wide range of materials to create functional components on 2d and 3D surfaces. This paper investigates the effect of ultrasonic current and atomizer flow on the print quality of the silver tracks, and establishes the best parameter combination for the material. For the ultrasonic current ranging between 0.3-0.6 A, an atomizer flow of 16 sccm prints homogenous, continuous and crack-free tracks. Optical microscope and SEM are used to ascertain the quality of printed lines. Aerosol jet capability of direct-write is also demonstrated on a curved dome shaped structure.

References

[1] Chua C K and Leong K F 2015 3D printing and additive manufacturing-principles and applications, *World Scientific*
[2] Lopes A J, MacDonald E and Wicker R B 2012 *Rapid Prototyping Journal* 18 129
[3] Goh G L, Ma J, Chua K L F, Agarwala S, Yeong W Y and Zhang Y P 2016 *Virtual and Physical Prototyping* 11 1
[4] Yang C, Zhou E, Miyanishi S, Hashimoto K and Tajima K 2011 *ACS Applied Materials and Interfaces* 3 4053
[5] Liu R, Shen F, Ding H, Lin J, Gu W, Cui Z and Zhang T 2013 *Journal of Micromechanics and Microengineering* 23 065027
[6] Tait J G, Witkowska E, Hirade M, Ke T H, Malinowski P E, Steudel S, Adachi C and Heremans P 2015 *Organic Electronics* 22 40
[7] Seifert T, Sowade E, Roscher F, WiemeM r, Gessner T and Baumann R R 2015 *Industrial & Engineering Chemistry Research* 54 769
[8] Hong K, Kim Y H, Kim S H, Xie W, Xu W D, Kim C H and Frisbie C D 2014 *Advanced Materials* 26 7032
[9] Liu R, Ding H, Lin J, Shen F, Cui Z and Zhang T 2012 *Nanotechnology* 23 505301

Experimental research to investigate the performance of bio coolant when turning of mild carbon steel

Tri Agus Susanto[1] and Rusdi Nur[1,2]

[1]Department of Mechanical Engineering, Politeknik Negeri Ujung Pandang, Makassar 90245, South Sulawesi, Indonesia
[2]Center of Materials and Manufacturing, Politeknik Negeri Ujung Pandang, Makassar 90245, South Sulawesi, Indonesia

E-mail: rusdinur@poliupg.ac.id

Abstract. Some literatures have been reported that the using bio coolant show better lubricating and cooling performances and reduce the occupational health risks associated with petroleum-oil-based coolant since they have lower toxicity. This paper investigates the effect the cutting conditions on the surface roughness through turning of mild carbon steel using dry, coolant and bio coolant. Measurement of surface roughness was conducted and then compared with the change of the cutting conditions. The relationship between surface roughness and cutting conditions was created in a curve for different of the cutting speed and coolant. The results indicate that the surface roughness was reduced when the speed of cutting is set to the highest level for all of coolant conditions (dry, coolant, and bio coolant) and constant of DOC and feed. The surface roughness had better performance using bio coolant than coolant conventional (mineral fluid).

1. Introduction

Nowadays, competitions in manufacturing industries continue to grow. In order to survive, the manufacturing industry should strive to have run its operations effectively and efficiently. Besides, it also consumer awareness of the importance of product quality also improved. The manufacturing industry had to maintain the quality of their products to fit the needs of consumers. Environmental issues (environment) as well as the safety and well-being are also factors that affect business activity. Increasing public awareness of environmental issues, safety, and health and government regulation caused the manufacturing industry will conduct its operations with due regard to environmental factors, occupational safety and occupational health.

AISI 1018 mild carbon steel has excellent weldability and produces uniform. AISI 1018 mild carbon steel offers a good balance of toughness, strength and ductility. Equipped with high mechanical properties, AISI 1018 hot rolled steel also include improved machining characteristics.
Special manufacturing controls used for surface preparation, chemical composition, rolling and heating processes. All these processes develop the highest quality products suitable for fabrication processes such as welding, forging, drilling, machining, cold drawing and heat treating.

Machining process was an important process in manufacturing industries [1]. Machining process was useful for removing unwanted material from the workpiece. Precise machining parameters needed to make the machining process implemented efficiently and effectively and to produce a quality product. One of the important parameters in the machining process is the use of cutting fluid. Machining process was resulting in high temperature that caused by friction between the cutting tool and workpiece. It was affect the dimension and surface machined. The high temperature in machining could be minimized by using a coolant.

Coolant and lubricants are usually employed in machining operations, including turning milling, drilling or grinding. The increased concern for environment and sustainability are pushing industry towards the new concept of sustainable production [2,3]. The use of liquid cooling was a simple way which can be used as coolant and lubricant. On the other hand, the use of coolant can cause environmental pollution and water for disposal [4,5]. Today, the using petroleum-oil based coolant is being debated for reasons of economy, health, and environment. Bio coolant and minimum quantity coolant have great potential, although still usage limited. Minimum quantity coolant has been used widely in the literature [6-9].

The researcher of Technical University of Denmark was stated that the use of bio coolant (e.g. vegetable oil) produce a better performance than other coolant [10]. The surface roughness is one of essential requirements to be considered in machining parts by customer [11]. This paper presents the influence speed of cutting and bio coolant on surface roughness.

The main goal of this experimental is to investigate the effect of the quantity of cutting fluid in turning of mild carbon steel under method of dry and wet (oil coolant and bio coolant). Other objective is to compare the parameters like workpiece quality of the machined surface under dry method and flood coolant method.

2. Bio coolant

Bio coolant, also known as bio-based coolant or bio-lubricants, are made from a variety of vegetable oils, such as rapeseed, canola, sunflower, soybean, palm, and coconut oils. The best application for bio coolants is in machinery that loses oil directly into the environment during use, and in machinery used in any sensitive areas, such as in or near water. Compared to petroleum-based coolant, use of bio coolant:

- Produces a cleaner, less toxic work;
- Environment and fewer skin problems for those working with machining;
- Offers better safety due to higher;
- Highly biodegradable;
- Costs less over the product's life-cycle;
- Due to less maintenance, storage and disposal requirements.

The increasing demand for biodegradable materials has opened the way for the use of vegetable oils as an alternative to petroleum-based ingredients polymer [12,13], especially in the operation of the machine. Awareness of the community on environmental issues has been continually evolving [14]. Vegetable-oils, especially canola [15] and rapeseed [16] are some of the more hopefully candidate as a base material for lubricants. They are more easily biodegradable and less expensive than synthetic base stocks. They frequently perform quite acceptable as a lubricant [17].

3. Experimental design

3.1. Cutting tools and materials

This paper was based on the experimental data which to assess the surface roughness in turning process. Machining tests were performed on a PL1000G lathe machine with 10kW of horse power and 7000 rpm of maximum speed. The cutting tool was ceramic insert (Sandvik) which the cutting parameters are shown in Table 1.

The material of experimental trials was AISI 1018 mild carbon steel round bar with 100 mm of diameter and 300 mm of length. The composition of AISI 1018 is shown in Table 2.

Table 1. The cutting parameters detailed

Parameters	Values
Speed of cutting (rpm)	500, 550, 700, 840, and 910
Depth of cut (mm)	0.5
Feed (mm/rev)	0.07
Coolant	Dry, Coolant and Bio Coolant

Table 2. The chemical composition of AISI 1018

Items	Composition
Carbon (C)	0.14 – 0.20 %
Iron (Fe)	98.81 – 99.26 %
Manganese (Mn)	0.60 – 0.90 %
Phosphorous (P)	≤ 0.04 %
Sulfur (S)	≤ 0.05 %

3.2. Experimental setup

The schematic of the experimental setup is shown in Fig.1. A Mitutoyo surface tester 301 was used to measure the surface machined for each turning test.

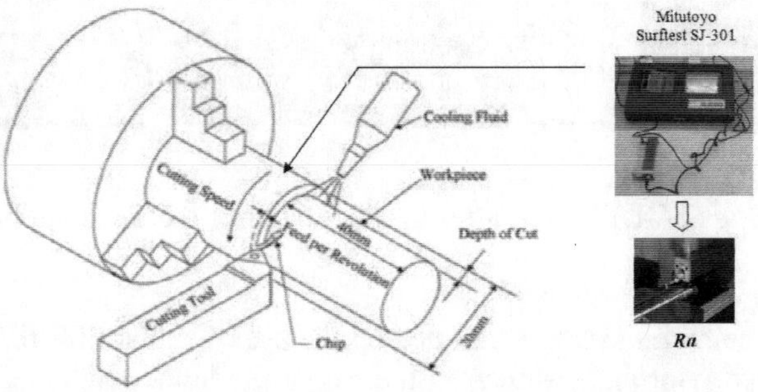

Figure 1. The schematic of experimental setup.

4. Results and discussion

Surface finish produced during machining has a direct impact on the condition of the tool. As the tool condition deteriorates, the surface roughness increases. It is accepted fact that the feed rate has a direct bearing on the surface finish of the machined work surface.

The value of the experimental surface roughness parameters was gained by stylus instrument and can be calculated theoretically, contemplating only the geometric component. The following equations [18, 19] can be used for the calculus of theoretical surface roughness parameters (Ra and Rt) in µm:

$$Ra = \frac{f^2}{32r} \times 1000 \tag{1}$$

$$Rt = \frac{f^2}{8r} \times 1000 \tag{2}$$

The experiment of turning on mild carbon steel shows that the surface roughness tends to decrease for higher speed of cutting. It can be shown in Fig.2 which shows that the surface roughness was proportional to speed of cutting also for using the coolant and bio coolant.

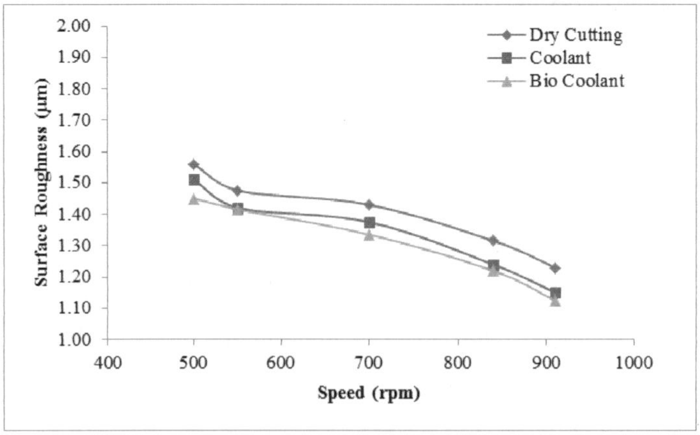

Figure 2. The graph of surface machined for different speed at feed of 0.07 mm/rev.

(a) Dry (b) Coolant (c) Bio coolant

Figure 3. The SEM views of flank wear after turning under dry, coolant and bio coolant at feed of 0.07 mm/rev

The decrease of surface roughness was affected by the use of coolant for the increasing speed. This result was confirmed by Xavior and Adithan in their investigation, which stated that coconut oil had greatest effect on the surface roughness and tool wear than compared to conventional mineral oil when turning of AISI 304 austenitic stainless steel with carbide tool [20]. Krisnha *et al.* studied the performance of SAE-40 and coconut oil when turning of AISI 1040 steel using cemented carbide [21]. It was reported that surface roughness initially minimized and then increased with increase in cutting speed at all the lubricating conditions and increased with increase in feed rate. They concluded that surface roughness decreased significantly with coconut oil-based nano-particle suspensions showed better performance compared to SAE-40. Another investigation was by confirmed by Nur *et al.* with different result. They present that surface roughness was only affected by changes in feed rate [22]. Sahin and Motorcu stated that the feed rate was main affecting factor on the surface roughness during turning of mild steel with coated carbide tools [23].

5. Conclusion

Lubricant and coolant are essential in minimizing the severity of processing contact between cutting tool and workpiece. Historically, the use of water was the main cooling materials due to the high of thermal capability and availability [24,25]. The aim of this investigation was to study the turning of mild carbon steel using coolant, bio coolant and dry coolant for the surface roughness as resulting performance. It concluded that bio coolant had better performance in surface roughness than compared to conventional mineral oil and dry coolant. The increment of speed will be decrease the surface roughness for all of coolant conditions.

Acknowledgments
Authors wishing to acknowledge Department of Mechanical Engineering Politeknik Negeri Ujung Pandang for generous access to submit this paper and also thank to Center for Materials and Manufacturing for supporting this research.

References
[1] Degarmo, Ernest P, Black J T, and Kohser R A 2003 Material and Processes In *Manufacturing 9th edition*. New Jersey. John Wiley & Sons
[2] Pusavec F, Krajnik P, and Kopac J 2010 *Journal of Cleaner Production* 18 (2) 174-184
[3] Zackrisson M. 2005 *Journal of Cleaner Production* 13 (1) 43-49
[4] Howes T D, Toenshoff H K and Heuer W 1991 *Ann CIRP.* 40(2) 623-629
[5] Bryne G and Scholta E 1993 *Ann CIRP.* 42(1) 471-474
[6] Davim J P, Sreejith P S, and Silva J 2007 *Materials and Manufacturing Processes* 22 45-50
[7] Thepsonthi T, Hamdi M, and Mitsui K 2009 *International Journal of Machine Tools & Manufacture* 49 156-162
[8] Sreejith P S 2008 *Materials Letters* 62 276-278
[9] Attanasio A, Gelfi M, Giardini C, and Remino C 2006 *Wear* 260 333-338
[10] De Chiffre L and Belluco W 2001 *Correlation Of Cutting Fluid Performance In Different Machining Operations* (Kgs Lyngby: Technical University Of Denmark)
[11] Astashev V K, Babitsky V I, and Kolovsky M Z 2000 *Dynamics and Control of Machines* (Springer) p 12-19
[12] Bisio, Attilio L, and Marino Xanthos 1995 *How to manage plastics waste* (Hanser Gardner Publications) How to Manage Plastics Waste (USA) 253
[13] Li F, Hanson M V, and Larock R C 2001 Polymer 42(4) 1567-1579
[14] Eichenberger and Hans F 1991 *Biodegradable hydraulic lubricant an overview of current developments in Central Europe.* (No. 910962. SAE Technical Paper)
[15] January 22,1993 *Operating hydraulics in "Green" fluids* (Machine Design) P73-77
[16] Flider F J 1995 INFORM 6(9) 1031-1035
[17] December 1994 *Vegetable oil based tractor lubricant* (Off-Highway Engineering) p 7-10
[18] Boothrpyd G and Knight W 1989 *Fundamentals of Machining and Machine Tools* (Marcel Dekker: New York) p 155-173
[19] Davim J P 2001 *Journal of Materials Processing Technology* 116 305-308
[20] Xavior M A and Adithan M 2009 *J. Mater. Process. Technol.* 209 900-909
[21] Krishna P, Vamsi Srikant R R and Rao D 2010 *Int. J. Mach. Tool Manuf.* 50 911-916
[22] Nur R, Noordin M Y, Izman S and Kurniawan D 2015 *Proceedings of the Institution of Mechanical Engineers, Part E: Journal of Process Mechanical Engineering* 1-8
[23] Sahin Y and Motorcu A R 2005 *Materials and Design* 26 321-326
[24] DIN ISO 14040ff 1998 Umweltmanagement-O¨kobilanzen-Festlegung des Ziels und des Untersuchungsrahmens sowie Sachbilanz (dt.Undengl.), Berlin, Deutschland
[25] McCoy J S 1994 *Introduction: tracing the historical development of metalworking fluids*, in: J.P.Byers (Ed.) (Metalworking Fluids, Marcel Dekker, NewYork) p 1-23

IOP Conf. Series: Materials Science and Engineering **191** (2017) 012029 doi:10.1088/1757-899X/191/1/012029

Numerical model of round rod in flat triangle pass in three-roll tandem rolling process based on mathematical analysis

PAN Lu[1,*], LIU XINYU[2]

[1]Department of Mechanical Engineering, Anhui Technical college of Mechanical and Electrical, China
[2]Department of Mechanical Engineering, Anhui Technical college of Mechanical and Electrical, China

E-mail: [*]ahjdpanlu@126.com

Abstract. In pass design of three-roll rolling process, accurate numerical model in deformation region is very important. Thus, the numerical model of round rod in flat triangle pass in three-roll tandem rolling process was proposed by mathematics analytic method in this paper. The parameters of numerical model, including arc length of the contact field, width of the contact field, rolling reduction(maximum, point) and rolling piece dimension at the exit, were established based on mathematical analysis.

1. Introduction

Three-roll mill is also called as Y-type mill, which is consisted of three dish-shape rollers, one of which is inclined to the other at 120 degree. Three-roll mill stands compactly, and temperature decrease is lower, which is benefit to control the rolling temperature [1]. The deformation in three-roll pass is equal, and spread is lower, which is less than 5%. In addition, the restriction of rolling piece is imposed from three directions by three rollers, which is propitious to low plastic, hard-deformed materials, such as titanium alloy, tungsten.

Conventional, deformation region in three-roll tandem rolling process was transferred to that of two-roll system, and mathematics model for pass design was the formulas written for two-roll system, which leaded to low-precision. Few studies for mathematics models have been carried out in three-roll sequence pass. Cao xifa proposed contact width in flat-triangle pass [2], J.H.Min, H.C.Kwon, Y.Lee, etc. proposed to use the mathematics model of two-roll in three-roll pass system [3-4].

In this study, considering that the spread of three-roll system is 1/3 of two-roll system, which was less than 5%, spread was ignored. The mathematics model was proposed by mathematics analytic method, including include contact arc length, contact width, contact area, rolling reduction(point, maximum), and cross-section area.

In this paper, formula derivation of mathematics model made the following assumptions: (i) ignore the influence of roller elastic flattening and rolled piece elastic deformation, (ii) ignore spread. And round rod in flat-round triangle sequence pass system was shown in Fig. 1.

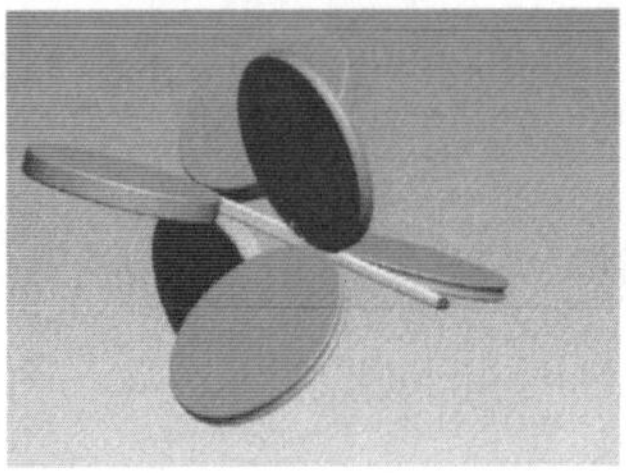

Figure 1. Round rod in flat-round triangle sequence pass system

2. Mathematics model

The parameters of round rod in flat triangle pass were shown in Fig. 2, The key parameters of round rod in flat triangle pass were shown in table 1.

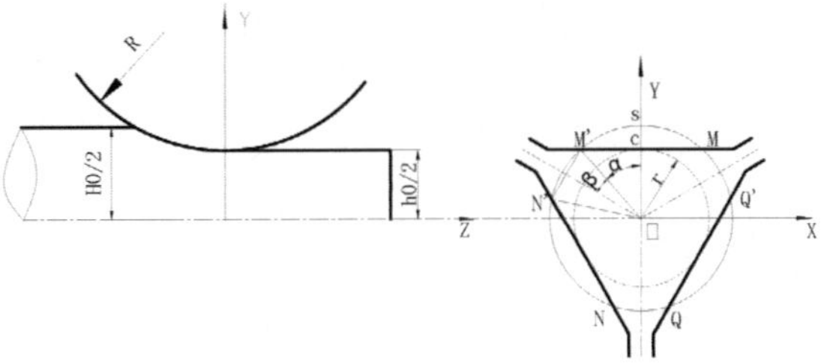

Figure 2. Geometric model of round rod in flat triangle pass

Table 1. Rolling parameters of round rod in flat triangle pass

Parameter	Value
Round rod diameter	H_0
Roller radius	R
Flat triangle Pass inscribed circle radius	r

Ignoring spread, the contact points is M and M', included angle between OM' and coordinate axis y is α.

Flat triangle pass curved surface equation in coordinate system XOY is calculated as Eq. 1:

$$\sqrt{[y-(R+r)]^2+z^2}=R \tag{1}$$

Round rod curved surface equation in coordinate system XOY is shown as Eq. 2:

$$x^2+y^2=(\frac{H_0}{2})^2 \tag{2}$$

2.1. Contact arc length

Ignoring spread in three-roll pass, the intersection curve between round rod curved surface and flat triangle pass curved surface can be calculated by solving the Eq.1 and Eq.2 simultaneously, as Eq. 3.

$$\begin{cases}\sqrt{[y-(R+r)]^2+z^2}=R \\ x^2+y^2=(\frac{H_0}{2})\end{cases} \tag{3}$$

Eq. 3 represents the curved boundary of deformation region of rolling piece. Overhead view curve from Z-axis of deformation region can be determined by setting Eq. 3 plane projection to x-z plane of eliminating variable y, as shown in Eq. 4. And then set Eq. 4 line projection to z-axis of eliminating variable x. Eq. 5 is the contact arc length in deformation region by definition of the contact arc length [4-5].

$$[\sqrt{(\frac{H_0}{2})^2 - x^2} - (R+r)]^2 + z^2 = R^2 \tag{4}$$

Contact arc length:

$$l = z = \sqrt{R^2 - [\frac{H_0}{2} - (R+r)]^2} \tag{5}$$

2.2. Contact arc width

As shown in Fig. 2, line MM' represents the contact width between rolling piece and roller. MM' is the intersection line between round rod curved surface and flat triangle pass curved surface of front view in x-y plane of CS XOY.

So, the positions of M, M' in coordinate system xoy are show as Eq. 6 and Eq. 7.

$$x_m = \sqrt{(\frac{H_0}{2})^2 - r^2}, y_m = r \tag{6}$$

$$x_{m'} = -\sqrt{(\frac{H_0}{2})^2 - r^2}, y_{m'} = r \tag{7}$$

Therefore, contact width MM' can be determined as Eq. 8.

$$k = 2 \times \sqrt{(\frac{H_0}{2})^2 - r^2} \tag{8}$$

2.3. Rolling reduction (point, maximum)

As shown in Fig. 2, line SC represents the maximum rolling reduction in deformation region, as shown in Eq. 9. And the reduction for every point in arc MM' can be calculated by Eq. 10.

$$\Delta h_{\max} = \frac{H_0}{2} - r \tag{9}$$

$$\Delta h_x = \sqrt{(\frac{H_0}{2})^2 - x^2} - r \tag{10}$$

Where, x represents the x-position in arc MM'

2.4. Rolling piece dimension at the exit

Round rod before rolled in flat triangle pass was shown in Fig. 3 (a). The shape of round rod at the exit of flat triangle pass has been reduced to sub triangular piece as Fig. 3 (b), ignoring spread. MM'N'NQQ' represents the profile of rolling piece rolled in flat triangle pass. M'N', NQ, Q'M are arc, and M'M, N'N, Q'Q are straight lines. So sub triangular piece dimensions are show in Table 2.

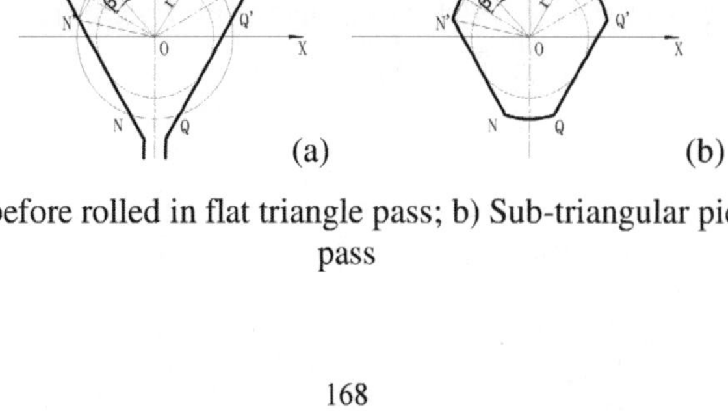

Figure 3. a) Round rod before rolled in flat triangle pass; b) Sub-triangular piece exiting flat triangle pass

Table 2. Rolling piece dimension for arc M'N', NQ, Q'M

Parameter	Center of arc	Radius of arc	Included angle
M'N', NQ, Q'M	(0,0)	$\frac{H_0}{2}$	$2\beta = 2\times[60° - \text{acos}(\frac{2\times r}{H_0})]$

2.5. Cross-section area and contact area

As shown in Fig. 2, are $M'SMM'$ are presents maximum cross-section area in deformation region, $S_h = \int_{x_{m'}}^{x_m} \Delta h dx = 2\int_0^{x_m} \Delta h dx$. Combining with Eq. 10, the mathematics model of S_h can be considered as Eq. 11.

$$S_h = 2\times[\frac{x_m}{2}\times\sqrt{(\frac{H_0}{2})^2 - x_m^2} + \frac{(\frac{H_0}{2})^2}{2}\times arsin(\frac{2\times x_m}{H_0}) - r\times x_m] \tag{11}$$

Mean rolling reduction in deformation region can be considered as $\overline{\Delta h} = \frac{S_h}{\text{k}}$, Where, k represents contact width and S_h represents cross-section area

As shown in Fig. 1, the geometric shape of deformation region can be reduced to rectangle. Therefore, contact area S_j can be calculated by multiplying the contact arc length l by contact width k, as shown in Eq.12.

$$S_j = 2\times\sqrt{R^2 - [\frac{H_0}{2} - (R+r)]^2}\times\sqrt{(\frac{H_0}{2})^2 - r^2} \tag{12}$$

3. Application of mathematics model in deformation region

In this paper, mathematics model of flat triangle-round triangle sequence pass system in three-roll tandem rolling process has been proposed, which has solved the pass design of three-roll tandem rolling process. In the case of the pass system for Φ25~Φ12 titanium alloy rod(TC4)product line, the parameters in deformation region was calculated with mathematics model. Titanium alloy rod (TC4) Φ25 was rolled to rod Φ12, and the eight-passes flat triangle-round triangle sequence pass system was shown in Fig.4. The entrance velocity of the flat triangle-round triangle sequence pass system was 1m/s, initial rolling temperature of rolling piece was 950°C.

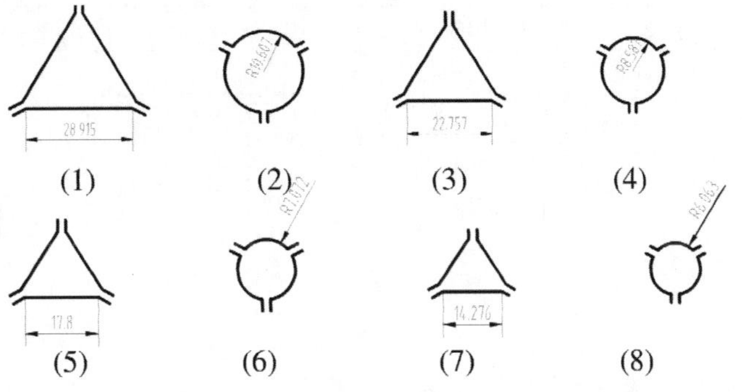

Figure 4. Φ25~Φ12 titanium alloy rod（TC4）eight-passes sequence pass system

3.1. Finite element analysis

Elastic-plastic finite element method was used in metal pressure process to predict shape and parameters during deforming process. And the result of FEM simulation can get high precision, and fits the practice results. Therefore, to prove the accuracy of mathematics model derivated in this paper, the eight-pass flat triangle-round triangle sequence rolling process of φ25~12 titanium alloy rod(TC4)product line was simulated with elastic-plastic finite element software, as shown in Fig. 5 [6-7].

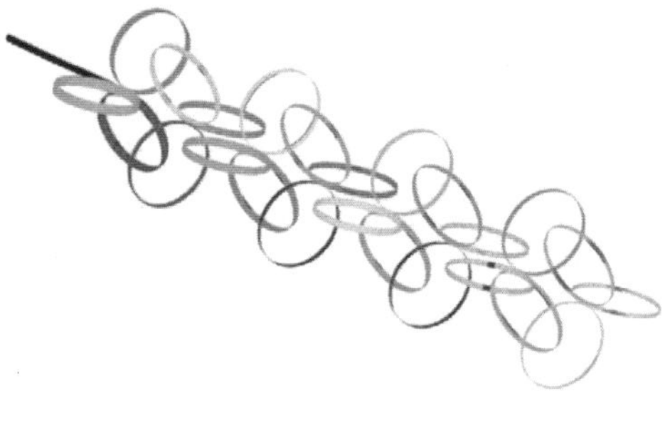

Figure 5. The finite element model of Φ25~Φ12 titanium alloy rod tandem rolling

3.2. Comparison of contact arc length between FEM simulation and mathematics model

As is shown in Table 3, error of contact arc length between FEM simulation and mathematics model was below 10%, and the contact arc length of mathematics model was smaller, as the result of ignoring the spread.

Table 3. Comparison between FEM simulation and mathematics model

Pass	contact arc length with simulation(mm)	contact arc length with Mathematics model(mm)	Error (%)
1	33.962	32.193	5.20
2	29.684	27.10919	8.67
3	31.755	29.08512	8.40
4	22.846	21.03437	7.92
5	26.073	25.33204	2.84
6	21.355	20.677	3.17
7	26.345	25.45635	3.373
8	22.786	20.87822	8.37

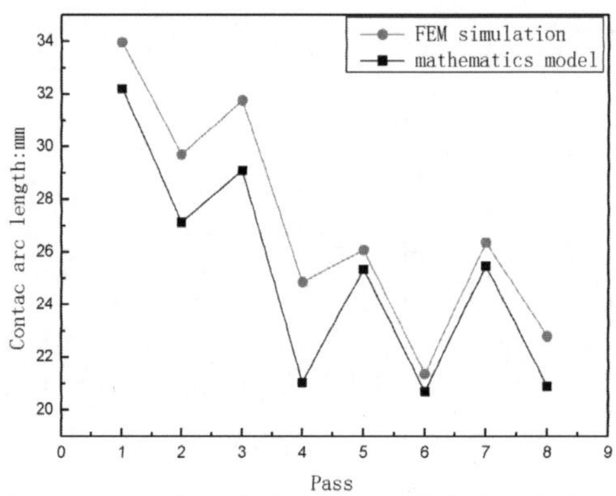

Figure 6. Comparison of contact arc length between FEM simulation and mathematics model

3.3. Comparison of contact width between FEM simulation and mathematics model

As is shown in Table 4, error of contact width between FEM simulation and mathematics model was mainly below 5%, and the contact width of mathematics model was smaller, as the result of ignoring the spread.

Table 4. Comparison between FEM simulation and mathematics model

Pass	contact width with simulation (mm)	contact width Mathematics model(mm)	Error(%)
1	17.95	17.4525	2.77
2	11.318	11.041	2.44
3	14.31	14.065	1.71
4	7.791	7.6535	1.76
5	11.836	11.29422	4.57
6	8.192	7.886542	3.72
7	10.417	10.08894	3.14
8	6.035	5.691089	5.69

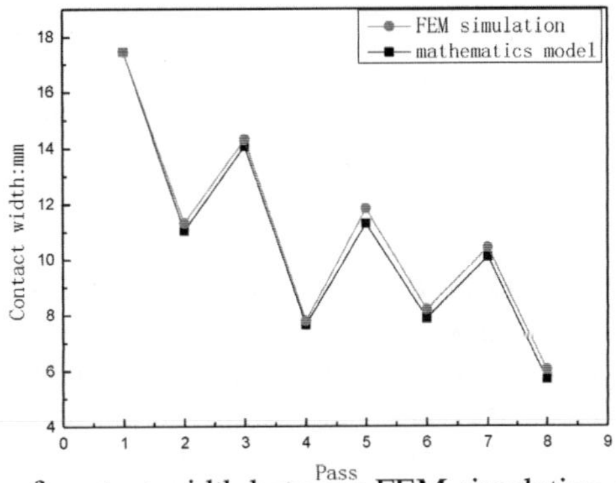

Figure 7. Comparison of contact width between FEM simulation and mathematics model

4. Summary

Mathematics model of round rod in flat triangle pass in three-roll rolling process was proposed in this paper, including arc length of the contact field, width of the contact field, rolling reduction(maximum, point) and rolling piece dimension at the exit. Therefore, Mathematics model of other pass system, such as flat triangle-spherical triangle pass system, spherical triangle-round triangle, can also be derivated by mathematics analytic method.

References

[1] C Xi-fa, *Steel rolling*. 4, 2 (1999).
[2] J.H.Min, H.C.Kwon, Y.Lee, J.S.Woo, Y.T.Im: *Journal of materials processing technology*. 140, 7 (2003).
[3] P Lu, S Mei-rong, H Qin-xue.*Taiyuan University of Science and Technology,* 2, 4 (2012).
[4] S Mei-rong, H Qin-xue, Q Jian-ping, *Rare Metal Materials and Engineering*, 6, 6 (2012).
[5] Mashiko. *Tran.ISIJ*, 28, 6(1988).
[6] S Mei-rong. *Taiyuan University of Science and Technology*, 5, 4 (2008).
[7] N Xing-long, W Guo-hong. *Rare metal materials and engineer*, 27, 5 (1988)

The limiting sheet diameter prediction model for cup-shaped part drawing process with diverse mould assemblage based on RSM

Zhefeng Guo and Wencheng Tang

Department of Mechanical Engineering, Southeast University, Nanjing 211189, China

E-mail: gdxtfj@163.com

Abstract. In order to ensure unbroken production in cup-shaped part deep drawing process with diverse mould assemblage, a prediction model of limiting sheet diameter based on response surface method (RSM) was presented. The simulated samples were used to fit the sheet diameter RSM prediction model, and the multiple determination coefficient of RSM was reached to 0.9. Setting the minimum sheet thickness to the critical value, the RSM will become a limiting sheet diameter prediction model. It shows a close agreement with FE simulated results through the test of examples. It can be concluded that presented RSM not only has higher efficiency, but also has higher accuracy. It will provide an early quality forecasting before actual deep drawing production.

1. Introduction

Sheet metal deep drawing is a widely used in manufacturing process, it is very suitable for mass production. Efficacious parameters for the defects of different types, which may appear in the forming parts can be divided into three main categories [1]: (1) material properties such as yield stress, work hardening coefficient, anisotropic coefficient, blank dimensions, and thickness of blank; (2) tool properties such as punch radius, die radius, and clearance; (3) process parameters such as blank holder force, friction coefficient, type and position of lubricant, strain rate, and pressure.

To achieve a desirable quality with a minimum cost, traditional process design has been performed in a time-consuming and costly way [2], usually involving a series of design modifications and test. As an efficient engineering tool, finite element (FE) simulation has become more and more prevalent, it allows us to precisely predict a forming process by detecting defects such as wrinkling and fracture in an early stage, thereby, reducing design and test costs to a considerable extent [3]. Utilization surrogate approximate model to map the relation between analysis results and input variables to replace the FE analysis is an efficient and convenient method, it cost less analysis times than direct simulation. Many kinds of surrogate model can achieve the approximation for the deep drawing process, such as artificial neural network (ANN) [4], radial basis function (RBF) [5] and response surface method (RSM) [6], et al.

In deep drawing process, the blank holder plays a key role in adjustment of metal flow in to the die cavity. Moreover, the quality of drawn parts is extremely affected by this flow [7]. Srirat, et al. [8] optimized the variable blank holder force (VBHF) trajectory and tools motion simultaneous by a sequential approximate optimization with RBF network. Ehsan Karajibani, et al. [9] investigated the formability of two layer (aluminum-st12 steel) sheets in the deep drawing process through numerical simulations and experiments. Jae-Jun Lee and Gyung-Jin Park [10] optimized the structural parameters and process parameters to avoid defects in sheet metal forming production. To observe above articles, it can be found that the research is insufficient for limiting sheet diameter in cup-shaped part deep drawing process.

At present, there are some moulds with diverse arc radius used in cup-shaped part drawing process. The diverse arc radius is the only difference during the same type moulds. Using the diverse exchange and assemblage of moulds, diverse cup-shaped parts can be produced. As we all know, in the deep drawing process, the required process conditions for different assemblage of mould are not the same. Users need to carry out a large number of repeated testing to determine the best drawing process for the eligible product, the efficiency is lower. Thus, it is necessary to develop a speedy and accurate prediction model to evaluate the drawing quality. In this paper, the precise FE simulation result was used to obtain the initial training samples, and then, based on the RSM surrogate model, the sheet diameter prediction model was established. Setting the minimum sheet thickness to the critical value, the RSM will become a limiting sheet diameter prediction model. Through the test examples, the RSM shows higher prediction accuracy. This prediction model can provide an early forecasting before actual deep drawing production, reducing the rate of flawed products and labor consumption of repeated testing, it is very helpful to improve the production efficiency and develop the numerical control system.

2. Limiting sheet diameter, RSM and material

2.1 Limiting sheet diameter

The limiting sheet diameter is a diameter of just before an edge crack occurs. In a drawing process, if the sheet diameter is more than the limiting sheet diameter, the drawing part will be occurred fracture defect. The assembly method of cup-shaped part in a deep drawing process is shown as figure 1.

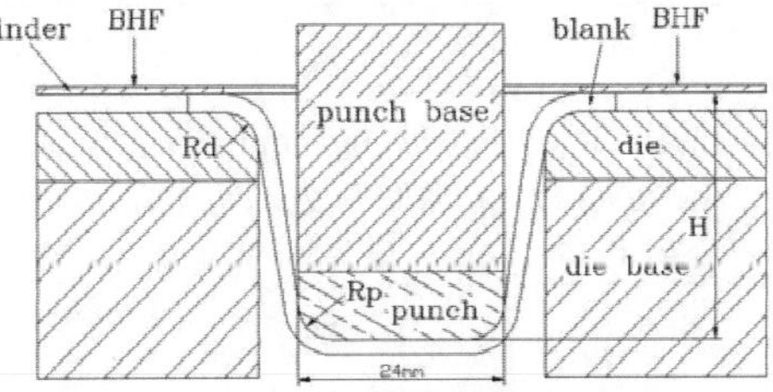

Figure 1. Schematic diagram of cup-shaped part deep drawing process.

The punch and die are installed on the bases respectively. In the deep drawing process, change the punch with diverse radius and initial sheet with diverse diameter, will obtain diverse cup-shaped part; change the die with diverse radius and blank holder force, will adjust the material flow condition, avoiding winkling and fracture. The combination of blank holder force (BHF), sheet diameter (D), punch radius (Rp) and die radius (Rd) in every group is corresponding to a drawing process; different drawing process will produce the parts with different quality. Thus, to develop a prediction model which can speedily and accurately evaluate the limiting sheet diameter, will reduce the repeated testing before drawing, it is significant to improve the production efficiency. In this paper, the BHF, D, Rp and Rd are considered as the research parameter, and their ranges and levels are shown in Table 1.

Table 1. Variable parameters with their levels.

levels	BHF(kN)	D(mm)	Rp(mm)	Rd(mm)
1	1	50	3	3
2	4	53	4	4
3	7	56	5	5
4	10	---	---	---

IOP Conf. Series: Materials Science and Engineering **191** (2017) 012030 doi:10.1088/1757-899X/191/1/012030

2.2 RSM

RSM provides an approximate relationship between the input variables x and their response y. In this paper, the approximate function of the response y is considered as second order polynomial:

$$y = \beta_0 + \sum_{i=1}^{k} \beta_i x_i + \sum_{i=1}^{k} \beta_{ii} x_i^2 + \sum_{i<j}^{k} \beta_{ij} x_i x_j + \varepsilon$$

In this paper, the accuracy of RSM was evaluated by multiple determination coefficients (R^2):

$$R^2 = 1 - \sum_{i=1}^{n} (y_i - \tilde{y}_p)^2 \Big/ \sum_{i=1}^{n} (y_i - \bar{y})^2$$

where n is the number of the samples, y is the response of RSM model, $\tilde{y}_c$ is the actual value of the sample outputs, and the $\bar{y}$ is the mean value of these actual outputs.

The more close to 1 the R^2 is, the higher the RSM prediction accuracy is.

2.3 Material and fracture criterion

In this paper, the low carton steel DC06 with 0.6 mm thickness was used to deep draw, the mechanical properties of material are obtained from a series of tensile tests and shown in Table 2, where E and υ are young's modulus and poisson's ratio respectively. In general, the true stress - true strain curve is fitted to exponential function type, in this paper, it was considered to the formula $\sigma = k(\varepsilon_0 + \varepsilon)^n$ ($\varepsilon_0 = 0.02$). Material anisotropic property uses the barlat'89 criterion [11,12].

Table 2. Mechanical properties of low carton steel DC06.

$E(GPa)$	υ	$k(MPa)$	n	r_0	r_{45}	r_{90}
207	0.28	530.0	0.24	2.21	1.77	2.76

In the deep drawing process, the fracture is a very important failure type. It is used to measure the drawing quality as the criterion. In this paper, the maximum permit thinning was used to judge and measure the fracture. Usually, a maximum thinning of 25% thickness is used as the critical fracture criterion [4]. Because the initial blank thickness is 0.6 mm, the maximum allowed thinning is 0.15 mm.

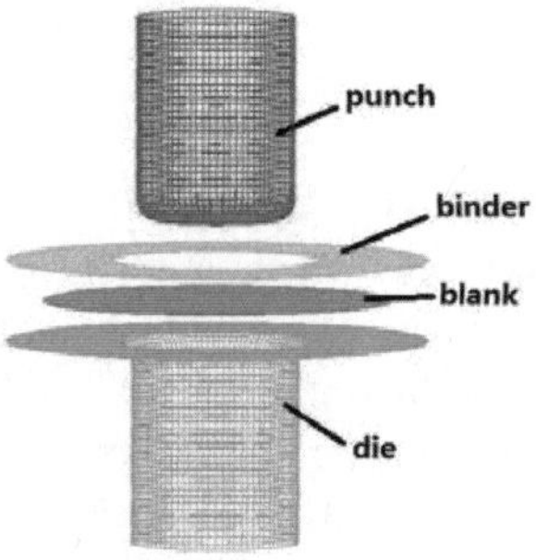

Figure 2. Mesh of the blank and drawing moulds.

3. FE model of deep drawing process

3.1 FE model

In this paper, the deep drawing process was simulated by the FE software Dynaform. The drawing depth was set up to 35 mm, in this drawing depth, if the fracture don't occurred, the final cup-shaped part will have not any flange section. The element type was chosen as a quadrilateral shell element (Belytschko-Tsay). The punch, die and binder were considered as rigid body, the drawing speed of the punch was set to 50 mm/s, friction coefficient was set to 0.1 as an ordinary scale of metal contact, and the mesh size of the parts was set between 0.5 and 1, with the mesh density in the arc region and planar region more dense and sparse, respectively. The mesh of blank and moulds was shown in figure 2, and the rest of the model parameters were set to the default values of the software.

3.2 Verification of FE model

For the verification of FE model, it is unnecessary to verify each FE model in the variable space of Table 1. In this paper, a set of process parameters within the variable space was chosen randomly to implement the FE simulation. When the BHF=10 kN, Rp=3 mm, Rd=3 mm, the FE simulated results for D=54.6 mm and D=54.3 mm were respectively shown as figure 3 (a) and (b).

It can be found that the minimum sheet thicknesses of D=54.6 mm and D=54.3 mm are 0.419 mm and 0.453 mm respectively. Therefore, the limiting sheet diameter must be a value between 54.6 mm and 54.3 mm, take the average value 54.45 mm as the FE simulated limiting sheet diameter of this drawing process. In the article of Zhang, et al[13], the drawing experiments with same process condition as the FE simulation have been done. The experimental limiting sheet diameter is 55.3 mm. The deviation between the EF simulated and experimental limiting sheet diameter is only 0.85 mm, the relative error is 1.5% at the same time. Thus, it is an indication that the FE model is capable to precisely simulate the drawing process under diverse process conditions.

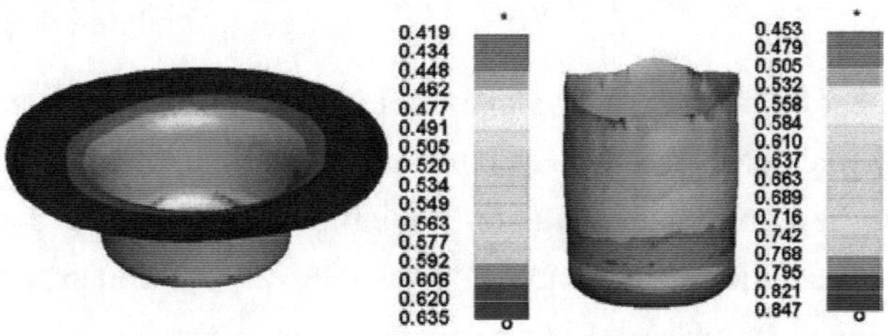

(a) (b)

Figure 3. FE simulated results for (a) D=54.6mm; (b) D=54.3mm.

4. RSM prediction model for limiting sheet diameter

4.1 Training samples

To further reduce the numbers of simulation, we implemented an efficient experimental design, which includes an orthogonal experimental design with 16 numbers and extra 24 random experiments, here only 40 samples need to simulate, and they are needful and sufficient to generate the required data for fitting RSM. In Table 3, the last column is the simulated minimum section thickness under each FE variable samples. If minimum thickness is less than 0.45 mm, the fracture has been occurred in deep drawing process.

Table 3. Experimental samples and their FE simulation results.

Num	BHF(KN)	D(mm)	Rp(mm)	Rd(mm)	t(mm)
1	1	50	3	3	0.522
2	1	53	4	4	0.523
3	1	56	5	5	0.501
4	1	53	3	3	0.505
--	--	--	--	--	--
39	10	56	4	4	0.277
40	10	56	5	3	0.047

4.2 RSM prediction model

Taking the BHF, Rp, Rd and final minimum sheet thickness t as the inputs, the initial sheet diameter D as the output respectively, the sheet diameter RSM prediction model was established by the samples of

Table 3. Then, the RSM is an approximate model for predicting the sheet diameter D. Fitted RSM function was shown as equation (1), the compared results between initial and predicted sheet diameter were shown as figure 4, and the statistical histogram of deviation was shown as figure 5 respectively.

$$D(BHF, Rp, Rd, t) = 88.58 - 3.95*BHF + 2.61*Rp + 0.46*Rd - 60.03*t + 0.015*BHF^2 - 0.278*Rp^2 + 0.023*Rd^2 - 52.56*t^2 - 0.098*BHF*Rp - 0.031*BHF*Rd + 7.956*BHF*t + 0.196*Rp*Rd + 0.214*Rp*t - 1.05*Rd*t \quad (1)$$

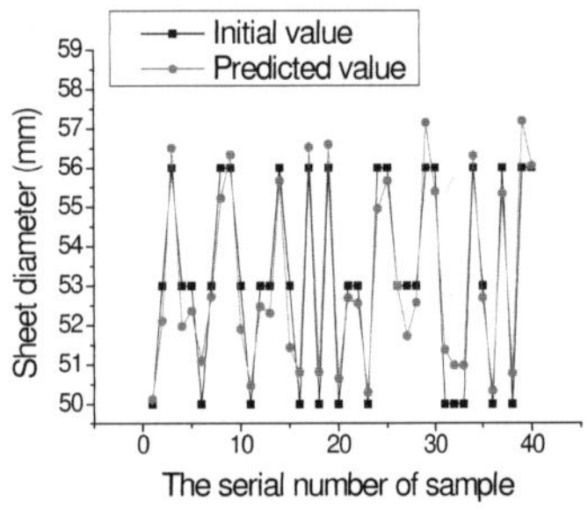

Figure 4. Initial and predicted sheet diameter.

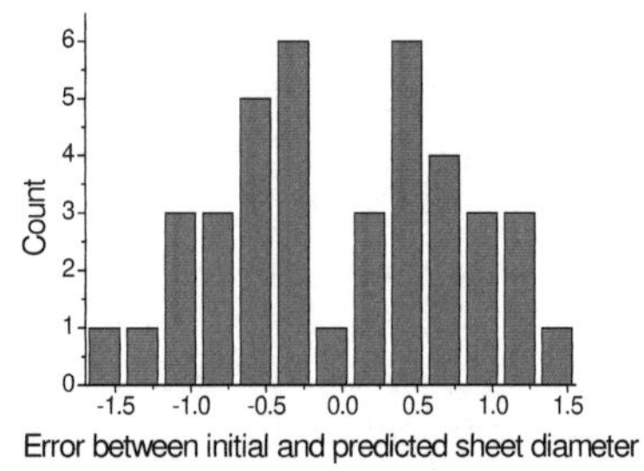

Figure 5. Statistical histogram of deviation between initial and predicted sheet diameter.

From the figure 4, it can be found that the maximum diameter deviation is 1.57 mm at the sample serial number 15, still in the acceptable range. Form the deviation statistical histogram of figure 5, it be clearly shown that there is only one diameter deviation more than 1.5 mm. Therefore, it can be declared that fitted RSM prediction model can provide a better approximate sheet diameter value.

R^2 of RSM was reached 0.9. The RSM is capable to map the complex and non-linear relation between sheet diameter and multi-parameter inputs. For the limiting drawing state, the sheet thickness has been reached to the crack edge. In this case, the minimum sheet thickness was equal to 0.45 mm. Therefore, setting the value of t in equation (1) to 0.45, the equation (1) will become a limiting sheet diameter prediction model. From this model, the user needs not to simulate the complex deep drawing process anymore by FE, the limiting sheet diameter can be obtained immediately when others input parameters are entered.

4.3 Test examples and discussion

In order to test the applicability and accuracy of limiting sheet diameter RSM, four groups of drawing process parameters were selected randomly, and their minimum sheet thickness, simulated and predicted sheet diameter were shown in Table 4.

Table 4. Process parameters and results of test examples.

Num	BHF(KN)	Rp(mm)	Rd(mm)	D(mm)/t(mm)		
				RSM	Unbroken	Broken
1	10	3	4	52.8/0.45	53.3/0.49	53.6/0.44
2	8	3	4	53.86/0.45	53.6/0.46	53.8/0.448
3	4	3	3	55.7/0.45	55.2/0.48	55.6/0.38
4	7	5	3	54.66/0.45	55.2/0.49	55.6/0.23

The procedure for obtaining the limiting sheet diameter on FE simulation as the following:

For one test example, FE was used to simulate two different deep drawing processes respectively, whose sheet diameter sizes must be different and very close. If one of the two sheet diameters is broken, and another one is unbroken, then, the limiting sheet diameter on FE simulation must be a value between these two sheet diameters. The comparisons were shown in figure 6.

From the figure 6, it can be found that the maximum limiting sheet diameter deviation is at the serial number 4, because the limiting sheet diameter is a value between the unbroken and broken sheet diameter, here, the average value 55.4mm was treated as the limiting sheet diameter, the deviation between predicted by RSM and simulated by FE is 0.74mm, the relative error is only 1.4% at this time. Thus, the limiting sheet diameter prediction RSM model has higher prediction accuracy. Because the RSM model has fast calculation speed and higher accuracy, it could provide a guidance to prevent fracture at the early stage of deep drawing process.

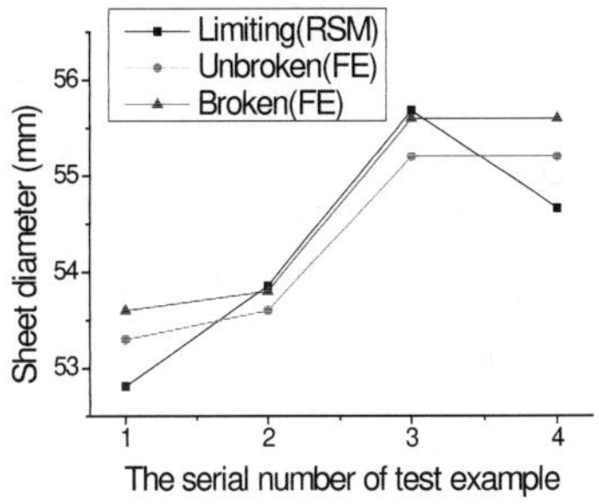

Figure 6. Drawing quality of initial and predicted sheet diameter.

5. Conclusions

In this paper, an efficient and accurate approximate model based on RSM was developed to predict the limiting sheet diameter in the cup-shaped part deep drawing process with diverse mould assemblage. It can be found that the limiting sheet diameter RSM prediction model in this paper not only has higher efficiency, but also has higher accuracy. It is very suitable to provide a guide for mould assemblage and parameter setting in the early stage of deep drawing process, and is also very suitable to be integrated in the numerical control system. In the future, we will use RSM model to predict the actual drawing production to improve the production efficiency.

Acknowledgements

The research was funded by the National Science and Technology Major Project of China (2012ZX0401031).

References

[1] Gantar G and Kuzman K 2002 *J Mater Process Tech.* 125 p 302-308
[2] Yi K, Choi K K, Kim N H and Botkin M E 2007 *Int J Numer Meth Eng.* 71 p 1483-1511
[3] Sun G Y, Li G Y and Li Q 2012 *Finite Elem Anal Des.* 59 p 76-90
[4] Manoochehri M and Kolahan F 2014 *Int J Adv Manuf Tech.* 73 p 241-249
[5] Kitayama S, Kita K and Yamazaki K 2012 *Int J Adv Manuf Tech.* 61 p 1067-1083
[6] Song J H, Huh H and Kim S H 2007 *J Eng Mater-T ASME.* 129 p 397-406
[7] Hosseini A and Kadkhodayan M 2014 *Int J Adv Manuf Tech.* 71 p 337-355
[8] Srirat J, Kitayama S and Yamazaki K 2012 *J Adv Mech Des Syst.* 6 p 1081-1092
[9] Karajibani E, Fazli A and Hashemi R 2015 *Int J Adv Manuf Tech.* 80 p 113-121
[10] Lee J J and Park G J 2014 *J Mech Sci Technol.* 28 p 605-619

[11] Barlat F and Lian K 1989 *Int J Plasticity.* 5 p 51-66

[12] Zhang D J, Cui Z S, Ruan X Y and Li Y Q 2006 *Comp Mater Sci.* 38 p 256-262

[13] Zhang W R, Cao Y G, Yang S L, Cheng L X, Deng P R and Yong X C 2014 *Forming and Stamping Technology.* 39 p 41-45 (in Chinese)

Adaptive laser focusing head design for the 3D free surface in parallel kinematic machining

K H YE[1] , B S Lee[2] and H W Choi[2]

1 Graduate School of Mechanical Design, Keimyung University, Daegu, Dalgubeol-daero 1095, Republic of Korea
2 Department of Mechanical and Automotive Design, Keimyung University, Daegu, Dalgubeol-daero 1095, Republic of Korea

E-mail: ykh9662@naver.com

Abstract. An adaptive laser head was designed for microscale patterning and welding applications. An optical design for laser head was performed by using ray tracing technique with commercial software. Setting the focal length of the lens at 100mm the calculated focus beam diameter was offset to be 10μm for the CFRP welding applications. To perform an auto-focusing calibration an LRF(Laser Range Finder) distance sensor was used to measure the distance to the target surface. Using a DC Motor with PID control loop, the distance was kept at constant between the laser head and the target material. In this paper, we propose an algorithm for auto-focusing calibration function and detail schematics of laser head design..

1. Introduction

Recently, many researches have been carried out new ways of machining using laser processing. In modern machine tool market it is required to shorten the time of the machining processes and improve productivity and quality of the product. Many efforts have been made in manufacturing process to combine a cutting process and a non-cutting process such as a laser [1-3]. The conventional machining processing requires tools which have friction between the tool and the material, and it takes up unnecessary load on the tool and material. In results, machining process is not energy efficient and sacrifices the precision of machining. However, when a non-contact tool is introduced, it is possible to process materials such as alloys and ceramics which are difficult to process by cutting regardless of hardness and rigidity [4].

In this paper, using a specially designed laser head we present a non-contact processing method, and propose a method for more precise processing using a laser head with an auto-focus correction. All the processes are designed to perform in a single machine without any additional equipment.

Figure 1. PKM (Parallel Kinematic Machines)

2. Laser head device development

2.1. Laser head design

In general, 5-axis machining is evaluated to be more efficient and flexible than 3-axis machining. It can be applied to various products such as aviation parts, blades, molds, etc., which cannot be machined in conventional 3-axis machining [5]. The laser head is optimally designed to be applied to a 5-axis processing machine in PKM (Parallel Kinematics Machine) Fig. 1.

Fig. 2 is an example of beam delivery system that can separate the visible light spectrum from the mixed spectrum of lights. In addition, focal length was set at 100mm and the optical design was optimized to flexibly work near 100mm. To minimize the chromatic aberration, achromatic lenses were used for various wavelengths. As a distance control hardware with DC Motor, the system was designed to perform an auto-focusing calibration and a pulley system was used to transmit the power from the motor to focusing unit. We modified several design alternatives and confirmed the possibility of prototyping using 3D printing. The final version was completed and the prototype was completed.

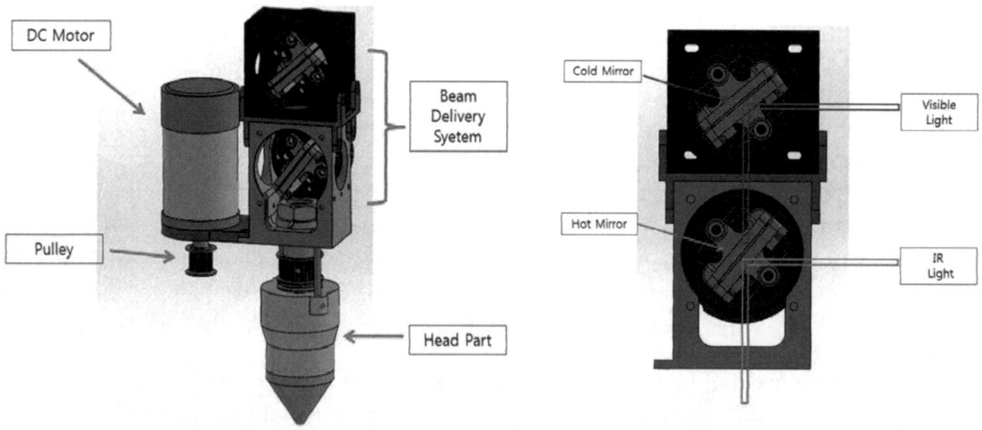

(a) 3D modelling of driving system **(b) Beam delivery system**

Figure 2. Laser Head Design and Beam Delivery System Diagram

In Fig. 3, a schematic diagram of distance measurement system was described and it uses an LRF based on the triangulation method. After measuring the distance from the optics to the target object by using the LRF, the predefined position of the laser head to the motor was controlled by using the analog data from LRF.

IOP Conf. Series: Materials Science and Engineering **191** (2017) 012031 doi:10.1088/1757-899X/191/1/012031

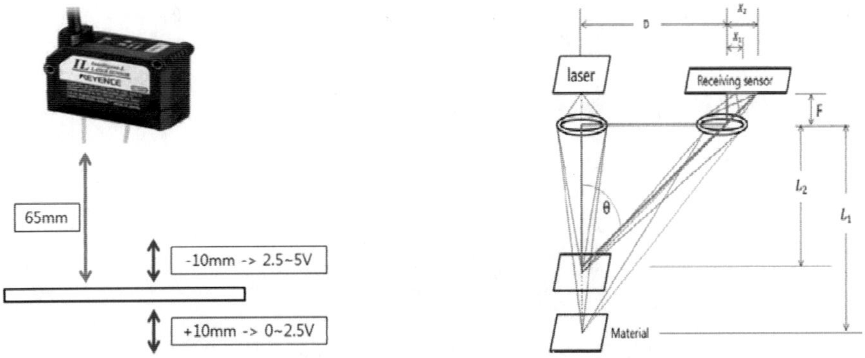

(a) Distance measurement range **(b) Theory of triangulation method**

Figure 3. Laser range finder using triangulation

To calculate the position of laser focusing head, it is necessary to define the size of the focus of the laser in the designed optical system. The spot size of the laser focused on the target material is approximately 13µm according to Equation 1, assuming that the Beam quality factor M^2 = 1, the laser wavelength λ = 1064nm, the focal length of the lens f = 100mm and the entrance pupil D = 10mm (Fig. 4).

$$\text{Spot size} = \frac{4M^2\lambda f}{\pi D} \quad (1)$$

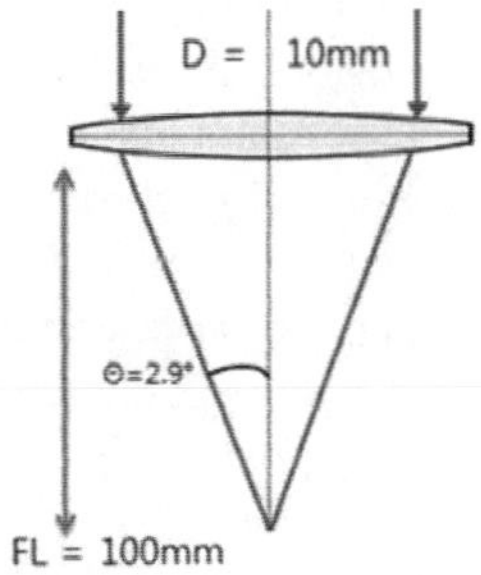

Figure 4. Laser beam spot size diagram

2.2. Position control algorithms

The laser head introduced in this paper should be compatible with the machine tool regardless of the noisy environment. Therefore, it is necessary to use minimum apparatus for the reduction of control time. Using the analog control data from the LRF, the PWM signal was generated for the Motor control which compensates the distance between the laser head and the target (Fig. 5). Based on the developed control method, it was proved that the system always keeps the distance at a certain distance neighboring the focal length. As shown in Fig. 6, the distance is always maintained at the predefined constant location and the current position data is transported along the direction of the laser head.

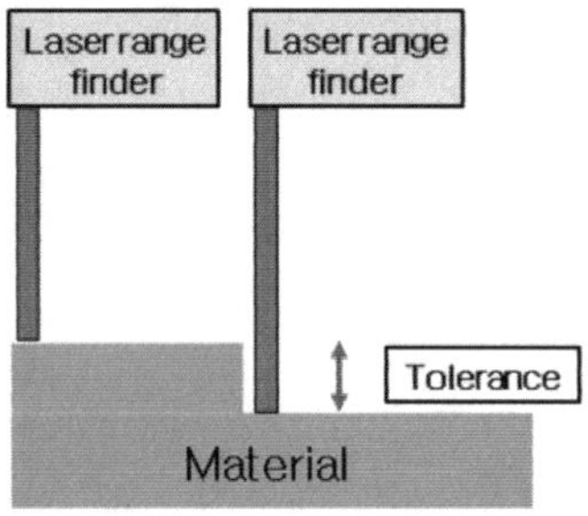

Figure 5. Step measurement using LRF

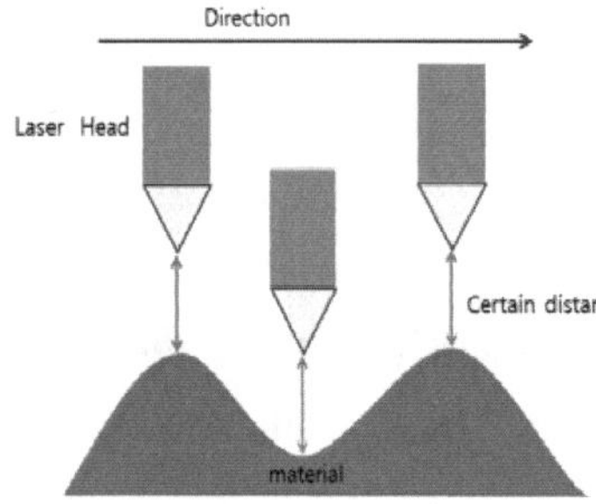

Figure 6. Diagram of position control

The feedback control algorithm measures the current location and compares it to the measured tolerance in LRF to reach the target value of "0" in real time. This control logic is applied to the embedded system, ARDUINO in Fig. 7. The ARDUINO is an open-source project controller that creates microcontroller-based kits for building digital devices and interactive objects with very compact control system design.

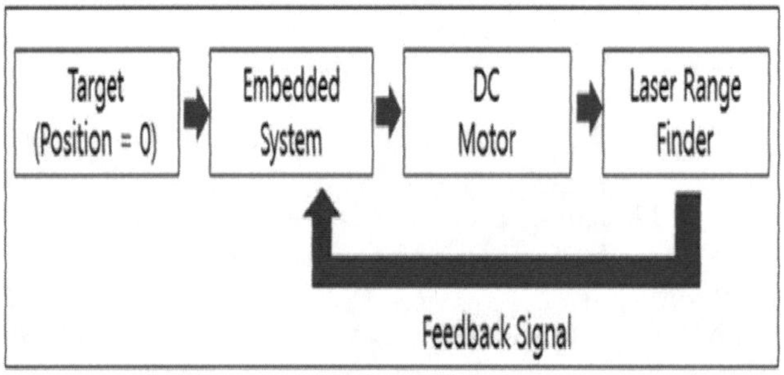

Figure 7. Positon Control Algorithm

3. Prototype productions and testing

The prototype laser head designed as shown in Fig.8 was completed. To maintain the focus of the optical system precisely, the backlash in all drive parts is designed to be close to zero. Based on precision machining with minimum backlash, the system response was proved to be fairly high which is suitable in this project. Even though there is some backlash, the gravity always helps to mesh the gap in the downward direction.

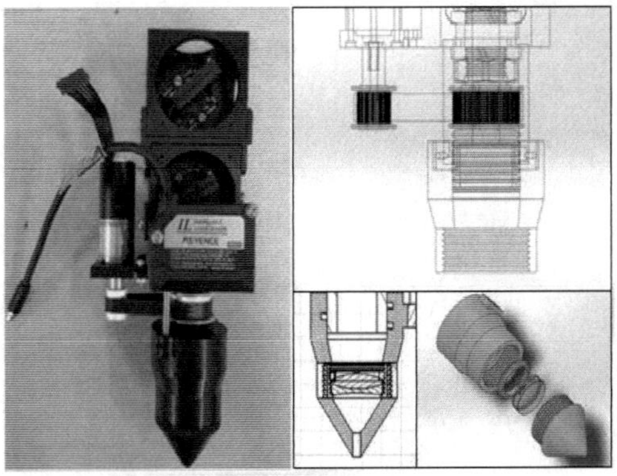

Figure 8. Prototype of laser head

For the optimum motor driving system design, we tested the stepping motor and a DC motor as the basic test motors. First, a stepping motor that moves 1.8 degrees per step was used for the motion. The designed laser head has a driving ratio of 1:2, and the lead of the head part screw is 3mm. In this case, as shown in Equation 2, the vertical travel distance of the head per pulse is about 7.5μm. The change in diameter of the beam per pulse was calculated to be about 0.76μm.

$$\frac{Distance}{Pulse} = \frac{deg/step}{360^\circ} * lead \quad (2)$$

Second, we tested a DC motor for faster response than the stepping motor in real time. As shown in Fig. 9, the response time between the output of the PWM to the DC motor and the analog data coming from the LRF in the embedded system was measured. As a result, the response time was measured to be about 10ms. It is a time that can be ignored when machine tool is transported with constant feed rate in real environment. Since typical traveling speed of PKM is 1,000mm/min, the system can move only 0.16mm in 10ms which may be negligible.

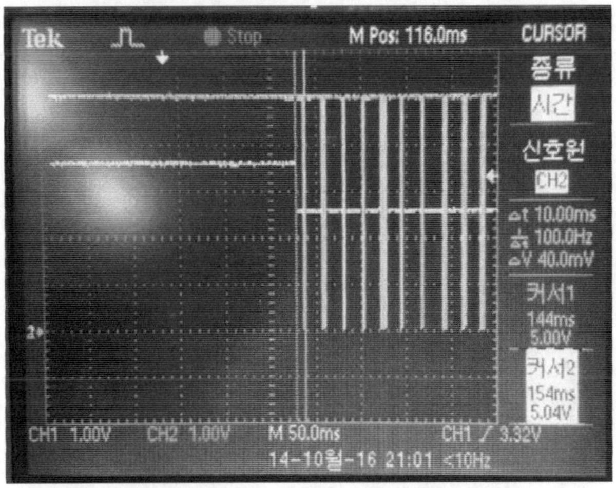

Figure 9. Reaction time of embedded system

When the analog data of the LRF reaches the desired target distance value of "0", the DC motor was slightly unstable or vibrating. By optimizing the PID controller parameters, it was possible to reduce the vibration of the DC motor near the target value. Tuning the PID gain value, the motor's fast response speed can be obtained while reducing the error value

IOP Conf. Series: Materials Science and Engineering **191** (2017) 012031 doi:10.1088/1757-899X/191/1/012031

4. Applied tests on machine tools

After completing the basic tests, we conducted test on the machine tool, which is a real environment. The experimental setup was constructed as shown in Fig. 10.

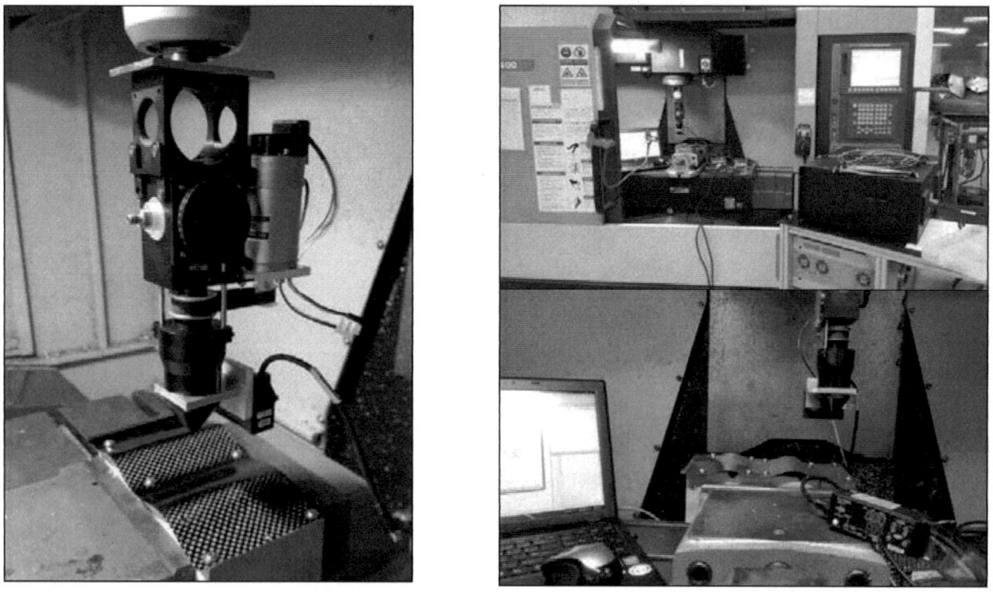

Figure 10. Test equipment setting

The diode laser used in the experiment has wavelength of 980nm and the maximum output power of 30W. In order to perform the auto-focusing function, an artificially bent specimen was prepared on the table of CNC system as shown in Fig. 11.

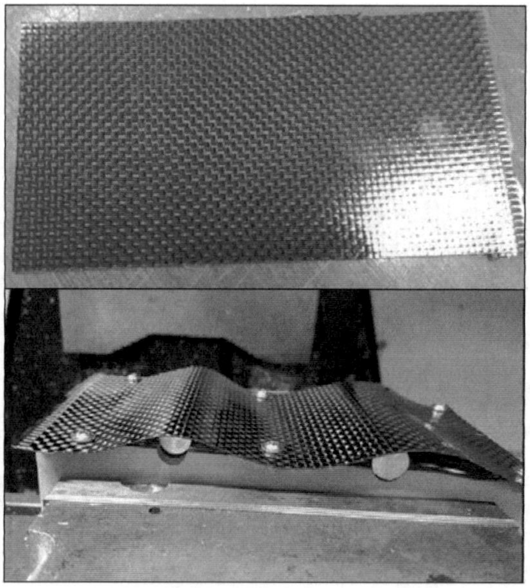

Figure 11. Curved specimens

The test performed with a range of laser output power of 5 to 10W on a CFRP work piece. The range of laser power selected so that the CFRP material is not bunt by the laser power. As a scanning speed, the feed rates of the machine tool were set as 1000, 2000 and 3000mm/min. From the visual inspection of the laser head motion, the auto-focusing function was normally functioning at the feed rate of 1000 and 2000mm/min. However, if the feed rate is increased at 3000mm/min, the speed of the DC motor tends to be unstable which cannot compensate the distance difference.

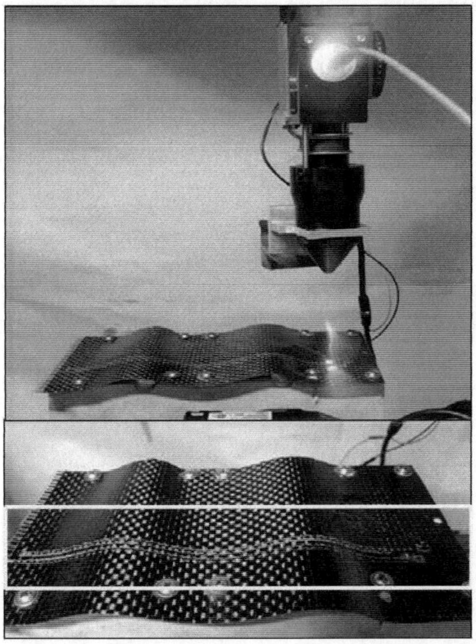

Figure 12. Auto focusing function test and sample results

Experiments using laser head with auto focusing function is depticted in Fig. 12. If the focal length of the laser does not match with the predefined value, defocusing occurs. Therefore, some unfocused parts will not burn which proves the unfocusing. However, the auto focusing function normally works, all the curved surfaces are burned.

5. Conclusion

The laser head was developed in this paper and a prototype of laser head was used for the experiment. However, instead of directly applying to the 5-axis machining centre, the test was first demonstrated with 3-axis machining centre. The proposed system will be applied to the 5-axis machine in near future. From this research, we found that it is possible to apply auto focusing correction to the conventional 5-axis machining tool.

The developed system is expected to provide the machine tool be more stable, flexible and precise with the non-contact advantages. By applying a laser heat source to a machining process of a general machine tool in a hybrid form, the time of the machining process can be shortened and the quality can be expected to be improved. In addition, the laser has advantages not only in processing but also in processes such as marking, texturing, welding, cutting, and heat treatment. If the system is equipped with auto focusing function, much more precise machining will be possible.

Acknowledgments

This research was sponsored by Korea Institute for Advancement of Technology (R0003919).

References

[1] Kim, Y., NA, S., Han, J., Shin, D., Lee, J. and Cho, H., "Interface design of laser hybrid machine based on machining center," Proc.of KSMTE, pp.295-295,2013.4

[2] Lee, J., Shin, D., Suh, J., Cho, H. and Kim, K., “Trends of Laser integrated Machine,” Journal of KSPE, 25(9),pp.20-26,2008

[3] Pyo. C., "Development of Auto Positioning Laser System by using Image Measurement Data," Journal of the KSMPE, 12(3), pp.36-40, 2013

[4] Choi, K., Kim, S. and Shin, K., "Development of Automated Non-contact Thickness Measurement Machine using a Laser Sensor," Journal of the KSMPE, 14(2), pp.51-58, 2015

[5] Kim, H., Wie, K. and Ahn, S., “Application of 5-axis Machining-laser Engraving, Molding Die Machining,” Journal of KSPE, 23(10), pp.25-31, 2009

IOP Conf. Series: Materials Science and Engineering **191** (2017) 012032 doi:10.1088/1757-899X/191/1/012032

The effect of the brake pad components to the some physical properties of the ecological brake pad samples

Mahmut Unaldi* and Recai Kus

Selcuk University, Technology Faculty, Department of Mechanical Engineering Konya, Turkey

Corresponding author: mahmutunaldi@gmail.com

Abstract. In this study, brake pad components in powder form such as miscanthus, cashew, alumina, phenolic resin and calcite, were used to manufacture ecological brake pads. In the brake pad applications trial and error method is often used to find the optimum proportion providing the best characteristics of the component in composite. The evaluation process became complicated and time consuming due to the number of components used for the manufacture of brake pad, randomly selected mixing ratios, many samples produced with trial and error method and a vast variety of results obtained from the measurements. Taguchi method was utilized in order to determine how the compositions of the brake pad components in the range of 5-20 wt. % had an effect on the properties of the brake pads. As a result of the experiments such as density, hardness, porosity and wear rate, the characteristics of the brake pad samples were more affected by the proportions of Miscanthus and phenolic resin in the mixture.

1. Introduction

Brake pad lining materials mostly made of five ingredients such as, reinforcement, lubricant, abrasive, binding agent and filler components. Each component has a specific function in composite structure such as, to adjust the balance friction and wear properties, increase strength, improve lifetime, porosity and noise. Different components and/or the proportion of the components in the brake pad formulation are reported in the literature that may affect the physical and mechanical properties of the brake pads to be developed [1-3].

Selection of components' types for the manufacture of brake pad samples is a difficult task because of affect the properties of brake pad samples. There are limited reports in the literature about the investigations of the effects of the manufacturing processes and components of brake pad materials to the mechanical and physical properties of brake pad materials. The physical and mechanical properties of brake pad materials also depend on manufacturing process parameters and the characteristics of the components used in production of brake pad [4].

Aleksendrić and Senatore [5] (2012) were investigated the effects of the manufacturing process such as moulding temperature, time and pressure values, heat-treatment temperature, and heat-treatment time to wear behaviour of the brake friction materials. They used artificial neural networks as an appropriate modelling technique for development of a model representing the effects of the manufacturing process on wear behaviour of the friction materials. In this article, moulding pressure was determined the most influential factor on wear behaviour of the brake friction materials and it is expressed that wear behaviour of friction materials were strongly affected by the manufacturing processes.

Kim, Kim and Jang [6] (2003), in their work, determined the best manufacturing parameters for 16 different manufacturing conditions via applying Taguchi method to manufactured brake pad samples comprising 15 ingredients. Hardness, porosity, friction coefficient and wear properties of brake pad samples were examined via Taguchi method and the best manufacturing parameters were suggested as moulding time 6 minutes, moulding temperature 225°C, moulding pressure 27MPa, heat treatment time 6 hours and heat treatment temperature 200°C for this particular formulation based on the S/N ratio and ANOVA analyses of tests results. It was expressed that there is no apparent relationship between the physical properties and tribological performance of a brake pads.

The effect of brake pad samples produced with different periwinkle shell particle size on the wear behaviour has been investigated by Amaren, Yawas and Aku [7] (2013). Pin on disk machine used to perform the wear tests under different test conditions such as periwinkle shell particle size, sliding speed, applied load and temperature. The results of the study showed that the wear rate increased with increasing the periwinkle particle size, sliding speed, load and temperature.

Pujari, Ramakrishna and Kumar [8] (2014) were discussed in detail about the applications and advantages of jute and banana fibre composites. They expressed that natural fibres are an alternative resource to synthetic fibres, as reinforcement for polymeric materials for the manufacture is cheap, renewable and environment friendly and when compared to glass fibres, natural fibres offer both reduction in density and cost.

Zaharudin, Talib, Berhan, Budin and Aziurah [4] (2012), designed the manufacturing parameters such as moulding pressure, moulding temperature and moulding time to analyse their effects on the physical and tribological properties of the friction materials according to Taguchi's L8 OA. The optimal parameters were determined as 50 tonnes moulding pressure, 150°C moulding temperature and 600 seconds moulding time. It was expressed that there is no relationship between the hardness and specific gravity or wear properties of the friction materials.

In this study, Taguchi method was utilized for determining the effect of the brake pad components and production parameters on the some properties of the brake pads. Therefore, reinforcement, lubricant, abrasive and binder components of the brake pad were chosen as the factors with four-levels and then experiments were conducted to $L_{16}(4^4)$ Taguchi orthogonal array design. 16 brake pad samples with different proportions were produced for experiments in accordance with $L_{16}(4^4)$ Taguchi orthogonal array design.

2. Materials and method

All powders were mixed with a 1000 W electrical mixer for 10 minute and then compacted at 170°C under a double acting hydraulic press machine. Brake pad samples were compacted with 175MPa for 10 minutes holding time. The production parameters were determined based on the understanding gained from the previous works and in accordance with literature. Finally, produced specimens were rubbed against an abrasive paper of grade 180, 320, 400 and 800 grit for make them ready to experiments. Specimens were produced in the mould which is designed by authors in Figure 1 and samples produced are dimensioned as diameter 12mm, height 25mm.

Density of produced specimens was calculated by a precision scales that can operate according to the Archimedes' principle (equation (1)) in accordance with the ASTM C-20 [7, 9] (Figure 2). Where ρ_w is density of pure water (22.5°C, ρ_w=0.9977 g/cm^3), ρ_s is sample density, (g/cm^3), and m_1 and m_2 are sample weight in air and water (g), respectively.

$$\rho_s = \frac{m_1}{m_1 - m_2} \times \rho_w \qquad (1)$$

Hardness values of all specimens were determined at Rockwell Hardness R scale (HRR) with 12.7mm spherical steel ball and 60kg.f the applied load by a hardness test machine (DIGIROCK-

RBOV). Five positions which were along the radial direction of the specimen were tested for each set of samples.

Porosity affects the properties of composite materials. Generally, for the same material, the lower the porosity is, the less the connected pores are. Thus, composite material has higher strength and thermal conductivity values and lower water absorption values. The porosity values of the brake pad samples were determined by equation (2) [10]:

$$P = (1 - \frac{\rho_E}{\rho_T}) x100 \quad (2)$$

where P is porosity (%), ρ is density (g/cm^3), E refers experimental and T refers theoretical.

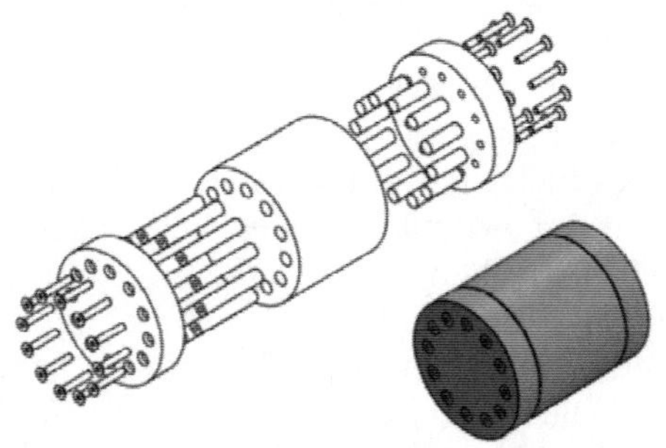

Figure 1. The mould used in the production of brake pad samples

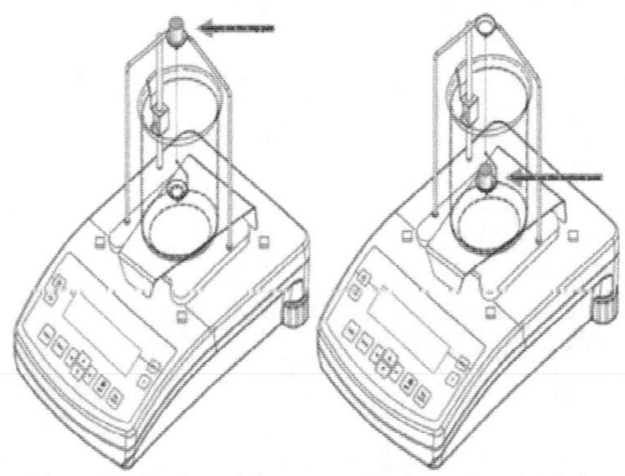

Figure 2. Density calculation kit according to the Archimedes' principle

Wear rate tests were carried out using a pin-on-disc type tribo-test machine. The pin-on-disc test apparatus with disc made of grey cast iron of hardness value 195HB (62.5kg.f, 5mm ball) and 180mm track diameter was used to investigate the dry sliding wear rate of the brake pad specimens in accordance with ASTM:D792 standard. The pin was pressed against the rotating cast iron disc of counter surface at a load of 1MPa, sliding speed of 1m/s and sliding distance of 1000m. All tests were conducted at room temperature. The differences in weight measured before and after tests give the weight loss of the samples. The equation (3) used to convert the weight loss into wear rate (W, mg/m). Where ΔW is the weight difference of the sample before and after the test (mg), S is total sliding distance (m) [11, 12].

$$W = \Delta W / S \quad (3)$$

The weight percentage of components such as miscanthus (A), cashew (B), alumina (C) and phenolic resin (D) used for production of ecological brake pad samples were chosen in the range of 5-20wt-% increased by 5wt-%. Calcite used as filler (in the range of 35-80wt-%) was preferred as a complement component for the brake pad mixture, because it had no effect on the characteristics of the brake pad sample as reported in the literature. 256 units brake pad samples can be manufactured considering these mixing ratios. But it is impossible because of the performing production, experimentation and evaluation processes of the brake pad samples. For these reasons, design of experiment is an important tool for designing processes and products. Instead of having to test all possible combinations like the factorial design, the Taguchi method tests pairs of combinations. Signal-to-Noise ratios (S/N) are used as the quality characteristics of choice on Taguchi Design. In the Taguchi method, the term 'signal' represents the desirable target for good products and the term 'noise' represents the undesirable value. The experimental observations are transformed into signal-to-noise (SN) ratios and the goal of the experiment is to improve the relationship between the signal factor and the response. The signal-to-noise ratio measures how the response varies relative to the nominal or target value under different noise conditions. SN ratios are available depending on the type of performance characteristics and can be characterized into three categories; Larger is better, Nominal is best and Smaller is better. The SN ratio category selected is important for the goal of the experiments. The parameter level which has the highest S/N value is the best result for the experiments regardless of the selected category of performance characteristics. Therefore, the optimal level of the parameters is the level with the greatest S/N value [13, 14].

According to the ANOVA results, in multiple linear regression analysis, R^2 is the regression coefficient (R^2>0.90) for the models, which indicate that the fit of the experimental data is satisfactory.

In this work, to evaluate the statistical data, the larger-is-better quality characteristic (equation (4)) was chosen for density and hardness and the smaller-is-better quality characteristic (equation (5)) was chosen for porosity and wear rate values of brake pad samples. The quality characteristic was chosen in accordance with the previously conducted studies made by Zaharudin, Talib, Berhan, Budin and Aziurah [4], Kim, Kim and Jang [6], Ibhadode and Dagwa [15]. These formulas were used to calculate SN ratio:

$$\text{Larger is better} \quad S/N = -10.\log[\Sigma(1/Y^2)/n] \tag{4}$$

$$\text{Smaller is better} \quad S/N = -10.\log(\Sigma Y^2/n) \tag{5}$$

Y= measured value,
n= Number of tests.

Table 1. The mean values of physical properties in accordance with Taguchi design.

	A	B	C	D	d	H	Pr	WR
1	1	1	1	1	1.459	105.2	0.15932	1.47E-06
2	1	2	2	2	1.485	104.9	0.1559	1.35E-06
3	1	3	3	3	1.583	109.4	0.15216	1.45E-06
4	1	4	4	4	1.563	112.3	0.17273	1.59E-06
5	2	1	2	3	1.528	105.6	0.16666	1.53E-06
6	2	2	1	4	1.491	104.6	0.16522	1.47E-06
7	2	3	4	1	1.489	111.7	0.21171	1.91E-06
8	2	4	3	2	1.527	106.9	0.23704	1.70E-06
9	3	5	3	4	1.563	108.9	0.16889	1.56E-06
10	3	2	4	3	1.543	111.5	0.16526	2.86E-06
11	3	3	1	2	1.408	100.9	0.22323	2.08E-06
12	3	4	2	1	1.407	104.7	0.25289	2.12E-06
13	4	1	4	2	1.538	108.1	0.17912	3.52E-06
14	4	2	3	1	1.46	105.8	0.21977	3.35E-06
15	4	3	2	4	1.435	102.9	0.21524	3.07E-06
16	4	4	1	3	1.401	95.5	0.24556	2.74E-06

d: Density (g/cm^3); H: Hardness (HRR); Pr: Porosity (%); WR: Wear Rate (cm^3/N.m)

3. Results and discussion

Experiments are conducted in accordance with Taguchi's L16 orthogonal array to determine the degree of influence of components and their ratios constituting brake pad samples to the physical properties of brake pad samples. Experiments were carried out three times and average values of experiments and mixing ratios of produced samples are given in Table 1. The resin was varying from 5 to 30wt% with interval of 5wt%.

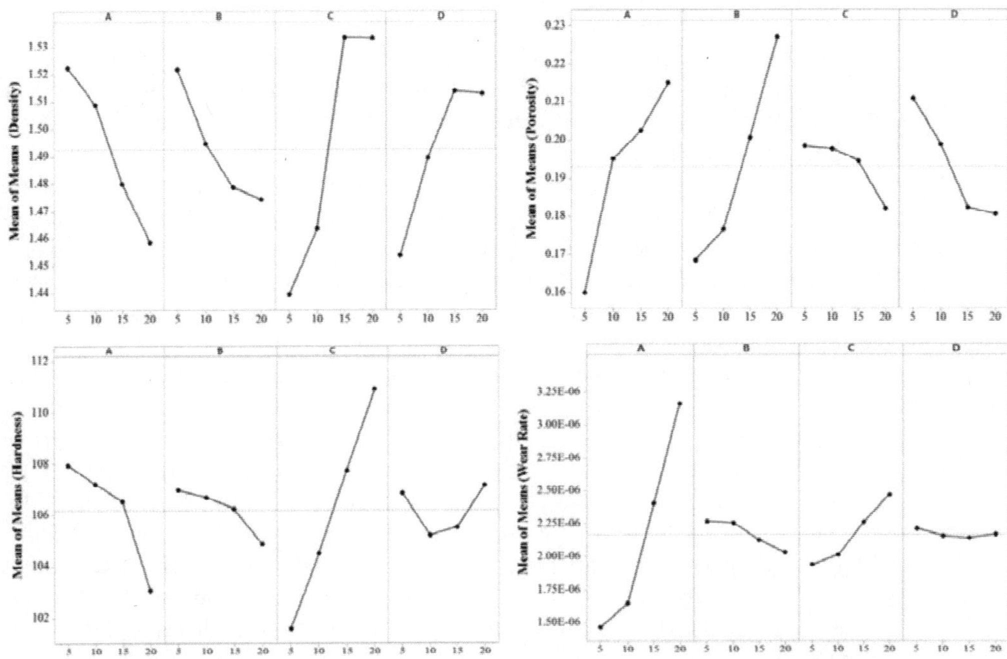

Figure 3. Mean value graphics of physical properties of brake pad samples

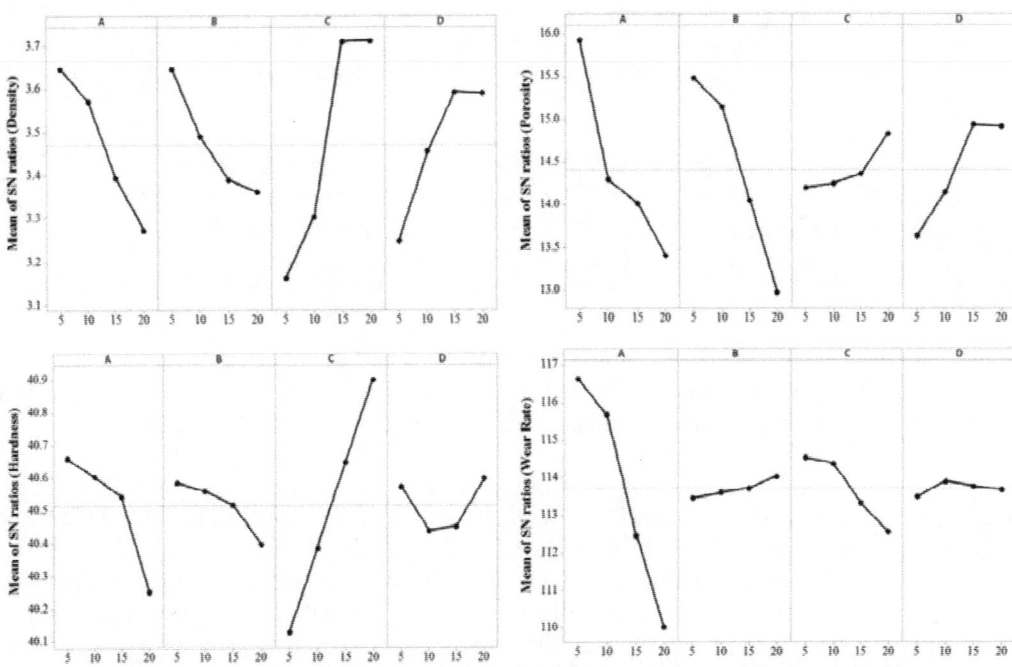

Figure 4. S/N value graphics of physical properties of brake pad samples

Mean value graphs that are obtained from the data in Table 1 are given in Figure 3. When these graphs are examined, miscanthus (A) and cashew (B) components exhibit similar behaviour for the density, hardness and porosity properties of brake pad samples. While miscanthus and cashew components' ratios in the structure of the brake pad samples increase, density and hardness values of the brake pad samples decrease.

While alumina (C) and phenolic resin (D) ratios increase, porosity values increase, and vice versa. The wear rates of the brake pad samples are increasing in proportion to the ratio of miscanthus and alumina. Cashew ratio is inversely proportional to the wear rates and it appears that the phenolic resin ratios have no impact on the wear rate. Referring to the mean value graphics we can make comments it this way. It could not be understood the effect of the brake pad components on the physical properties of the brake pad samples by investigation of the mean value graphics. For this reason, SN ratio graphs and ANOVA tables should be examined.

When the line is horizontal on the mean of SN ratio graphs, then there is no main effect present. When the line is not horizontal, then there is a main effect present. Different levels of the factor affect the characteristic differently. The greater the differences in the vertical position of the plotted points are, the greater the magnitude of the main effect is. By comparing the slopes of the lines, we can compare the relative magnitude of the factor effects. The S/N value graphics plotted to determine the degree of influence of the brake lining components and ratios to the physical properties such as density, hardness, porosity and wear rate of the brake linings are shown in Figure 4 and the ANOVA results are given in Table 2-5. The ANOVA table shows the relative importance of the brake pad components affecting the physical properties of brake pad samples. This can be evaluated from the contribution (Cont.%) in percentage on the total variation indicating their degree of influence on the result. F test is used to observe that the design parameters have a major effect on the quality characteristics. In the analysis, the F-ratio is traditionally used to determine the significance of a factor. P (probability of significance) values give the statistically significant factors affecting the property. The factors whose P values are less than 0.05% (95% confidence levels) are generally regarded as statistically significant. Based on this criterion, it is found that miscanthus ratio is not statistically significant for wear rate and compression strength values [16-18].

Table 2. ANOVA for density SN ratios

Source	Seq SS	Adj. MS	Cont. (%)	F	P
A	0.3431	0.1143	18.69	11.62	0.037
B	0.1968	0.0655	10.72	6.67	0.077
C	0.9525	0.3174	51.88	32.26	0.009
D	0.3141	0.1046	17.11	10.64	0.042
Residual Error	0.0295	0.0098			

Table 3. ANOVA for hardness SN ratios

Source	Seq SS	Adj. MS	Cont. (%)	F	P
A	0.3964	0.1321	20.49	9.21	0.05
B	0.0823	0.0274	4.25	1.91	0.304
C	1.3328	0.4442	68.89	30.98	0.009
D	0.0802	0.0267	4.37	1.86	0.311
Residual Error	0.043	0.0143			

Table 2-5 suggest that alumina, miscanthus, phenolic resin and cashew components are significant parameters for density values of brake pad samples and the contribution values in percentage 51.88, 18.69, 17.11 and 10.72, respectively. Alumina, miscanthus and phenolic resin components' P values are less than 0.05% for density property.

Hardness values of brake pad samples are more affected by alumina and miscanthus components. Because alumina and miscanthus have 68.89, 30.98 and 20.49, 9.21 values of the contribution values (%) and F values, respectively. Alumina and miscanthus components' P values are less than 0.05% for hardness property. Cashew, miscanthus and phenolic resin are more effective components because their contribution values are 43.35, 38.92 and 13.51, respectively, and these components have less than 0.05% P values for porosity property. Miscanthus is a more effective component than others for wear rate property of brake pad samples because of its higher contribution value and F value. According to

ANOVA results, miscanthus and alumina have an effective influence on all the physical properties of brake pad samples whereas others are not so effective. R-Sq values of SN Ratios of density, hardness, porosity and wear rate of brake pad samples are 98.4, 97.8, 98.5 and 99.2 respectively.

Table 4. ANOVA for porosity SN ratios

Source	Seq SS	Adj. MS	Cont. (%)	F	P
A	13.883	4.6275	38.92	26.44	0.012
B	15.465	5.1551	43.35	29.45	0.01
C	0.9818	0.3273	2.75	1.87	0.31
D	4.8199	1.6066	13.51	9.18	0.051
Residual Error	0.5251	0.175			

Table 5. ANOVA for wear rate SN ratios

Source	Seq SS	Adj. MS	Cont. (%)	F	P
A	110.34	36.78	89.94	107	0.001
B	0.721	0.24	0.59	0.7	0.611
C	10.273	3.424	8.37	10	0.045
D	0.325	0.108	0.27	0.32	0.815
Residual Error	1.027	0.342			

4. Conclusion

Ecological brake pad samples were produced the first time using miscanthus as reinforcing component of brake pad. As a result of this study which investigated the degree of influence production parameters and the components constituting the brake pad samples on the physical properties was obtained as follows:

When miscanthus component ratio in brake pad samples increase, density and hardness values of brake pad samples decrease and porosity and wear rate values of brake pad samples increase.
When cashew component ratio in brake pad samples increase, only porosity values of brake pad samples increases and others decrease.

When alumina component ratio in brake pad samples increase, only porosity values of brake pad samples decrease.

When phenolic resin component ratio in brake pad samples increase, density and hardness values of brake pad samples increase and porosity values of brake pad samples decrease. Phenolic resin component has no effect on wear rate values.

The conclusion reveals that density and hardness properties are most affected by alumina component of brake pad samples. Porosity property is most affected by miscanthus and cashew components of brake pad samples. Wear rate property is most affected by miscanthus component of brake pad samples.

Acknowledgement
The data in this paper were benefited from Ph.D. thesis studies of Mahmut UNALDI. The authors acknowledge the support provided by the Scientific Research Projects Coordinatorship of Selçuk University Contract No: 13101024.

References

[1] Y. Lu, Composites Science and Technology, 66 (2006) 591-598.
[2] H. Jang, K. Ko, S.J. Kim, R.H. Basch, J.W. Fash, Wear, 256 (2004) 406-414.
[3] T.R. Jaafar, M.S. Selamat, R. Kasiran, Selection of Best Formulation for Semi-Metallic Brake Friction Materials Development, in: K. Kondoh (Ed.) Powder Metallurgy2012, pp. 30.
[4] A. Zaharudin, R. Talib, M. Berhan, S. Budin, M. Aziurah, International Journal of Mechanical and Materials Engineering, 8 (2012) 83-88.
[5] D. Aleksendrić, A. Senatore, Journal of Composite Materials, 46 (2012) 2777-2791.
[6] S.J. Kim, K.S. Kim, H. Jang, Journal of Materials Processing Technology, 136 (2003) 202-208.
[7] S.G. Amaren, D.S. Yawas, S.Y. Aku, Results in Physics, 3 (2013) 6.
[8] S. Pujari, A. Ramakrishna, M.S. Kumar, International Journal of Current Engineering and Technology, (2014) 121-126.

IOP Conf. Series: Materials Science and Engineering **191** (2017) 012032 doi:10.1088/1757-899X/191/1/012032

[9] M. Maleque, A. Atiqah, R. Talib, H. Zahurin, International Journal of Mechanical and Materials Engineering, 7 (2012).
[10] K. Alaneme, International Journal of Mechanical and Materials Engineering, 7 (2012) 96-100.
[11] V.S. Aigbodion, U. Akadike, S.B. Hassan, F. Asuke, J.O. Agunsoye, Tribology in Industry, 32 (2010) 12-18.
[12] U.D. Idris, V.S. Aigbodion, I.J. Abubakar, C.I. Nwoye, Journal of King Saud University-Engineering Sciences, 27 (2015) 185-192.
[13] J. Benson, 1st International Conference on Prevention and Infection Control (ICPIC), Geneva, Switzerland, 2011.
[14] Anonymous, Design of Experiment.
[15] A. Ibhadode, I. Dagwa, Journal of the Brazilian Society of Mechanical Sciences and Engineering, 30 (2008) 166-173.
[16] F. Ficici, M. Durat, M. Kapsiz, Journal of the Brazilian Society of Mechanical Sciences and Engineering, 36 (2014) 653-659.
[17] R. Rama, G. Padmanabhan, International Journal of Engineering Research and Applications (IJERA), 2 (2012) 192-197.
[18] S. Basavarajappa, G. Chandramohan, J.P. Davim, Materials & design, 28 (2007) 1393-1398.

IOP Conf. Series: Materials Science and Engineering **191** (2017) 012033 doi:10.1088/1757-899X/191/1/012033

Static analysis for magneto-electro-elastic plates based on the scaled boundary finite element method

Pengchong Zhang[1], Jun Liu[1,2,3] and Gao Lin[1]

[1]School of Hydraulic Engineering, Faculty of Infrastructure Engineering, Dalian University of Technology, China
[2]State Key Laboratory of Coastal and Offshore Engineering, Dalian University of Technology, China
[3]State Key Laboratory of Structural Analysis for Industrial Equipment, Dalian University of Technology, China

E-mail: zhangpengchong2014@126.com

Abstract. The scaled boundary finite element method (SBFEM) and the precise integration algorithm (PIA) are utilized to analyze the extended displacement field in clamped or simple-supported magneto-electro-elastic plates produced by external transverse loadings. There are no limitation on boundary conditions and types of external forces. Only the in-plane dimensions are divided into 2D elements. By introducing a set of scaled boundary local coordinates, 3D governing partial differential equations are converted into the second order ordinary differential matrix equation. By means of the internal nodal force, a first order ordinary differential equation is obtained and its general solution is a matrix exponential. The PIA is introduced to calculate the matrix exponential and any desired accuracy can be obtained. Finally, several numerical examples are provided to validate the versatility of the proposed technique.

1. Introduction

In the recent decade, magneto-electro-elastic plates due to exceptional nature to convert one form of energy into another have been drawn considerable discussions. With the increase in the usage of magneto-electro-elastic plates in engineering practices, an excellent understanding of their structural behavior are needed.

Over the past few years, mechanics of magneto-electro-elastic plates have received considerable research effort. Phoenix et al. [1] developed a finite element model based on the Reissner mixed variational theorem to analyze the static and dynamic analysis of magneto-electro-elastic plate problem. Built on the third-order shear deformation theory, Moita et al. [2] proposed a finite element model to investigate the static and free vibration problem of thin and thick magneto-electro-elastic plates. Liu [3] made use of the Kirchhoff hypothesis and classical thin-plate theory to study the bending problem of the magneto-electro-elastic rectangular plate subjected to certain type of applied loads acting on the top or bottom surfaces.

The purpose of this paper is to introduce the scaled boundary finite element (SBFEM) developed Wolf and Song [4] and Song and Wolf [5] to explore the static response of magneto-electro-elastic plates. The SBFEM method has its own attractive features. Recently, this method is extended to discuss the plate bending problem [6, 7]

The paper is outlined as follows. The solution procedure of the fundamental equations in SBFEM is presented in section 2. Subsequently, three numerical examples are provided to validate verify the proposed approach as well as to highlight its performance in Section 3.

2. Solution procedure

In this section, only the necessary equations are introduced and further details are listed in Ref. [6, 7]. The 3D geometry of a magneto-electro-elastic plate is shown in figure 1.

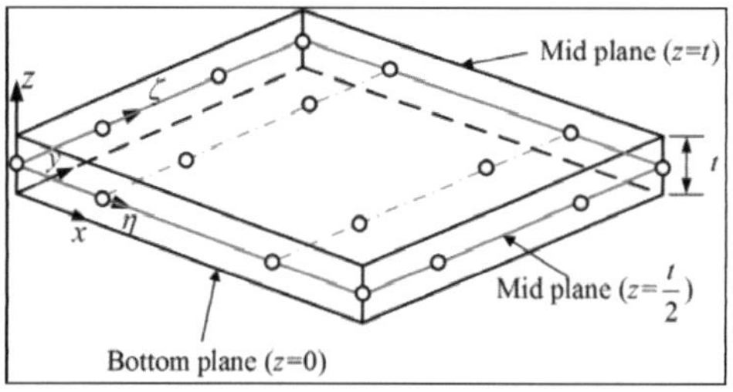

Figure 1. The 3D geometry of a magneto-electro-elastic plate.

With the internal nodal force, the matrix exponential can be expressed as

$$\begin{Bmatrix} \{u_B\} \\ \{F_B\} \end{Bmatrix} = \begin{Bmatrix} \{c_1\} \\ -\{c_2\} \end{Bmatrix}$$
$$\begin{Bmatrix} \{u_T\} \\ \{F_T\} \end{Bmatrix} = \exp(-[Z]t) \begin{Bmatrix} \{c_1\} \\ \{c_2\} \end{Bmatrix} \tag{1}$$

The PIA proposed by Zhong et al. [8] is utilized to solve the matrix exponential. The total thickness t is divided into 2^N mini-layers of equal thickness.

$$\{X_T\} = \exp\left(-[Z]t/2^N\right)^{2^N} \{c\} = \exp\left(-[Z]\tau\right)^{2^N} \{c\} = \mathbf{T}^{2^N} \{c\} = \bar{\mathbf{T}}\{c\} \tag{2}$$

Since τ is an extremely small thickness, $\mathbf{T}$ is calculated as

$$\mathbf{T} = \exp(\mathbf{H}\tau) \approx \mathbf{I} + \mathbf{T}_a \quad \mathbf{H} = -[Z]$$
$$\mathbf{T}_a = \mathbf{H}\tau + \frac{1}{2!}(\mathbf{H}\tau)^2 + \frac{1}{3!}(\mathbf{H}\tau)^3 + \frac{1}{4!}(\mathbf{H}\tau)^4 \tag{3}$$

The estimation of $\bar{\mathbf{T}}$ is performed in the following way:

$$\bar{\mathbf{T}} = (\mathbf{I} + \mathbf{T}_a)^{2^N} = \left[(\mathbf{I} + \mathbf{T}_a)^{2^{N-1}}\right]^2 = \mathbf{I} + \mathbf{T}_r^N \tag{4}$$

in which

$$\mathbf{T}_r^i = 2\mathbf{T}_r^{i-1} + \mathbf{T}_r^{i-1} \times \mathbf{T}_r^{i-1} \quad (i = 1, 2, ..., N) \quad \mathbf{T}_r^0 = \mathbf{T}_a \tag{5}$$

Substituting equation (4) into equation (1) and grouping the extended displacement field and external forces, the stiffness equation can be acquired.

$$[S] \begin{Bmatrix} \{u_B\} \\ \{u_T\} \end{Bmatrix} = \begin{Bmatrix} \{F_B\} \\ \{F_T\} \end{Bmatrix} \tag{6}$$

in which

$$[S] = \begin{bmatrix} \bar{\mathbf{T}}_{12}^{-1}\bar{\mathbf{T}}_{11} & -\bar{\mathbf{T}}_{12}^{-1} \\ \bar{\mathbf{T}}_{21} - \bar{\mathbf{T}}_{22}\bar{\mathbf{T}}_{12}^{-1}\bar{\mathbf{T}}_{11} & \bar{\mathbf{T}}_{22}\bar{\mathbf{T}}_{12}^{-1} \end{bmatrix} \tag{7}$$

By applying boundary conditions, the extended displacement field of the magneto-electro-elastic plate is obtained.

3. Numerical examples

In this section, all examples are conducted with a material $BF_{50\%}$, whose material properties are listed in Table 1 of Ref. [9].

3.1. Verification

Comparisons provided by Alaimo et al. [9] are under condition. From table 1, it is obvious that the present results agree well with that from the reference.

3.2. Circular plate

It is meaningful to investigate the deformation of a circular magneto-electro-elastic plate as illustrated in figure 2 under uniform electric displacement load on the top plane. Thickness to length ratios t/l=0.01, 0.02 and 0.05 are selected, in which the characterized length is $l=2R$. Clamped boundary conditions are given at the edge of the circular plate. The central transverse displacements under different kinds of external loadings are demonstrated in table 2. Figure 3 delineates variations of the electric and magnetic potential along the normalized thickness. From all tables and figures, a conclusion can be drawn that the thickness-to-length ratio greatly influences the distribution of electric and magnetic potential.

Table 1. Results of a square magneto-electro-elastic plate.

Element order	t/l=0.01	t/l=0.02	t/l=0.05
4	-1.574×10^{-9}	-1.072×10^{-2}	9.106×10^{-6}
6	-1.570×10^{-9}	-1.069×10^{-2}	9.033×10^{-6}
8	-1.567×10^{-9}	-1.065×10^{-2}	8.818×10^{-6}
10	-1.560×10^{-9}	-1.045×10^{-2}	8.228×10^{-6}
Ref. [9]	-1.507×10^{-9}	-9.820×10^{-3}	8.365×10^{-6}

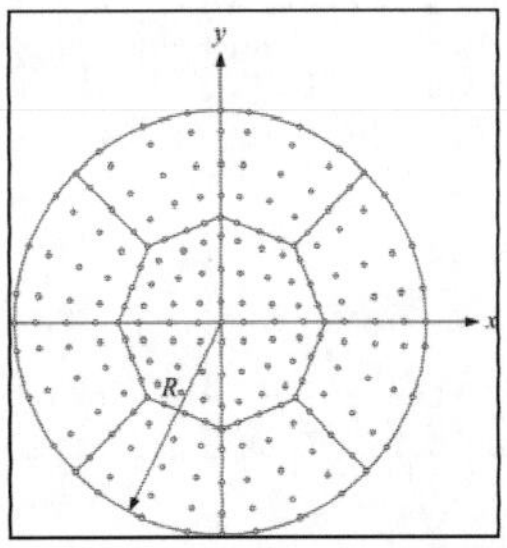

Figure 2. A circular plate meshed with fourth order spectral elements.

Table 2. Central transverse displacement for the circular plate.

Element order	t/l=0.01	t/l=0.02	t/l=0.05
4	1.287×10^{-13}	-7.982×10^{-12}	4.366×10^{-10}
6	1.687×10^{-13}	-1.505×10^{-11}	4.592×10^{-10}
8	2.204×10^{-13}	-1.545×10^{-11}	4.822×10^{-10}
10	2.238×10^{-13}	-1.612×10^{-11}	5.220×10^{-10}

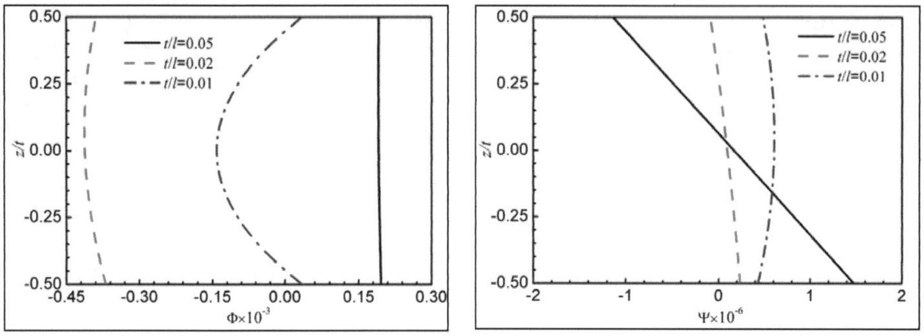

Figure 3. Variations of the electric and magnetic potential.

3.3. Square plates with four holes

In this example, a square plate with four holes are selected as a model as demonstrated in figure 4 under a sinusoidal mechanical load on the top plane. Simply supported boundary conditions are prescribed as all four outer edges while the inner edges are free. Additionally, zero electric potential and magnetic potential are applied on both the upper and lower plate surfaces. Table 3 reveals deflections of an open-hole square plate at the center. Variations of the electric and magnetic potential are displayed in figure 5. Similar findings on variations of the deflection in the previous example can also be found in Table 3. It is obvious that central transverse displacements in the case of t/l=0.05 are largest. Moreover, curves of the electric potential and magnetic potential are approximately parabolic in figure 5.

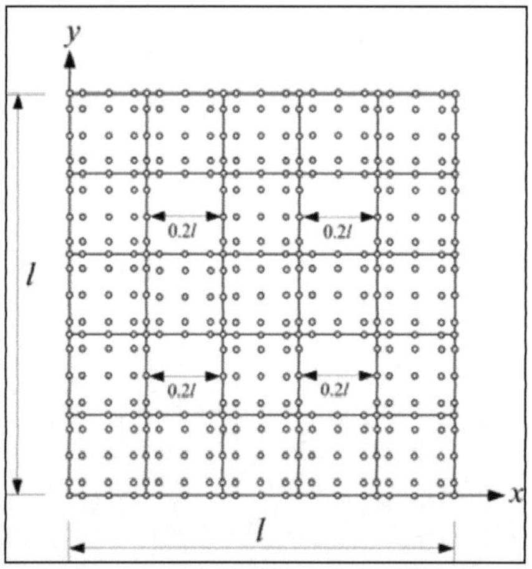

Figure 4. An square plate model with open-holes.

Table 3. Deflections of an simply supported square plate at the center.

Element order	t/l=0.01	t/l=0.02	t/l=0.05
4	8.838×10^{-9}	1.318×10^{-7}	1.041×10^{-6}
6	9.719×10^{-9}	1.457×10^{-7}	1.155×10^{-6}
8	9.888×10^{-9}	1.475×10^{-7}	1.165×10^{-6}
10	9.954×10^{-9}	1.483×10^{-7}	1.169×10^{-6}

IOP Conf. Series: Materials Science and Engineering **191** (2017) 012033 doi:10.1088/1757-899X/191/1/012033

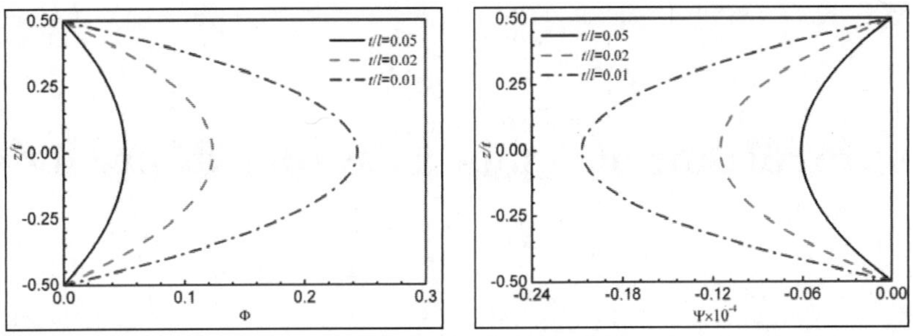

Figure 5. Through-thickness variations of the electric and magnetic potential.

4. Conclusion

Based on the SBFEM and PIA, the magnitude and distribution of the transverse deflection, electric potential and magnetic potential in a magneto-electro-elastic plate is discussed in the paper. Several numerical examples are given to verify the proposed approach. In the circular plate, a conclusion can be drawn that the thickness-to-length ratio plays an important role in determining the distribution of electric and magnetic potential. As to the open-hole square plate, curves of the electric potential and magnetic potential are approximately parabolic induced by mechanical loadings.

References

[1] Phoenix S S, Satsangi S K and Singh B N 2009 *J Sound Vib* 324 798-815
[2] Moita J M S, Soares C M M and Soares C A M 2009 *Compos Struct* 91 421-26
[3] Liu M F 2011 *Appl Math Model* 35 2443-61
[4] Wolf J P and Song C 2000 *Comput Struct* 78 191-210
[5] Song C and Wolf J P 2000 *Comput Struct* 78 211-25
[6] Man H, Song C, Gao W and Tin-Loi F 2012 *Int. J. Numer. Methods Engineering* 91 491-515
[7] Man H, Song C, Xiang T and Tin-Loi F 2013 *Int. J. Numer. Methods Engineering* 95 331-60
[8] Zhong W X, Lin J H and Gao Q 2004 *Int. J. Numer. Methods Engineering* 60 11-25
[9] Alaimo A, Milazzo A and Orlando C 2013 *Comput Struct* 129 120-33

IOP Conf. Series: Materials Science and Engineering **191** (2017) 012034 doi:10.1088/1757-899X/191/1/012034

Closed loop oscillating heat pipe as heating device for copper plate

Patrapon Kamonpet*, and Waranphop Sangpen

Department of Mechanical Engineering, Chiang Mai University, Thailand

patraponk@gmail.com

Abstract. In manufacturing parts by molding method, temperature uniformity of the mold holds a very crucial aspect for the quality of the parts. Studies have been carried out in searching for effective method in controlling the mold temperature. Using of heat pipe is one of the many effective ways to control the temperature of the molding area to the right uniform level. Recently, there has been the development of oscillating heat pipe and its application is very promising. The semi-empirical correlation for closed-loop oscillating heat pipe (CLOHP) with the STD of ±30% was used in design of CLOHP in this study. By placing CLOHP in the copper plate at some distance from the plate surface and allow CLOHP to heat the plate up to the set surface temperature, the temperature of the plate was recorded. It is found that CLOHP can be effectively used as a heat source to transfer heat to copper plate with excellent temperature distribution. The STDs of heat rate of all experiments are well in the range of ±30% of the correlation used.

1. Introduction

In manufacturing parts by molding method, temperature uniformity of the mold holds a very crucial aspect for the quality of the parts. Poor designed molds may lead to non-uniform temperature of the mold cavity and core plates which lead to low quality parts and part rejection. Studies have been carried out in searching for effective method in controlling mold temperature. Using of heat pipe is one of the many effective ways to control the temperature of the molding area to the right uniform level. Standard parts of heat pipe are cataloged in many cutting edge mold making companies. Recently, there has been the development of oscillating heat pipe (OHP) and its application is very promising. However, most applications of OHP, nowadays are concentrated in energy management of air conditioner, refrigerator, heat exchanger, and in computer application. There are not many studies of OHP in industry application such as molding. Since molding industry has been growing rapidly, i.e. in plastics injection molding alone, the growth was anticipated to be increased by 40% within 5 years [1]. In injection molding, rapid cooling and heating technique of cavity and core plates is of important. There are two main categories of heating system used, surface heating and volume heating. Surface heating, as its name implied, is heating of the cavity surface using different heating method such as induction heat and infrared. Whereas in volume heating, channels are drilled and cartridge heaters are placed evenly in the core side as shown in Fig. 1(a), where the center plate is the heating plate and the right-hand plate is the cooling plate. Fig. 1(b) shows another design where water and heater could be placed in the same channel.

The objective of this paper is then to adapt the volume heating concept by replacing the cartridge heater and water with CLOHP. By this adjustment, the heating and cooling process can be combined using just CLOHP and the system is much more simplified as there is no leakage problem since CLOHP is a closed system with no moving parts. The basic characteristics of OHP is a capillary tube meandered into equally space-wise turns as shown in Fig 2. Inside of the tube is evacuated and partly filled with working fluid. It can then be placed in between the mold plate to transfer heat into the cavity with a more uniform temperature distribution. Moreover, the tube can be bent in any required shape (Fig.3). For preliminary study, this research is concentrated on only heating of the copper plate to the set surface temperature. The temperature distribution was then studies and reported as the standard deviation value of the set surface temperature. For comparison, the same experiment but using cartridge heater was also carried out. The results were as well compared with the result studied with P-20 of the previous study.

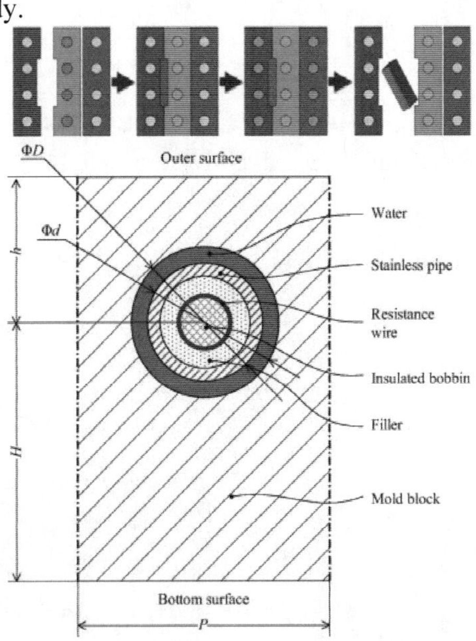

Figure 1. Rapid Mold Heating/Cooling Process [2, 3]

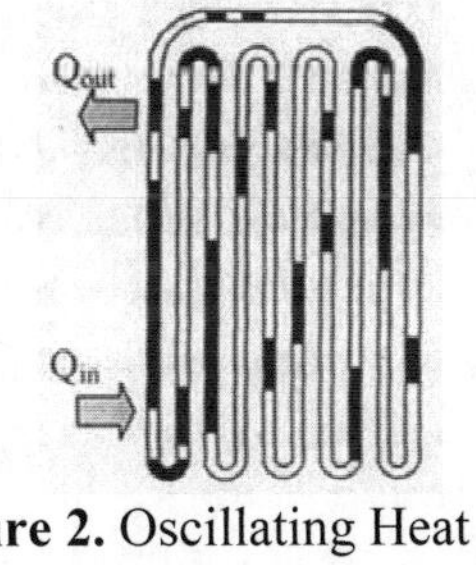

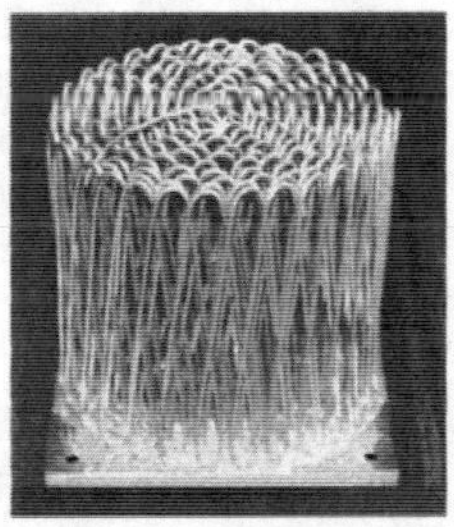

Figure 2. Oscillating Heat Pipe **Figure 3.** OHP in Commercial Used [4]

Khandekar et al. [5] proposed the semi-empirical correlation which determine the maximum heat transfer achievable for CLOHP (filling ratio of 50%) with the STD of ±30%. The correlation together with other dimensionless terms required in the design are as shown in Eq. 1 to Eq. 5 and Fig. *4*.

$$\dot{q} = \left(\frac{\dot{Q}}{\pi D_i \cdot N \cdot 2L_e} \right) = 0.54\,(\exp(\beta))^{0.48}\, Ka^{0.47}\, Pr_{liq}^{\;0.27}\, Ja^{1.43}\, N^{-0.27} \tag{1}$$

$$B_o = \frac{g(\rho_l - \rho_v) D_i^2}{\sigma} \tag{2}$$

$$Ja = \frac{h_{fg}}{C_{pl} \Delta T_{sat}^{e-c}} \tag{3}$$

$$Pr_{liq} = \left(\frac{C_{pl} \mu_l}{k_l} \right) \tag{4}$$

$$Ka = fRe^2 = \frac{\rho_l (\Delta P)_l D_i^3}{\mu_l^2 L_{eff}} \tag{5}$$

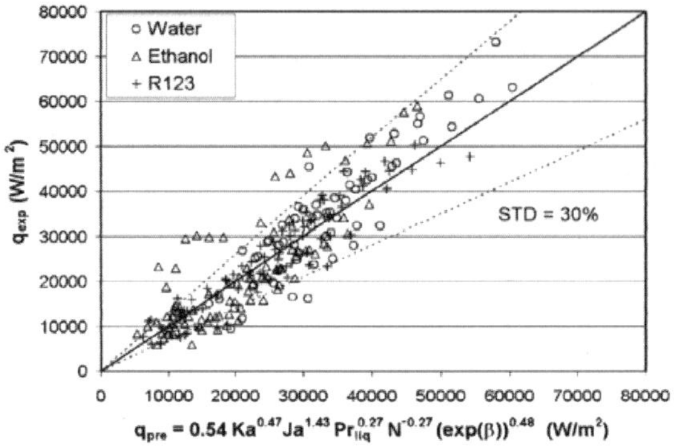

Figure 4. Semi-empirical correlation of CLOHP [5]

2. Experimental setup and analysis

CLOHP was designed for 300 Watts of heat transfer using Eq. 1 to Eq 5. Choosing water as working fluid, the number of turns needed is 12 with the inner diameter of 2 mm. Fig. 5 shows dimensions of CLOHP used in this study. Fig. 6 shows the experimental setup. Fig.7 shows locations of 16 thermocouples. Fig. 8 shows the recorded temperature from the experiment with standard deviation (SD) at each point. The CLOHP was placed at 5 mm and 10 mm from the surface of the copper plate. The surface temperature of interest in this study was 80°C -130°C, with the increment of 10°C.

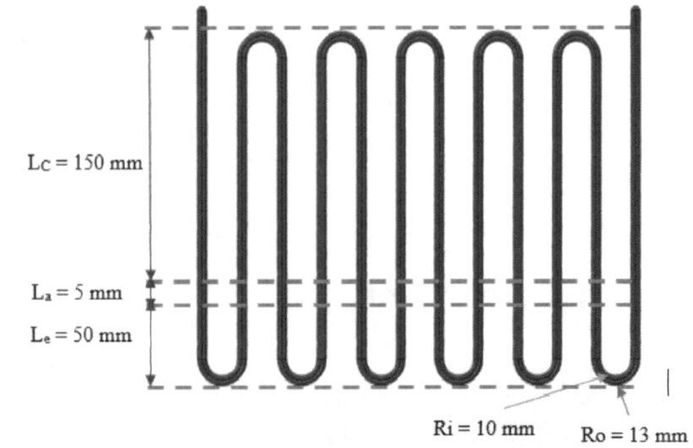

Figure 5. Dimensions of CLOHP

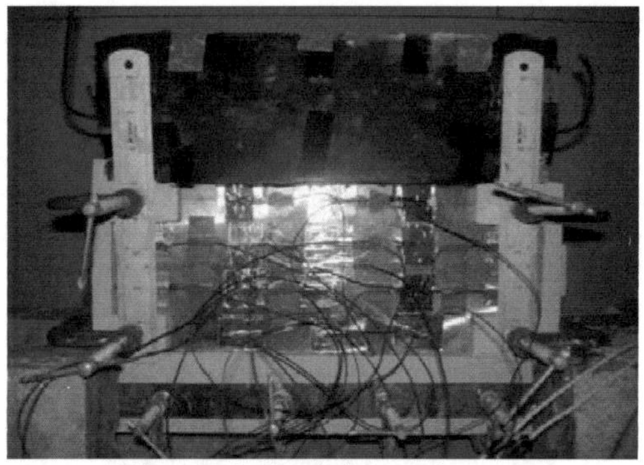

Figure 6. Experimental setup

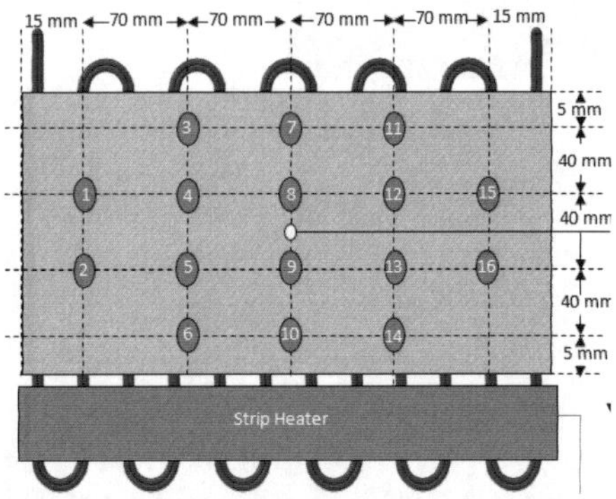

Figure 7. 16-location of thermocouples

COND. SIDE

T / SD	3	7	11	CLOHP
	80.2 ±0.163	80.0 ±0.115	79.6 ±0.216	
1	4	8	12	15
77.6 ±0.187	80.4 ±0.173	80.3 ±0.067	80.0 ±0.257	77.7 ±0.248
2	5	9	13	16
79.2 ±0.177	81.0 ±0.139	80.8 ±0.065	80.6 ±0.209	80.0 ±0.261
	6	10	14	
	80.9 ±0.099	81.0 ±0.094	81.1 ±0.150	

EVAP. SIDE

Figure 8. Recorded temperature with SD value

When the system was at steady state, the temperatures of 16 locations were recorded for 60 minutes. With the data from 3 replications, the temperature at each location was reported as mean temperature with SD value which was calculated by Eq. 6.

$$SD = \sqrt{\frac{\sum_{i=1}^{N}\left(X_i - \bar{X}\right)^2}{N}}. \tag{6}$$

Where SD is the standard deviation at each location
X is the measured temperature
$\bar{X}$ is the average measured surface temperature
N is sampling size

However, when refer to surface temperature of the whole plate, the standard deviation (STD) is used, where $\bar{X}$ is set surface temperature and N is sampling size. However, for comparison, when 2 set of temperature distribution results are considered, the relative standard deviation (%RSD), as in Eq. 7, was used.

$$\%RSD = \frac{STD}{\bar{X}} x100 \tag{7}$$

3. Result and discussion

The designed CLOHP can heat the copper plate to the desired set temperatures with acceptable deviation. The results of set surface temperature of 100 °C will be used as a representative for the data at other set temperature as well since they all show the same trends. Fig. 9 shows 16-location of surface temperature of copper plate at set surface temperature of 100 °C. The deviation (STD) of the whole plate from the set surface temperature was ±0.28 °C. The deviations (SD) at each 16-location are also shown on the right side of the figure. It can be clearly seen that the temperature distribution of the copper plate with CLOHP as the heat source is rather uniform. Fig. 9 also shows the result in the case of using cartridge heater (HR) at the same surface temperature for 16-location as well. The deviation in this case was ±0.62 °C. From Table 1, it can clearly be seen from the temperature deviation that the copper plate with CLOHP as the heat source shows good temperature distribution than those with cartridge heater (HR).

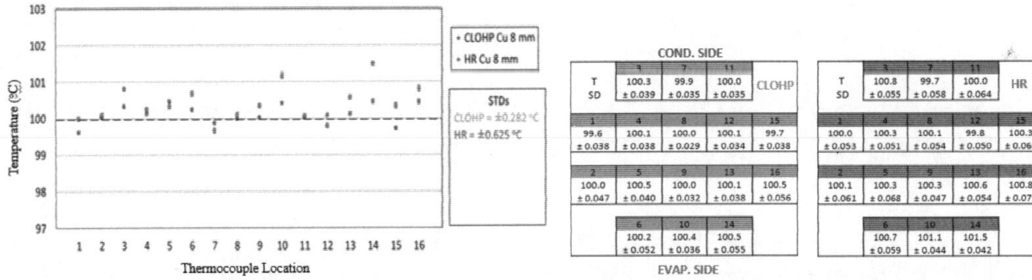

Figure 9. 16-location surface temperature of the copper plate at set temperature of 100 °C with the deviation (SD) at each location and deviation temperature of the whole plate (STD), comparison between using CLOHP and HR as the heat source

Table 1. STD of surface temperature at 5mm and 10mm depth.

Set Surface Temperature (°C)	STD (± °C) at 5mm Depth		STD (± °C) at 10mm Depth	
	CLOHP	HR	CLOHP	HR
80	0.51	0.64	0.50	0.64
90	0.45	0.59	0.52	0.59
100	0.28	0.62	0.47	0.62
110	0.25	0.58	0.43	0.58
120	0.22	0.48	0.48	0.48
130	0.24	0.53	0.55	0.53

From the result above, it is found that the CLOHP can transfer heat to the copper plate with better temperature distribution than HR, consequently, the discussion hereafter will be concentrated on using only CLOHP as the heat source and comparing the temperature distribution of the copper plate to those of P-20 plate from previous study [6]. Table 2 compares the result of the STD values of the surface temperature of both plate using CLOHP at 5mm and 10mm deep from the surface of the plate. The STD values of the copper plate, in all cases, are less than those on the P-20 plate. This suggests that the temperature distribution of the plate using CLOHP as the heat source with copper plate is more uniform than those with P-20. Fig. 10 and Fig. 11 compare the effect of specific heat and thermal conductivity on the RSD, respectively. Copper which has lower specific heat shows less RSD than P-20 which has higher specific heat. Lower specific heat material requires less thermal energy to raise the temperature of material of the same mass up to one degree, as a result the temperature distribution is undoubtedly more uniform. When considering the heat conductivity, copper shows less RSD than P-20. Material with higher thermal conductivity allows thermal energy per unit area to transfer and increase its temperature much faster. These results are in agreement with the study of Wang, G. et. al. [7]

Table 2. STD of surface temperature at 5mm and 10mm depth.

Set Surface Temperature (°C)	STD (± °C) at 5mm Depth		STD (± °C) at 10mm Depth	
	Copper	P-20	Copper	P-20
80	0.51	1.03	0.50	0.63
90	0.45	0.98	0.52	0.89
100	0.28	0.99	0.47	0.86
110	0.25	0.97	0.43	1.07
120	0.22	0.96	0.48	0.91
130	0.24	1.08	0.55	0.78

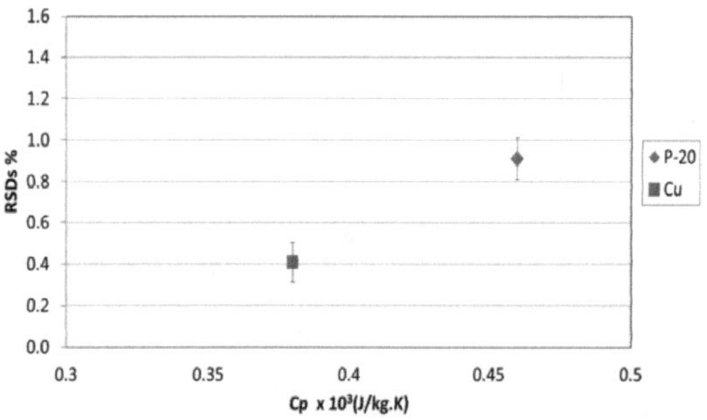

Figure 10. Effect of Specific Heat on RSD

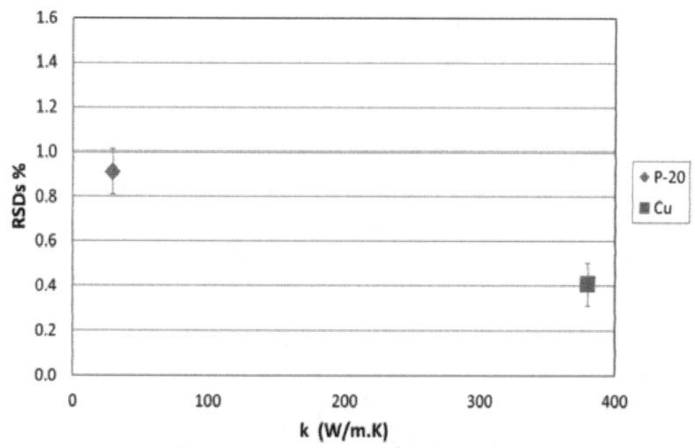

Figure 11. Effect of Thermal Conductivity on RSD

Table 3 shows the STD of the all the experiments. It is found that the STD of all the experiments are well in the range of the STD of the correlation which is ±30.0%. Consequently, the correlation can be well used for this design.

Table 3. STD of the heat transfer rate from experiment

Depth	STD of the Heat Transfer Rate	
	Copper (%)	P-20 (%)
5 mm	29.48	25.87
10 mm	9.41	6.76

4. Summary

CLOHP can be effectively used as a heat source to transfer heat to copper plate with much better temperature distribution. The STD of surface temperature from CLOHP is ±0.22°C to ±0.55°C while those used HR is ±0.48°C to ±0.64°C. Copper plate yields better temperature distribution than P-20. The STD of surface temperature of copper plate is ±0.22°C to ±0.55°C while those of P-20 plate is ±0.63°C to ±0.1.08°C. The STDs of heat rate of all experiments are well in the range of ±30% of the correlation used.

References

[1] http://www.grandviewresearch.com/industry-analysis/injection-molded-plastics-market
[2] https://knowledge.autodesk.com/support/moldflow-flex/learn-explore/caas/CloudHelp/cloudhelp/2015/ENU/MoldflowInsight360/files/GUID-2D69A8E5-36D6-4B22-8675-ADAF2707B403-htm.html.
[3] G. Wang, et. al., Development and evaluation of a new rapid mold heating and cooling method for rapid heat cycle molding, International Journal Of Heat and Mass Transfer, 78, 99-111, (2014).
[4] Y. Mishing, in: Diffusion Processes in Advanced Technological Materials, edtied by D. Gupta Noyes Publications/William Andrew Publising, Norwich, NY (2004), in press.
[5] Khandekar, S., Charoensawan, P., Groll, M., and Terdtoon, P., Closed-Loop Pulsating Heat Pipe Part B: Visualization & Semi-Empirical Modeling. Applied Thermal Engineering, 23, 2021-2033, (2003).
[6] P. Kamonpet, P. SakulchangSatjaThai, Temperature Uniformity of Heated Mold Plate by Oscillating Heat Pipe, International Conference in Mechatronics and Mechanical Engineering, September 15-16, Singapore, (2015).
[7] G. Wang, G. Zhao, H. Li, and Y. Guan, Thermal Response Simulation and Mold Structure Optimization for Rapid Heat Cycle Molding Processes with Steam Heating and Electric Heating, Material and Design, 31, 382-395, (2010).

IOP Conf. Series: Materials Science and Engineering **191** (2017) 012035 doi:10.1088/1757-899X/191/1/012035

Recovery of uranium from low concentration leach liquor of acid in-situ leaching

Zhenqian Wen[1,2] and Yuqing Niu[1]

[1] Beijing Research Institute of Chemical Engineering and Metallurgy, CNNC, Beijing 101149, China;
[2] School of Chemistry and Chemical Engineering, Central South University, Changsha 410083, China

E-mail: 87024942@qq.com

Abstract. For problems of environment protection and cost in uranium recycle, the process flows of ion exchange and Eluex, which recovered uranium from low concentration leach liquor of acid in-situ leaching, were studied. Although the flow sheet of ion exchange process was simple, the Eluex process had an advantage over it due to large quantity of effluent and high processing cost in ion exchange process by comparative studies.

1. Introduction

In northern China, because of low uranium amount per square meter, uranium grade is generally lower (uranium grade is around 0.01%). The surrounding rock has good permeability (permeability coefficient is from 7 m/d to 10 m/d) and low content of carbonate. The process of acid in-situ leaching was employed in this ore block. However, the concentration of uranium in the leach liquor was low (15-30 mg/L). So the difficult problem, which concentrated uranium from low concentration solution, should be urgent to be resolved.

To get the high quality of uranium precipitation product and reduce the consumption of NaOH during the process of hydrometallurgy, ion exchange process and Eluex process had been employed to recovery uranium from acid in-situ leaching solutions. Comparisons of technique, economy and environmental protection between these two methods showed that Eluex process is more suitable for uranium recovery from the low concentration leach liquor in acid in-situ leaching. Finally, the comparatively good process and parameters were determined, and the ideal effect was acquired too.

2. Ion exchange process

Ion exchange process [1-4] is the main technology used in recovery of uranium from in-situ leaching solution, and it generally includes procedures of uranium adsorption from leach liquor, uranium elution from saturation resin and resin transformation. In China, the process of adsorption with fixed bed and elution with moving bed is employed by many uranium milling plants. However, this method usually causes the problems of high cost and environmental protection. In this study, only adsorption and elution process of fixed bed was discussed.

The pH value of the acid leaching solution is 1.62 and ORP is 315mV. The concentrations of U, Ca^{2+}, Mg^{2+}, Fe^{2+}, Al^{3+}, SO_4^{2-} and Cl^- were 22,465,413,480,277,8453 and 488mg/L respectively. Strong base anionic resin named D1 was used as adsorption material. The adsorption with three columns in series, which are fixed beds, is employed. And feeding mode is upward. Adsorption temperature is from 10 to 20 °C with contact time 6 minutes. When uranium concentration of leach liquor was 20-30 mg/L, uranium saturation capacity of resin is 11-15 mg/mL, the breakthrough volume is about 600 bed volumes, and the ratio of saturation volume and breakthrough volume is about 1.5.

The acidic NaCl solution is employed to elute U from the saturation resin in a single fixed bed column. The eluent solution contains 90 g/L of NaCl and 5 g/L of H_2SO_4.The elution contact time is about 45-60 minutes. When the bed volume of eluate reaches to 9, the uranium concentration of eluate can drop to 50 mg/L. If the bed volume of eluate reaches to 3, the uranium concentration of eluate is highest. The OK liquor can be obtained from the 2nd to 5th BV. When uranium concentration in leach liquor is 25-30 mg/L, the saturation U capacity of resin is 10-13 mg/mL. Uranium concentration in the OK liquor can be maintained between 4.5 to 5.5 g/L, while the average consumption of NaCl is about 4.63 t/(t·U) and the elution rate is higher than 97%.

After elution of uranium loaded resin, the barren resin will be converted to Cl type. So it is necessary to convert the resin from the chloride form to the sulfate form before returned to the adsorption cycle. After uranium adsorption, effluent contains 8-10 g/L of SO_4^{2-}, which can be used to convert the resin.

The conversion results show that at early stage of the transformation (0.25-0.5 BV), Cl^- concentration in the liquor is relatively high and such liquor can be used for elution. In the next 5 BV of transformation solution, Cl^- concentration becomes low enough to discharge into evaporation pond to prevent Cl^- from accumulation in the leaching system.

The primary product of yellow cake is precipitated by the method of circulation aging from OK liquor. The precipitation reagent NaOH is put slowly into the OK liquor, and the pH value is adjusted to 7.0-7.5. The agitation is carried for 20-30 minutes, and the time of settlement aging is 12 hours. After multi-precipitations, the alkali consumption of 1 ton uranium product is relatively high. The average amount of alkali is 1.33 t, while the U concentration in the mother liquor is about 5 mg/L.

3. Eluex process

The Eluex process is a combined technology which makes the processes of ion exchange and solvent extraction to link together. Firstly uranium is extracted by the process of ion exchange from leach liquor, and then the saturation resin is eluted by sulfuric acid. Secondly the uranium is extracted by the process of solvent extraction from the acidic eluate, and then uranium and sulfuric acid are separated from each other. Finally, the uranium is concentrated at high magnification.

The uranium is extracted by process of ion exchange from leach liquor, and the saturation resin is eluted by 100 g/L of sulfuric acid. Under the condition of contact time with 40 minutes, uranium content of barren resin is less than 0.1 mg/(ml·wsr), and uranium concentration of effluent is less than 0.1mg/ml after 9 BV of elution. Uranium concentration of OK liquor is 2-4g/L.

During the procedure of sulfuric acid elution, the anionic uranium complexes in the resin are displaced by the rule of mass action due to high concentration of anionic bisulfate in eluent. The complexes of $UO_2(SO_4)_2^{2-}$ and $UO_2(SO_4)_3^{4-}$ in the eluate are dissociated partly, and the resultants of UO_2SO_4 and UO_2^{2+}, which are not easy to be adsorbed by the resin with HSO_4^- type, are formed. Elution rate is dependent on the concentrations of exchangeable anion and diffusion of two kinds of exchanging anion due to simplicity of mass action. So the longer contact time and large amount of eluents are needed during the elution.

As feed solution of solvent extraction [5-7], the OK liquor of elution is extracted by the organic phase of 5% P204, 10% TBP and sulphonated kerosene. The size of the mixing chamber of mixer-settler is 1m×1m×1.5m. The side ratio of the mixing chamber and settler chamber is 1:3. The flow ratio of organic phase and aqueous phase is 1:2-1:4. The operating temperature is 18℃. After 4 stages of countercurrent extraction, the uranium concentration in the organic phase can reach to about 6g/L, and the uranium concentration in the raffinate is below 0.01g/L.

The loaded organic phase was handled by 2 stages of scrubbing and 3 stages of stripping. The flow ratio of organic phase and aqueous phase containing 110g/L of Na_2CO_3 in scrubbing process was 10:1, and the flow ratio of organic phase and aqueous phase in stripping process is 5:1-7:1. The operating temperature is 20℃. Stripping results indicated that under the condition of 6 g/L of uranium concentration in loaded organic phase, the uranium concentration of OK liquor can reach to 50 g/L by 3 stages of stripping.

4. Comparison of 2 processes

4.1. Process flow

Compared with the processes of ion exchange and Eluex, the procedure of solvent extraction is increased in the Eluex process, which is more complicated.

4.2. Comparison of environmental benefit

4.2.1. Effluent quantity of acidic NaCl elution process

1) Effluent of resin transformation

To avoid the accumulation of Cl^- in leach solution, the amount of transformation effluent is totally 5 BV. If it is calculated by the 12 mg/mL of saturation capacity, the effluent quantity is 417 m^3/t U.

2) Effluent quantity of discharged sodium sulfate

Generally, the production capacity of sodium sulphate is 3.9 t/tU in the process of acidic NaCl elution. Assuming that the solution will be discharged by precipitation of the mother liquor when the concentration of Na_2SO_4 is 100g/L, the effluent quantity is 39 m^3/t U.

4.2.2. Effluent quantity in eluex process

1) Raffinate discharge

The barren organic phase contains Na^+, of which 5% P204. The concentration of Na is around 0.149mol/L. The Na^+ enters into aqueous phase by exchanging with UO^{2+} and H^+ during extraction process and then sodium sulphate is produced. Uranium capacity of saturated organic phase during Eluex process is 6.16g/L, therefore, the yield of sodium sulphate is 1.74t Na_2SO_4/t U. Assuming that the solution will be discharged when the concentration of sodium sulphate reaches to 100g/L, the effluent quantity is 17.4m^3/t U.

2) Effluent quantity of discharged sodium carbonate

During the recirculation of stripping feed solution in the preparation procedure of mother liquor, sodium carbonate will be accumulated by the addition of sodium bicarbonate. If the amount of sodium bicarbonate is 0.21 t $NaHCO_3$/t U, then the accumulation amount of sodium carbonate is 0.26t Na_2CO_3/t U. Assuming that the concentration of sodium carbonate is 110 g/L, the effluent quantity of discharged carbonate discharge is 19.8m^3/t U.

The effluent quantity in ion exchange process is much more than that in Eluex process, which is 456 m^3/t U compared with 19.8m^3/t U in total. By utilization of Eluex process, the effluent quantity is reduced by above 95% of it.

4.3. Comparison of reagent cost

4.3.1. Reagent consumption of ion exchange process. The primary reagent consumption includes sulphuric acid, sodium hydroxide and sodium chloride.

1) The consumption of H_2SO_4

The sulfuric acid concentration in the eluent is 5g/L. In the condition that the uranium concentration is 5g/L, the theoretical acid consumption is 1t H_2SO_4/t U.

2) The consumption of NaOH

The consumption of sodium hydroxide occurs in the procedure of precipitation, includes neutralization of sulphuric acid and precipitation of uranium. The actual average consumption is 1.33t NaOH/t U.

3) The consumption of NaCl

The reason why sodium chloride consumes during production is that transformation solution and effluent of sodium sulphate are discharged. The actual average consumption is 4.63t NaCl/t U.

4.3.2. Reagent consumption in Eluex process. The primary reagent consumption in this process consists of the inorganic and organic reagent. The inorganic reagent includes sulphuric acid, sodium hydroxide and sodium bicarbonate, and organic reagent includes P204, TBP and kerosene, etc.

1) The consumption of H_2SO_4

The primary consumption of sulphuric acid occurs in the elution and extraction process. During the elution process, the decreasing portion of sulphuric acid enters the barren resin, and then it is returned to absorption process, and recycles in the process of in-situ leaching, which has no acid consumption. The sulphuric acid consumption in extraction is 0.38t H_2SO_4/t U.

2) The consumption of organic reagent

The primary consumption of organic reagent occurs in the extraction process. If the dissolution loss of organic phase is calculated after oil removal, the concentration of organic phase in the raffinate is less than 20mg/L and about 15% loss.The consumption of organic reagent is 0.66kg/t U.

3) The consumption of $NaHCO_3$ and Na_2CO_3

In stripping process, besides sodium carbonate consumption in initial operation, the sodium bicarbonate is added into mother liquor to prepare the stripping feed solution after recirculation, only the sodium bicarbonate is consumed, and the consumption of it is 1.36t $NaHCO_3$/t U.

4) The consumption of NaOH

The primary consumption of sodium hydroxide occurs in precipitation. Since the uranium concentration of stripping solution is high, the alkaline consumption is low, which is 0.95t NaOH/t U.

Table 1. Consumption comparison of two processes

Reagent type	Unit price (CNY/t)	Acid sodium chloride elution		Eluex process	
		Consumption $(t\ (t\cdot U)^{-1})$	Cost $(CNY(t\cdot U)^{-1})$	Consumption $(t\ (t\cdot U)^{-1})$	Cost $(CNY\ (t\cdot U)^{-1})$
H_2SO_4	700	1	700	0.38	266
NaCl	900	4.63	4167	—	—
NaOH	3000	1.33	4000	0.95	2850
TBP	25000	—	—	0.0034	85
P204	20000	—	—	0.014	280
kerosene	9000	—	—	0.176	1584
$NaHCO_3$	2000	—	—	1.36	2720
sum	—	—	8867	—	7785

Calculating by the current market price of reagent, the cost of these two processes is shown in table 1. From this table, it can be seen that the total cost per tone of uranium is 8867 CNY in ion exchange process, while 7785 CNY in Eluex process.

5. Conclusion

The two process flows of acidic NaCl elution and Eluex are employed to recovery uranium from low concentration leach liquor of acid in-situ leaching. Although there is no operation of extraction and stripping in the process of acidic NaCl elution, the Eluex process flow is more economical and environmental due to transportation operation of barren resin, large amount of Na_2SO_4, large quantity of effluent and high consumption of reagent in the process of acidic NaCl elution.

The further comparative work, includes investment of equipment, facility and construction as well as operation staff, should be considered in the future.

Reference

[1] Xu C, Xie G, He T, Yu X and Li X 2009 Uranium recovery from leach solution in in-situ leaching of uranium with ion-exchange adsorption *Journal of Chongqing University*, 32 1398-1402.

[2] Cai P, Tan K, Shi W, Hu e and Huang A 2010 Experimental study of ion exchange and

IOP Conf. Series: Materials Science and Engineering **191** (2017) 012035 doi:10.1088/1757-899X/191/1/012035

neutralization on acid in situ leach uranium mining waste water *Modern Mining*,11 21-3.

[3] Niu Y, Han Q and Wang Z 2003 Recovering uranium from the sulphuric acid leaching liquor containing more chlorides with anion-exchange resin *Uranium Mining and Metallurgy*, 22 84-7.

[4] Zhang X, Fu C, Xiao J, Zhou Y, Zhang Y and Tan Y 2011 Deeply purifying and recovery uranium from the uranium mine water by ion exchange *Chem Industry and Eng Progress*,30 126-9.

[5] Merritt R C 1971 *The extractive metallurgy of uranium* (Colorado: Johnson Publish Company) p209.

[6] Jia X, Li P, Chen T, Huang Y, Liu H, Liu Z and Wen Z 2015 Extraction and recovery of uranium in vanadium extraction process by P204 *Uranium Mining and Metallurgy*, 34 107-10.

[7] Edwards R I and Riele W A 1983 *Handbook of Solvent Extraction* (New York: Wiley Interscience) P723-31.

IOP Conf. Series: Materials Science and Engineering **191** (2017) 012036 doi:10.1088/1757-899X/191/1/012036

Impact of solvent extraction organics on adsorption and bioleaching of *A. ferrooxidans* and *L. ferriphilum*

Yu Hualong[1] and Liu Xiaorong[1]

[1]School of Materials Science and Engineering, Shanghai Institute of Technology, Shanghai, 201418 China.

E-mail: yuhualoong@hotmail.com

Abstract. Copper solvent extraction entrained and dissoluted organics (SX organics) in the raffinate during SX operation can contaminated chalcopyrite ores and influence bioleaching efficiency by raffinate recycling. The adsorption and bioleaching of *A. ferrooxidans* and *L. ferriphilum* with contaminated ores were investigated. The results showed that, *A. ferrooxidans* and *L. ferriphilum* cells could adsorb quickly on minerals, the adsorption rate on contaminated ores were 83% and 60%, respectively, larger than on uncontaminated ores. However, in the bioleaching by the two kinds of acid bacterias, contaminated ores presented a lower bioleaching efficiency.

1. Introduction

Bioleaching-Solvent Extraction-Electrowinning technology (BL-SX-EW) for copper has been widely used due to its advantages of low costs and environmental friendly [1]. However, low efficiency is a key problem in bioleaching process. Many studies have shown that various factors greatly affect bacterial growth and the bioleaching process [2]. Among which, the adsorption of bacterial on minerals could be a more important one [3].

On the other hand, SX organics lost in the raffinate in forms of entrainment and dissolution, could contaminate bioleaching system through the recycling use of the raffinate [4]. Studies have shown that the LIX extraction reagent is harmful to the bacterials [5]. When exposed to different extraction compounds, the bacterial activity and leaching ability of acidophilic bacteria were different. *A. thiooxidans* and *A. ferrooxidans* were confirmed to be more sensitive than *S. thermosulfidooxidans* and *Alicyclobacillus spp* [6]. A biochemical mechanism was speculated for the inhibition effect by SX organics [7]. Most of the studies were from the view of bacteria growth and bioleaching efficiency. The organics could contaminated minerals and bacterial by means of adsorption, which could finally impact the leaching efficiency. Thus, we studied the adsorption of *A. ferrooxidans* and *L. ferriphilum* on the contaminated chalcopyrite ores by SX organics, and its following bioleaching process, which are of theoretical and practical significance.

2. Materials and methods

2.1. Materials

2.1.1. Minerals and extraction regents. The low-grade chalcopyrite ore (0.2% Cu, 3.4% Fe, and 5.1% S) used in the experiments was provided by the Dexing Copper Group Company, Jiangxi, China. The copper extraction reagent, LIX 984N, was obtained from the BASF Chemical Co., Ltd. (Shanghai, China). Commercial 260# kerosene was used as the diluent.

2.1.2. Bacteria and cultivation. Mackintosh (MAC) medium[8] containing 0.132 g/L $(NH_4)_2SO_4$, 0.1 g/L KCl, 0.025 g/L $MgCl_2$•$6H_2O$, 0.027 g/L KH_2PO_4, and 0.147 g/L $CaCl_2$•$2H_2O$, pH 2.0 was used as the culture medium. The inorganic salts were analytical grade. *A. ferrooxidans* and *L. ferriphilum* (provided by the Biofilm Center, University of Duisburg-Essen) was adapted to 10% low-grade chalcopyrite ore in the MAC medium. The cultures were incubated in Erlenmeyer flasks at 30 °C with shaking at 180 r/min in a constant-temperature air-bath shaker (ZHWY-2102, Zhicheng Analytical Instrument Manufacturing Co., Ltd, Shanghai).

2.2. Methods

2.2.1. Solvent extraction. MAC medium was mixed with an isopycnic organic phase in an agitating tank. After stirring for 10 min at 400 r/min at room temperature, the MAC medium, which contaminated by SX organics, was separated by a funnel. Low-grade chalcopyrite ore (washed with 75% ethanol and rinsed with deionized water several times) was added to 100 mL of uncontaminated or contaminated MAC medium. The mixtures were shaken for 2 h in Erlenmeyer flasks in air-bath shakers at 30 °C and 180 r/min. The contaminated ore powders were filtered and air dried in a dust-free environment.

2.2.2. Microorganism adsorption. Microorganism adsorption on the minerals was investigated using the contaminated mineral, with inoculation of 10% (v/v) culture. A microscope was used to examine the bacterial cells. The adsorption rate was calculated as follows. Cells adsorbed on uncontaminated minerals were used as a control.

$$r_e = \frac{n_i - n_t}{n_i} \times 100\% \tag{1}$$

where r_e is the adsorption rate, and n_i and n_t are the initial number of cells and the number of cells at time t, respectively.

2.2.3. Bioleaching experiments. Bioleaching experiments were carried out in 250 mL erlenmeyer, containing 150 mL MAC medium and 15 g contaminated/uncontaminated minerals in a shaker at 200rpm, 30°C . Each medium were inoculated with 5% (v/v) of bacterial culture (5×10^7 cells/ml). Deionized water was supplemented for the evaporation losses.

3. Result and discussion

3.1. Adsorption of A. ferrooxidans and L. ferriphilum on chalcopyrite ores
Figure 1 reflected the adsorption of *A. ferrooxidans* and *L. ferriphilum* on low-grade chalcopyrite ores. As Figure 1a shown, the adsorption on contaminated low grade chalcopyrite ores by SX organics had an obvious improvement, the maximum rate was nearly 83%, double of that on uncontaminated ores. In Figure 1b, the adsorption behavior of *L. ferriphilum* presented a same trend as *A. ferrooxidans*. Cells of *L. ferriphilum* adsorbed fast on minerals in first 5 minutes and be stable after then. The adsorption capacity (60%) on contaminated ores was larger than on uncontaminated ores (45%).

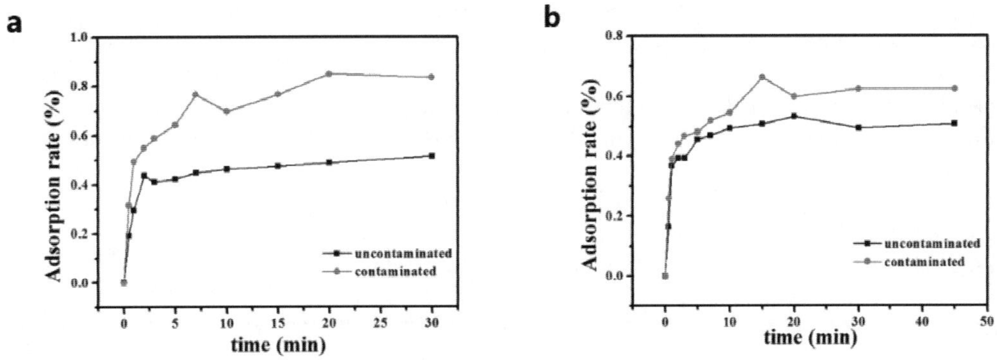

Figure 1. Adsorption of *A. ferrooxidans* (a) and *L. ferriphilum* (b) on contaminated/uncontaminated chalcopyrite ores

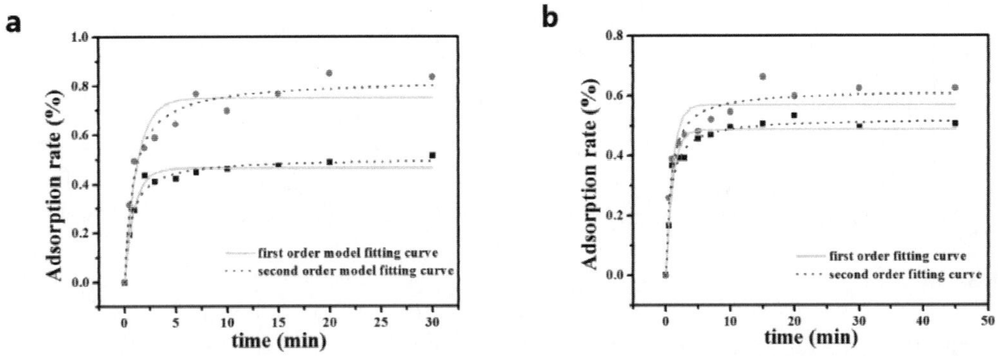

Figure 2. The fitting kinetic curves of the adsorption of *A. ferrooxidans* (a) and *L. ferriphilum* (b)

Table 1. Related parameters of fitting kinetic curves

		first order model			second order model		
		k1	r1	R2	k2	r2	R2
A.f	**uncontaminated**	1.036	0.466	0.969	2.942	0.504	0.976
	contaminated	0.810	0.751	0.911	1.377	0.822	0.966
L.f	**uncontaminated**	0.954	0.487	0.946	2.657	0.523	0.966
	contaminated	0.944	0.571	0.897	2.128	0.618	0.959

It could be inferred that the SX organics adsorbed on minerals surface promote the adsorption of *A. ferrooxidans* and *L. ferriphilum* cells. This could be explained by the hydrophobic interaction of the bacteria and minerals. SX organics contain the major componds of LIX984N and sulfonated kerosene. LIX984N was made up by hydrophilic polar groups and lipophilic nonpolar group [9], which was a surfactant and could adsorb on minerals surface to enhance the hydrophobic. On the other hand, the cytomembrane of gram-positive bacterias contains some proteins and lipids, which were related to the adsorption on minerals [10]. Many studies have indicated that the bacterial trend to adsorb on the hydrophobicity surface [11], and it was certified again in Figure 1. The adsorption process were fitted by absorbing kinetic equations of the first order and second order in Figure 2. Comparing the data in Table 1, the second-order kinetic equation fitted better for the cells adsorption.

3.2. Bioleaching of A. ferrooxidans and L. ferriphilum on chalcopyrite ores

Since the adsorption of bacterials on mineral surface is a prerequisite for further interactions during bioleaching process. A higher adsorption rate of cells means a more intimate contact between them and will lead to a higher leaching efficiency. However, as shown in Figure 3, the bioleaching efficiency of contaminated and uncontaminated low-grade chalcopyrite ores by *A. ferrooxidans and L. ferriphilum*. It is obvious that the efficiency of contaminated minerals was smaller than that of the uncontaminated ores, the highest rate was 15% and 25% for *A. ferrooxidans*, 7.5% and 6.8% for *L. ferriphilum*. Higher cells adsorption does not result in higher bioleaching efficiency. SX organics improved the cells adsorption but inhibited the final leaching efficiency.

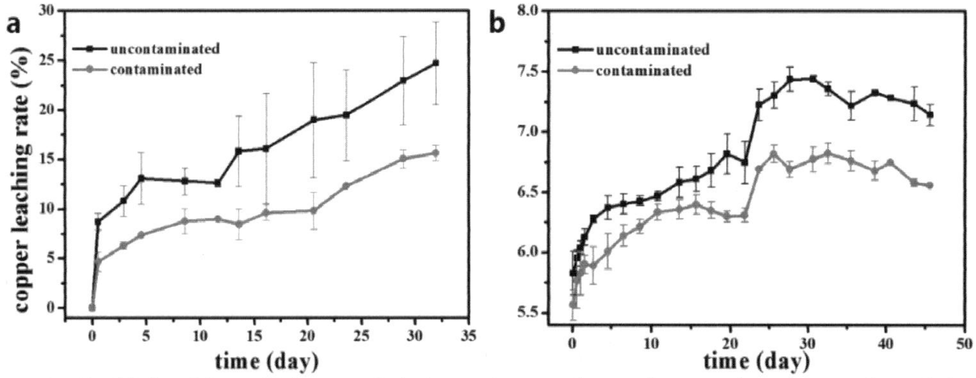

Figure 3. The bioleaching efficiency of chalcopyrite ores by *A. ferrooxidans* (a) , *L. ferriphilum* (b)

4. Conclusions

A. ferrooxidans and L. ferriphilum could adsorb quickly on minerals and the adsorption rate were 83% and 60% respectively on contaminated minerals, larger than on uncontaminated of 42% and 45%. The second-order kinetic equation fitted better for the adsorption.

The efficiency of contaminated minerals was smaller than that of the uncontaminated ores, the highest rate was 15% and 25% for *A. ferrooxidans*, 7.5% and 6.8% for *L. ferriphilum.*

Higher cell adsorption does not result in higher bioleaching efficiency. SX organics improved the cells adsorption but inhibited the final leaching efficiency.

Acknowledgments

Authors wishing to acknowledge the financial support of the National Nature Science Foundation of China (51474150).

References

[1] Tan S N and Chen M 2012 Hydrometallurgy 120 94.
[2] Devasia P and Natarajan K A 2010 Int J Miner Process 94 139.
[3] Dong Y, Lin H, Xu X and Zhou S 2013 Hydrometallurgy 140 47.
[4] Watling H R 2006 Hydrometallurgy 84 108.
[5] Dopson M, Sundkvist J and Börje L E 2006 Hydrometallurgy 81 213.
[6] Davis-Belmar C S I, Demergasso C and Rautenbach G 2012 Hydrometallurgy 130 139.
[7] Liu X, Li H, Liu Y, Su C and Sand W 2015 Hydrometallurgy 157 213.
[8] Mackintosh M E 1978 J. Gen. Microbiol 105 218.
[9] Diao M, Tuan A H, Elena T, Stephen M and Anh V Nguyen 2015 Colloid Surface B 132 280.
[10] Panda S, Parhi P K, Pradhan N, Mohapatra U B, Sukla L B and Park K H 2012 Hydrometallurgy 124 119.
[11] Lu J and Dreisinger D 2013 Hydrometallurgy 137 17.

IOP Conf. Series: Materials Science and Engineering **191** (2017) 012037 doi:10.1088/1757-899X/191/1/012037

In-situ high temperature X-ray diffraction study on the phase transition process of polymetallic sulfide ore

Guangshi Li[1], Hongwei Cheng[1,2], Xiaolu Xiong[1], Xionggang Lu[1,2], Cong Xu[1], Changyuan Lu[1], Xingli Zou[1,2] and Qian Xu[1,2]

[1]School of Materials Science and Engineering, Shanghai University, Shanghai 200072, People's Republic of China
[2]State Key Laboratory of Advanced Special Steel, Shanghai University, Shanghai 200072, People's Republic of China

E-mail: hwcheng@shu.edu.cn; luxg@shu.edu.cn

Abstract. The phase transformation of the polymetallic sulfide ore is quite complicated, especially the variety and diversity of the roasted intermediate. The mineralogy properties of the ore particles were characterized by XRF, ICP and SEM-EDS. The thermal stability of the complex ore was investigated using TG-DSC under different atmosphere (pure N_2 and 1% O_2–N_2). The phase changes and intermediates were revealed by the in-situ XRD in pure N_2 and 1% O_2-N_2 atmospheres with the temperature range from 25 to 800 °C. These results indicated that the crystal transformation of the pyrrhotite from monoclinic to hexagonal at 320 °C, and the pentlandite and chalcopyrite were decomposed into monosulfide solid solution in pure N_2. While the pentlandite and chalcopyrite were completely oxidized into $M^*_xFe_{3-x}O_4$ (M*=Fe, Ni, Cu) and Fe_2O_3, which is also the final product of the ore sample oxidized under 1% O_2–N_2 atmosphere.

1. Introduction

The pyrometallurgical process of polymetallic sulfide minerals is relatively energy efficient compared with hydrometallurgical extraction, which mainly due to the comprehensive utilization of the exothermic oxidation of sulfide ores. The oxidation process of metal sulfides is particularly important, besides, it is of much complexity and variability. Terkel Rosenqvist [1] has investigated and mapped the phase equilibria in roasting and smelting complex sulfide, but these systems were relatively simple compared with the nature sulfide minerals. Later, J. G. Dunn [2] has studied the oxidation behavior of natural pentlandite using thermogravimetry (TG) and differential thermal analysis (DTA). The oxidation process of the synthetic pentlandite under the 973K isothermal condition could be divided into three steps: firstly pentlandite oxidized into Fe_2O_3, $Ni_xFe_{3-x}O_4$, SO_2, $NiSO_4$, and NiO; secondly, a nickel-rich phase and a sulfur-rich phase were generated; Also, $Ni_xFe_{3-x}O_4$ and NiO were observed as the final oxidation products [3, 4]. Lately, the oxidation mechanism of nickel concentrate was investigated using TG-DTA and X-ray diffraction (XRD) by Dawei Yu [5, 6]. The thermodynamic properties of chalcopyrite were performed by B. R. Conard [7] and the phase transition of chalcopyrite under the oxidative roasting process was investigated by Bayer.G [8]. S.Prasad [9] summarized the alternative processes to treat chalcopyrite and pointed out that the roasting of chalcopyrite was an important step for the extraction of copper. By adsorbing and integrating the previous research results, this paper mainly studied the application of the in-situ XRD and simultaneous thermal analysis (STC), probing into the temperature programmed reactions of the complex nickel sulfides and the mechanism of sulfides transition during the pure N_2 and 1% O_2–N_2 atmosphere respectively. In-situ XRD pattern provides unique information about important properties of phases under controllable non-ambient conditions [10-14]. The application of in-situ techniques may lead to considerable advances in understanding of phase decomposition and oxidation reactions in pyrometallurgical and hydrometallurgical processes [15-17].

2. Experiments and characterization

2.1. Sample preparation and elements analysis

The collected sulfide nickel ore sample (Sinkiang, China) was dried at 100 °C for 24h and ground into fine powder (-200 mesh) using an agate mortar. The fine powder was carried out for the identification and quantification of chemical elements by the X-ray fluorescence (XRF) analysis (XRF-1800, SHIMADZU LIMITED). Furthermore, the inductively coupled plasma atomic emission spectrometry (ICP-AES) was measured to accurate analysis of several main elements on a PERKINE 7300DV.

2.2. SEM-EDS characterization

The cross section view of raw ore sample was prepared in an epoxy mount and polished. Subsequently the mineralogical characterization was performed by a tungsten filament scanning electron microscope (SU-1510, Hitachi, Japan), and the elements distribution was characterized by the energy-dispersive X-ray spectroscopy (EDS) mapping features within the TEAMTM EDS Analysis System in a SU-1510.

2.3. Thermal analysis

The effect of different atmospheres on the decomposition and oxidation processes of the fine powder sample has been studied over the temperature range of 25-1000 °C using the STC (the simultaneous application of TG and DSC to one and the same sample in a single instrument.) in a NETZSCH STA 449 F3 Jupiter.

2.4. In-situ XRD

In-situ high temperature XRD data were collected using a Bruker D8 Advance diffractometer with a Bruker TC-DOME temperature stage and a platinum-rhodium heating strip with TCPU1 temperature controller, in dynamic pure N_2 and 1% O_2–N_2 atmosphere respectively. Cu-Kα radiation was used for scans of 20 to 70° 2θ with a step size of 0.02° 2θ, a dwell time of 2 seconds/step and a heating and cooling rate of 5°C/min. Diffraction patterns were collected at room temperature, then every 50 °C between 300 °C and 600 °C and every 100 °C between 600 °C and 800 °C on heating with a holding time of 30 minute prior to data collection on reaching the measurement temperature. The fine mineral powder sample (-200 mesh) was deposited on a platinum foil substrate for heating, which was placed on the Pt-Rh heating strip.

3. Results and discussion

3.1. The mineralogy properties of polymetallic sulfide ore

Table 1. Elemental analysis (wt%) of the mineral sample using XRF and ICP.

	Fe	S	O	Si	Cu	Ni	Mg	Al	Ca	Na	Co
XRF	40.12	26.64	15.32	5.96	3.58	2.35	2.29	1.94	0.94	0.57	0.16
ICP	43.31	--	--	3.15	3.13	2.33	0.81	0.93	0.57	0.27	0.14

The chemical composition of the complex mineral fine powder sample are showed in table 1 using XRF and ICP-AES methods. The comparison of the data shows a good agreement between the XRF analysis results and those obtained by ICP, although there is some deviation in some certain elements (Mg, Al, Si,). The major elements of this mineral are iron, nickel and copper. As a result, this is a complex nickel sulfide ore with rich iron.

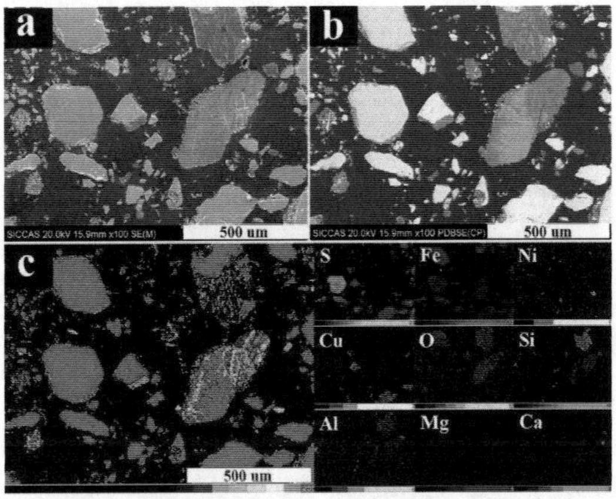

Figure 1. a) SE and b) BSE image of the sulfide ore particles, and c) the element distribution of the mineral particles' cross section.

Figure 1 a and b show the secondary electron (SE) and backscattered electron (BSE) micrograph of the cross sections of the ore grains. The elements distribution of the complex ore particles is shown in Figure 1c. It is clearly revealed that the large particles mainly consist of iron sulfide, iron oxide or silicon-magnesium oxide. And the nickel-iron sulfide and copper-iron sulfide were distributed sporadically in smaller granule. The complex sulfide nickel ore is composed of pyrrhotite, pentlandite, chalcopyrite, magnetite and gangue phases, which is detected by the XRD pattern in figure 2b.

3.2. The thermal stability of the complex ore sample in different atmosphere

In order to evaluate the effects of the atmosphere on the thermal stability of mineral, the TG-DSC measurements were performed in pure N_2 atmosphere and 1% O_2–N_2 atmosphere respectively for comparison. As shown in figure 2a, an obvious weight loss peak appeared in both atmospheres, which indicated that there is at least one weight-loss reaction in the temperature range of 450-800 °C. For further study on the thermal events, the analysis of the DSC curves revealed that there are only two endothermic peaks under pure N_2 atmosphere. The endothermic reaction at 320 °C without any weight changes, indicated that there may be a structure transition. The weak endothermic peak at 480 °C may be a certain decomposition reaction of one or more phases, as the TG curve appeared a slight weight loss at the temperature range from 450 °C to 800 °C under the pure N_2 atmosphere. In 1% O_2–N_2 atmosphere, the DSC curve is much more complicated. The large exothermic peak broadened in a temperature range from 400 °C to 650 °C, this might be the effect of a series of oxidative reactions by micro-aerobic environment. And the two endothermic peaks appeared later at 680 °C and 710 °C, as a result, it may be the decomposition of two or more kinds of intermediate products.

The final products, which were measured by TG-DSC test under different atmospheres, were investigated by XRD shown in figure 2b. The tested sample consist of hexagonal pyrrhotite ($Fe_{1-x}S$) and magnetic ferrite ($M^*_xFe_{3-x}O_4$, M*=Fe/Ni/Cu) under the pure N_2 atmosphere. However, the nickel and copper sulfides were not detected, one of the reason is that such small amounts of the nickel and copper phases that it could not be detected by XRD. But there is also the possibility, but unsubstantiated and discussed using in situ XRD later, that the major elements (Ni/Cu) might solubilize in hexagonal pyrrhotite and magnetic ferrite. Hematite (Fe_2O_3) and magnetic ferrite ($M^*_xFe_{3-x}O_4$) were the main phases of the TG-DSC measured residuals in 1% O_2–N_2 atmosphere. It indicated that the nickel and copper have been oxidized into nickel ferrite and copper ferrite respectively, and it was confirmed by previous work [5, 6, 9, 11].

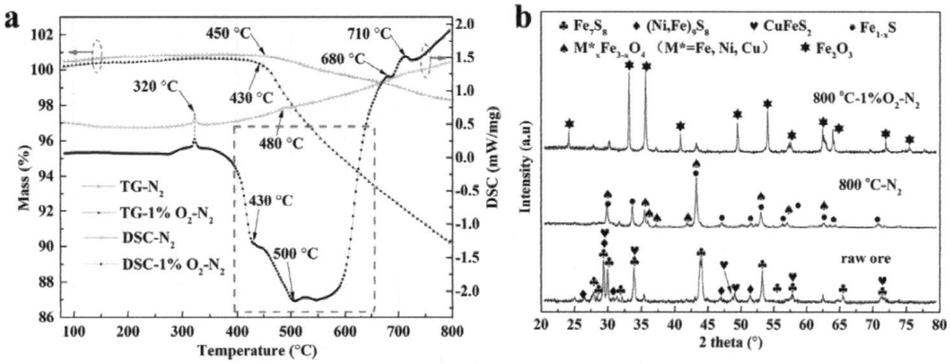

Figure 2. a) TG-DSC curves of the nickel mineral at 10 °C/min in the different atmosphere condition (1% O_2–N_2 and pure N_2 atmosphere); b) the XRD patterns of the residuals of TG-DSC measured samples compared with the raw ore powder sample.

3.3. The phase transition process of main phases

In order to investigate the phase transition process of the complex ore in detail, the in-situ XRD was used to study the reactions during the high temperature under different atmospheres. Figure 3a shows that pentlandite was stable below 450 °C, then turned into a monosulfide solid solution ($(Ni_{0.8}, Fe_{0.2})_{0.96}S$) at 500 °C, and finally decomposed into monosulfide (NiS), which was confirmed by Zhu et al [3]. However, chalcopyrite was more stable than pentlandite, and partially decomposed into bornite (Cu_5FeS_4) beyond 600 °C under pure N_2 atmosphere, as shown in figure 3b. The monoclinic pyrrhotite (Fe_7S_8) transformed into hexagonal structure from 350 to 400 °C, and then grew into $Fe_{1-x}S$ (hexagon) beyond 400 °C in pure N_2 atmosphere. The changes of the pyrrhotite diffraction peak and interplanar spacing with the changes of temperature were further investigated in figure 4. It is obvious that the variations of different interplanar spacing (Δd_i) were non-negative values, besides there were two maximum values at 300 °C and 400 °C, which results from the crystal lattice volume expansion and structure transition of iron-sulfide phase with the rise in temperature during pure N_2 atmosphere. And the maximum value at 300 °C was reported due to the magnetic transition which accompanied by a structural transformation from the monoclinic structure to hexagonal structure by Anthony V. Powell [18]. From the transition of pentlandite and chalcopyrite, it is possible to suggest that the crystal lattice volume expansion of pyrrhotite at 400 °C may be due to the monosulfide solid solution (Ni-Cu-Fe-S) formed by the migration of nickel and copper atoms.

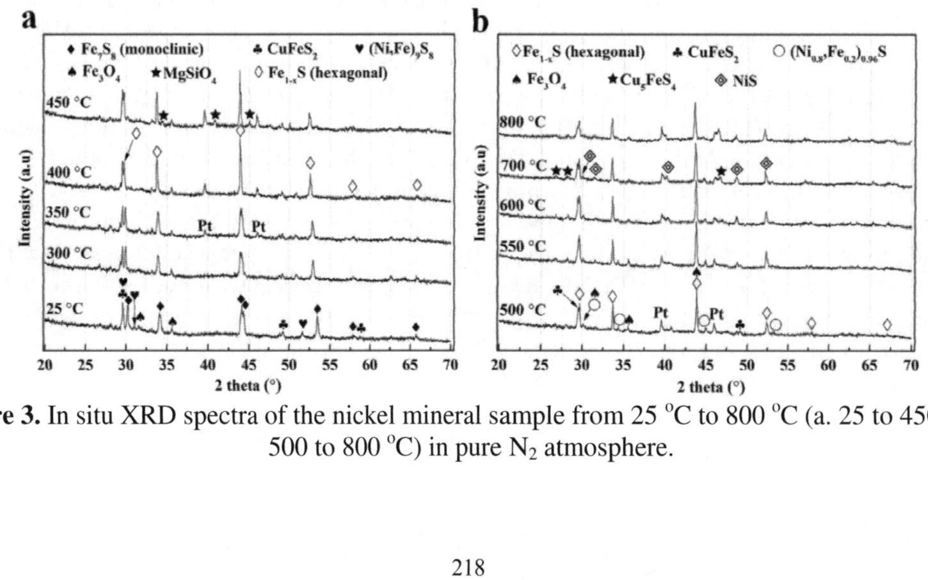

Figure 3. In situ XRD spectra of the nickel mineral sample from 25 °C to 800 °C (a. 25 to 450 °C, b. 500 to 800 °C) in pure N_2 atmosphere.

Figure 5a revealed that the structural transformation of pyrrhotite also appeared at around 350 °C, however the magnetic ferrite increased with the disappearance of the pentlandite and chalcopyrite at temperatures above 450 °C under the 1% O_2–N_2 atmosphere. As shown in figure 5b the hematite was formed during the oxidation of pentlandite and chalcopyrite. As was reported by previous' work, nickel and copper atoms migrated to the crystal lattice of magnetic ferrite and transformed into solid solution lastly [5, 6, 19, 20].

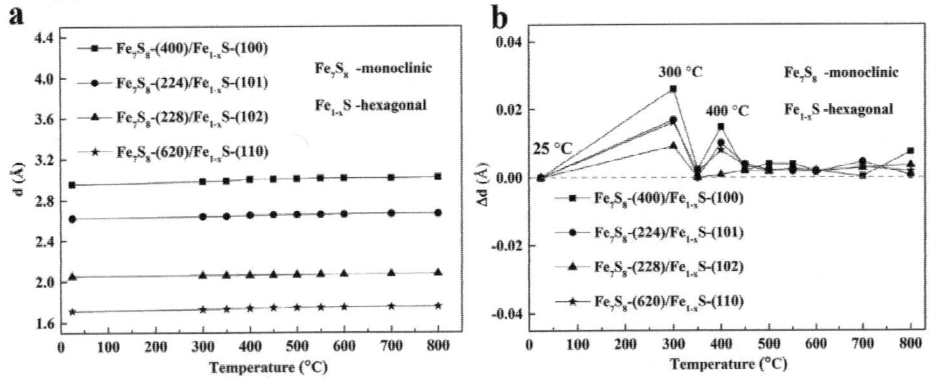

Figure 4. The calculating crystalline interplanar spacing (a) and lattice space variation gradient (b) of the Fe-S phase as a function of temperature in the pure N_2 atmosphere.

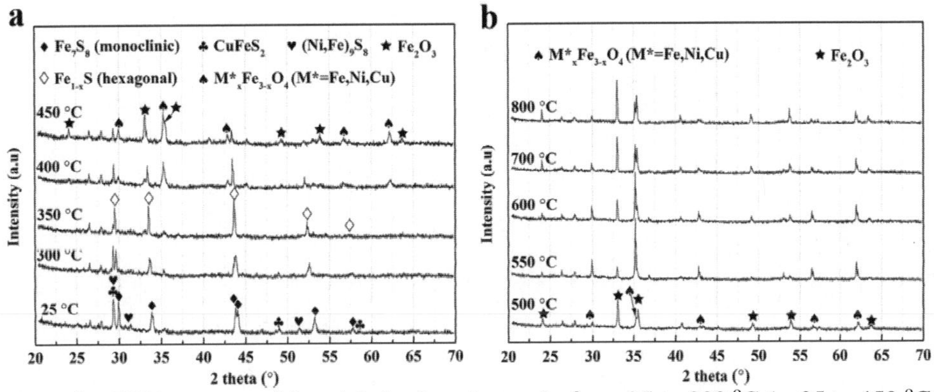

Figure 5. In situ XRD spectra of the nickel mineral sample from 25 to 800 °C (a. 25 to 450 °C, b. 500 to 800 °C) in pure 1% O_2–N_2 atmosphere.

4. Conclusion

The mineralogy properties of the complex sulfide ore were further investigated by means of XRF, ICP, SEM-EDS and XRD. The thermal stability tested by TG-DSC suggests that the ore sample is more active by micro-aerobic environment than that under pure N_2 atmosphere as the temperature rises. A further study of the phase transition was performed using the in-situ XRD in different environment. It is demonstrated that the structural transformation from the monoclinic to a hexagonal structure appeared in pyrrihotite phase with temperatures over 300 °C, while the pentlandite and chalcopyrite turned into monosulfide solid solution with temperature increasing under pure N_2 atmosphere. In 1% O_2–N_2 environment, the pentlandite and chalcopyrite transformed into magnetic ferrite solid solution ($M^*_xFe_{3-x}O_4$, M*=Fe/Ni/Cu) and Fe_2O_3, as a result of the migration of nickel and copper. Meanwhile, the pyrrihotite was gradually oxidized into Fe_2O_3 by temperatures above 400 °C. From these results, it is possible to illustrate the phase transition process of the complex sulfide nickel ore and it may be a fundamental one for the new process of nonferrous metallurgy in controllable products with gentle conditions.

Acknowledgments
This work was financially supported by the Grant 2014CB643403 from the National Key Program for Basic Research of 973 Program.

References

[1] Rosenqvist T 1978 *METALL MATER TRANS.* B 9 337-51.
[2] Dunn J G and Kelly C E 1980 *J THERM ANAL CALORIM.* 18 147-54.
[3] Zhu H H, Chen J, Deng J X, Yu R B and Xing X R 2012 *METALL MATER TRANS.* B 43 494-502.
[4] Zhu H H, Deng J X, Chen J, Yu R B and Xing X R 2014 *J MATER CHEM.* A 2 3008-14.
[5] Yu D W, Zhu M Q, Utigard T A and Barati M 2013 *MINER ENG* . 54 32-8.
[6] Yu D W and Utigard T A 2012 *THERMOCHIM ACTA.* 533 56-65.
[7] Conard B Sridhar R and Warner J 1980 *J CHEM THERMODYN.* 12 817-33.
[8] Bayer G and Wiedemann H G 1992 *THERMOCHIM ACTA.* 198 303-12.
[9] Prasad S and Pandey B 1998 *MINER ENG.* 11 763-81.
[10] Cahill C L Benning L G Barnes H L and Parise J B 2000 *CHEM GEOL.* 167 53-63.
[11] Majuste D Ciminelli V S T Eng P J and Osseo-Asare K 2013 *HYDROMETALLURGY.* 131-132 54-66.
[12] Pienack N and Bensch W 2011 *Angew Chem Int Ed Engl.* 50 2014-34.
[13] Walton R I and O'Hare D 2000 *CHEM COMMUN.* 23 2283-91.
[14] Xia F O'Neill B Ngothai Y Peak J Tenailleau C Etschmann B Qian G Brugger J Studer A Olsen S and Pring A 2010 *J APPL CRYSTALLOGR.* 43 511-19.
[15] Webster N A S Madsen I C Loan M J Knott R B Naim F Wallwork K S and Kimpton J 2010 *J APPL CRYSTALLOGR* 43 466-72.
[16] Webster N A S Pownceby M I Madsen I C Studer A J and Kimpton J A 2014 *POWDER DIFFR.* 29 S54-8.
[17] Webster N A S Pownceby M I Madsen I C Studer A J Manuel J R and Kimpton J A 2014 *METALL MATER TRANS.* B 45 2097-105.
[18] Powell A V Vaqueiro P Knight K S Chapon L C Sánchez R D 2004 *PHYS REV* B. 70 0144-15.
[19] Siriwardane R V Poston Jr J A Fisher E P Shen M S and Miltz A L 1999 *APPL SURF SCI.* 152 219-36.
[20] Yu D W Utigard T A and Barati M 2013 *METALL MATER TRANS.* B 45 653-61.

IOP Conf. Series: Materials Science and Engineering **191** (2017) 012038 doi:10.1088/1757-899X/191/1/012038

The effects of different bending techniques on corrosion resistance and nickel release of superelastic orthodontic NiTi archwires

N Rujeerapaiboon[1], N Anuwongnukroh [1], S Dechkunakorn[1] and M Jariyaboon[2]

[1]Orthodontic Department, Faculty of Dentistry, Mahidol University, Bangkok, 10400, Thailand
[2]Chemistry Department, Faculty of Science, Mahidol University, Nakhon Pathom, 73170, Thailand

E-mail: kiko_pink@hotmail.com

Abstract. Bending superelastic NiTi archwire is indicated in some stages of orthodontic treatment. The difference in bending techniques may affect corrosion resistance and nickel release. The purpose of this study was to investigate the corrosion resistance and nickel release after different bending techniques of NiTi archwires. Preform-curved NiTi archwires were used as a template for bending and used as a control group. 0.016×0.022 inches superelastic NiTi archwires were bent to curve-shape by cold bending, DERHT bending and cold bending then DERHT technique. Potentiodynamic polarization technique was used to measure corrosion behavior of the wires. Corrosion potential (E_{CORR}), corrosion density (I_{CORR}), and breakdown potential of each wire were determined. In addition, the amount of nickel release in the solution after the test was inductively coupled plasma mass spectrometry (ICP-MS). Although, the results showed that E_{CORR} and I_{CORR} were not statistically significantly different among all groups, the difference in breakdown potential and nickel release were observed. Similar corrosion resistance and nickel release were presented in the preform-curved NiTi archwires, cold bending, and cold bending then DERHT group. The DERHT bending group showed the lowest breakdown potential and highest nickel release.

1. *Introduction*

Superelastic nickel titanium (NiTi) archwires are widely used in clinical orthodontics because of their superelasticity properties, giving continuous and light force transmitted to the dentition over a long period of activation. The constant stress allows the design of orthodontic archwires to communicate through the brackets for leveling and alignment in the stages of tooth movement [1].

The best technique of bending NiTi archwires has not been clarified. There are two ways to alter the shape of the archwires. Heat treatment, for example direct electrical resistance heat treatment (DERHT), allows shape setting of superelastic wires. Without DERHT, NiTi archwires can be shaped by mechanical force (cold bending) until it reaches to permanent deformation. Both techniques have been recommended for using in orthodontic practice [2, 3].

Heat treatment at 500 °C ranged from minutes to an hour can alter the surface structure of NiTi archwires. Therefore, after heat treatment, the oxide layer in the form of TiO_2 will increase. These titanium oxides are important for protection of wire surface and increase the corrosion resistance. This indicates that heat treatment may produce protective surface and favor of good biocompatibility of NiTi archwires [4, 5].

On the other hands, mechanical bending can affect the surface of NiTi archwires. The influence of surface roughness might be a reason for varying corrosion tendency. The corrosion resistance is higher in the smooth surface of the NiTi archwires [6, 7].

During orthodontic treatment, orthodontic metallic appliances are prone to corrosion or degradation in the oral environment. This degradation of metal has been demonstrated to adversely affect biocompatibility, which considered a concern for clinicians due to the long period of orthodontic treatment. The Corrosion of NiTi archwires raises a major concern because of their high nickel content (47% to 50% nickel) [8, 9]. These individualized reports have indicated that insertion of NiTi alloys may occasionally lead to formation of rashes, swelling, and painful erythematous lesions in oral and labial mucosa [10].

Unfortunately, there is not much study about different modes of wire bending on corrosion resistance and nickel release. Thus the purpose of this study was to investigate the corrosion resistance and nickel release after different bending techniques of superelastic orthodontic NiTi archwires.

2. Material and methods

2.1. Sample preparation

0.016×0.022 inch superelastic NiTi archwires (Highland, USA) were used in this study and divided into 4 groups with different bending techniques as illustrated in Table 1.

Table 1. Wire bending techniques

Group	Condition	Performing methods
1	As-received	Preform-curved NiTi archwires
2	Retrieve	Cold bending by mechanical force
3	Retrieve	DERHT bending using Bender soarer-X at heat level 4
4	Retrieve	Cold bending then DERHT using Bender soarer-X at heat level 4 for 3 seconds

2.1.1. Group 1. 0.016×0.022 inch NiTi preform-curved NiTi archwire (Highland, USA) was used as a control. The template was made from preform-curved NiTi archwire with acrylic resin.

2.1.2. Group 2. 0.016×0.022 inch NiTi plain archwire was bent by mechanical force from thumb and index finger. The archwire was curved until the shape was similar to preform-curved NiTi archwires (group 1) by using the template as a guide. Permanent deformation occurred after bending.

2.1.3. Group 3. 0.016×0.022 inch NiTi plain archwire was bent by using DERHT (Bender soarer-X equipment) at level 4 (Figure 1). The pliers from DERHT were used to sweep the archwires. The archwire was permanently curved until the shape was similar to preform-curved NiTi archwires (group 1) by using the template as a guide.

2.1.4. Group 4. 0.016×0.022 inch NiTi plain archwire was bent by mechanical force from thumb and index finger. The archwire was curved until the shape was similar to preform-curved NiTi archwires (group 1) by using the template as a guide. Once the permanent deformation achieved, the wire was DERHT (Bender soarer-X equipment) at level 4 for 3 seconds.

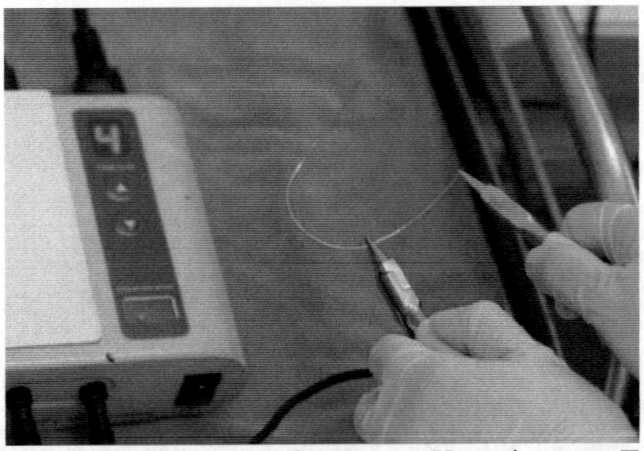

Figure 1. DERHT bending using Bender soarer-X equipment, Tomy Incor, Japan

2.2. Surface morphology examination
All groups were examined surface morphology with scanning electron microscope & energy-dispersive X-ray microanalysis type (SEM-EDS, JSM5410LV, JEOL LTD) at 400x and 800x magnification and operating voltage at 20 kV.

2.3. Potentiodynamic polarization testing
The curved ends of the archwires were cut to a length of 30 mm. All samples were cleaned with ethanol in an ultrasonic bath for 5 minutes and dried for 20 seconds. The corrosion behaviors were examined by measuring potentiodynamic polarization measurements (Autolab PGSTAT302N - High Performance, The Netherlands). Modified Fusayama artificial saliva with pH 5.3 was used as the electrolyte. The compositions of saliva were listed in Table 2. Reference and counter electrodes were Ag/AgCl and Pt, respectively.

Table 2. Composition of Modified Fusayama artificial saliva

Compound	Concentration (g/L)
KCL	0.4
NaCl	0.4
$CaCl_2.2H_2O$	0.906
$Na_2H_2PO_4.2H_2O$	0.690
$Na_2S.9H_2O$	0.005
Urea	1

The wire samples were immersed in the electrolyte and were maintained at 37°C. Open circuit potentials (OCP) were run for 30 minutes. After that, anodic polarizations were run at a scan rate of 10 mV/min. Corrosion potential (E_{CORR}), corrosion density (I_{CORR}) and breakdown potentials of each group were determined. The experiments were performed 4 ties for each group.

2.4. Nickel release testing
The amount of nickel release in solution after potentiodynamic polarization test was determined by inductively coupled plasma mass spectrometry (ICP-MS, ELAN-6000) at Faculty of Science, Mahidol University.

2.5. Statistical analysis
E_{CORR}, I_{CORR} and breakdown potentials were analyzed by Kruskal-Wallis Test and *a Mann-Whitney U*

Test for the post hoc test at 95 % confidence interval (α = 0.05). All statistical analyses were performed with IBM SPSS statistics 20 software.

3. Results

3.1. Surface morphology examination

Scanning electron micrographs of each group at 400x and 800x magnification were presented in Figure 2 and 3, respectively

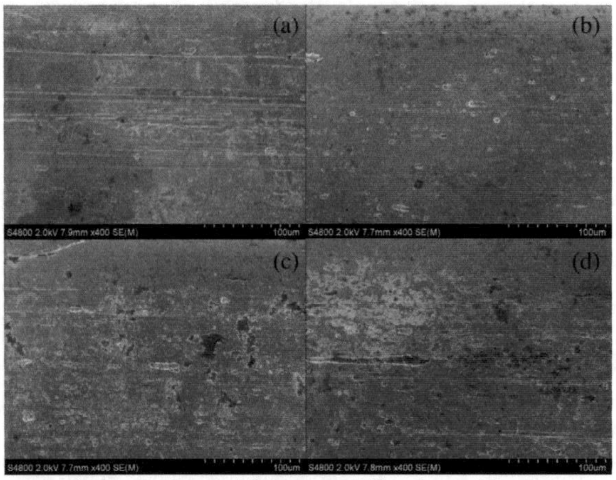

Figure 2. Scanning electron micrographs at 400x magnification: (a) group 1, (b) group 2, (c) group 3, and (d) group 4.

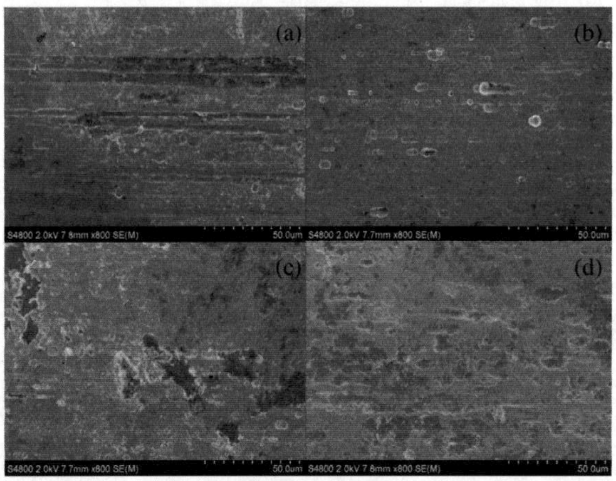

Figure 3. Scanning electron micrographs at 800x magnification: (a) group 1, (b) group 2, (c) group 3, and (d) group 4.

3.2. Potentiodynamic polarization testing

Polarization curves were shown in Figure 4. The passive current density of group 4 was lower than that of others. (It should be noted that the experiment was assumed surface area as 1 cm^2, however actual exposed area was smaller approximately of 0.11 cm^2.)

The means and standard deviations for E_{CORR}, I_{CORR}, and breakdown potentials for each group were shown in Table 3. E_{CORR} and I_{CORR} were similar for all groups. Breakdown potential of group 3 was statistically significantly less than that of the others (0.76±0.10 V, (Ag/AgCl)) ($P<0.005$, α = 0.05). There were no statistically significantly different between group 1 and group 2 as shown in Table 4

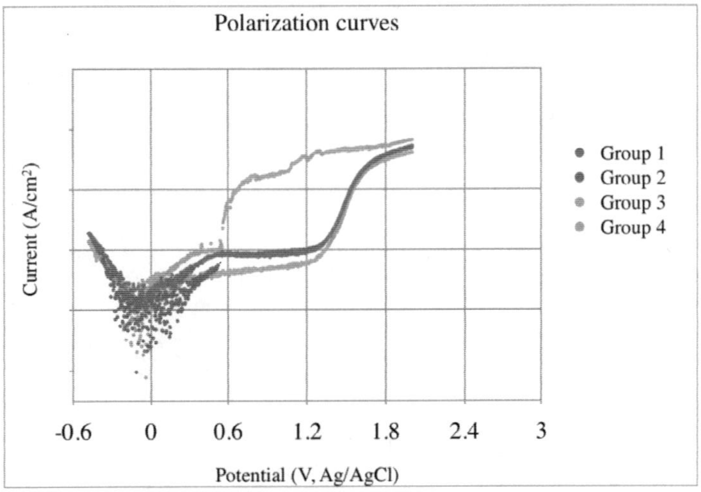

Figure 4. Polarization curves of all groups

Table 3. E_{CORR}, I_{CORR}, and breakdown potentials for each group

Group	E_{CORR} (V, Ag/AgCl)	I_{CORR} (A/cm^2)	Breakdown potential (V, Ag/AgCl)
1	-0.09±0.02	$1.59x10^{-7}$±8.5 $x10^{-10}$	1.33±0.00
2	-0.15±0.02	1.30 $x10^{-7}$±5.4 $x10^{-9}$	1.35±0.01
3	-0.16±0.01	1.38 $x10^{-7}$±3.05 $x10^{-9}$	0.76±0.10
4	-0.09±0.01	1.18 $x10^{-7}$±4.3 $x10^{-9}$	1.28±0.00

Table 4. Significantly different among four groups of breakdown potentials

Group	1	2	3	4
1	-	NS	*	*
2	NS	-	*	*
3	*	*	-	*
4	*	*	*	-

*Significantly different at P<0.005 and α = 0.05.

3.3 Nickel release testing

In Figure 5, the amount of nickel release in each group was shown. The amount of nickel release of group 3 was higher than that of others (269±11.0 μ g/L). There were similarities in the amount of nickel in group 1, groups 2 and group 4. The mean nickel release was presented in Table 5.

IOP Conf. Series: Materials Science and Engineering **191** (2017) 012038 doi:10.1088/1757-899X/191/1/012038

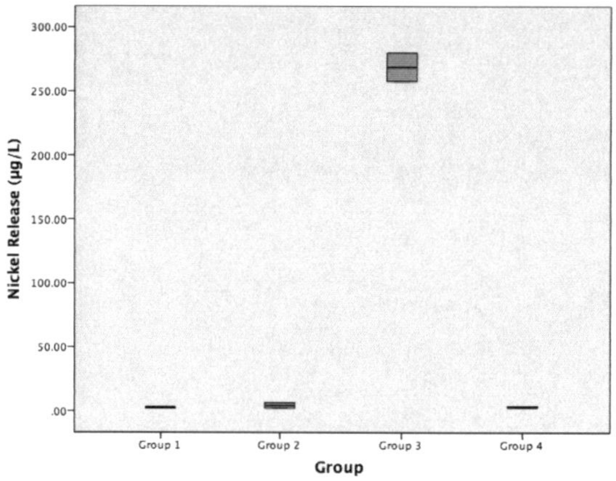

Figure 5. The amount of nickel release in all groups

Table 5. The amount of nickel release

Group	Mean amount of Nickel (μ g/L)	SD
1	2.65	0.65
2	4.35	2.25
3	269.00	11.0
4	3.45	0.35

4. Discussion

As a chair-side procedure, the manipulation of NiTi archwires can be achieved mechanically through cold bending and forming from the heat sources [2, 3].

In orthodontics, DERHT is one of the methods to alter NiTi mechanical properties. DERHT transformed electric current to high temperature and induced restructuring of the lattice. However, the temperature received on the archwire was not as high as heat treatment by furnace in air or salt bath in the laboratory [11]. In our study, the temperature received on the archwire in DERHT bending and cold bending then DERHT group was 42.83±1.96° C and 210.33±6.96°C respectively.

The temperature used as a chair side heat treatment in orthodontics treatment should not reach 500°C. Considering the mechanical properties, heat treatment will increase hardness and unloading force of the wire. Moreover, transitional temperature range (TTR) will decrease. Thus, NiTi archwire will increase its superelasticity. However, at a very high temperature, heat treatment will deteriorate these mechanical properties [5].

From SEM, the surface of DERHT bending group was rougher than that others. In Figure 3, DERHT bending group showed deep and wide spread area of corrosion as a result of the process of wire bending. The DERHT-pliers (Figure 1) were used to sweep the archwire until the shape was set. During sweeping, the beak of the pliers could scratch the wire surface. On the other hands, thumb and index fingers were used to sweep the archwire in cold bending and cold bending then DERHT group. This resulted in smoother surface than DERHT bending group and was similar to preform-curved NiTi archwire.

The breakdown potential in our study ranged from 0.76-1.35 V (Ag/AgCl). The breakdown potential of preform-curved NiTi archwire was 1.33 V (Ag/AgCl). This was not significantly different in cold bending group (1.35 V (Ag/AgCl)). Iljima previously studied the corrosion resistance of NiTi archwire. The breakdown potential of as-received wire was similar to our study [12].

Heat treatment was applied in 2 groups: DERHT bending group and cold bending then DERHT group. The heat treatment had an effect on corrosion behavior by affecting on surface structure. The oxidization on NiTi surface depended on the temperature of the heat treatment. The increase in temperature would increase in protective oxide layer [13, 14].

The passive current density was a measure of a protective oxide film. The more negative passive current density, the more oxide film was presented. However, once the passive current density was more positive, this oxide layer was breakdown. Thus corrosion occurred [15]. In our study, cold bending then DERHT group received highest temperature in heat treatment. From polarization curve (Figure 4), cold bending then DERHT group had lowest passive current density. This implied that this group had greater protective oxide layer.

Recent studies from Shabalovskaya and Firstov found that the temperature that suitable for the formation of titanium oxide layer (TiO_2) was 500°C [4, 9]. At this temperature, it would produce a smooth protective oxide surface, nickel-free and favor of good biocompatibility of NiTi [4, 5]. The increase in temperature would increase oxide layer leading to greater corrosion resistance. This oxide layer serves two purposes. Firstly, it increases the surface layer stability and protects the surface from corrosion. Secondly, it creates a physical and chemical barrier against nickel oxidation by modifying the oxidation pathways of the nickel [13].

The amounts of nickel release of preform-curved NiTi archwire, cold bending, and cold bending then DERHT group were similar (2.65, 4.35, and 3.45 μ g/L respectively). Hwang studied Ni release from simulated fixed orthodontic appliance and found nickel release of 17.03 μ g/L, which was greater than our study. The amount of nickel release increased in the first 3 months after that nickel release decreased or stopped after the certain time [16]. However, the result of nickel release in this study cannot be compared with that of Hwang because of different methodology.

In our study, DERHT bending group presented the lowest breakdown potential (0.76±0.10 V (Ag/AgCl)) and highest amount of nickel release (269±11.0 μ g/L). This may be because of the process of bending the archwire by using DERHT-pliers, which could damage the surface of archwire, resulting in surface roughness. The increase in surface roughness will decrease in corrosion resistance [6, 7]. Moreover, the temperature received on wire during DERHT bending was relatively low. Once the heat treatment is below 300°C, the surface of NiTi will be modified in a different way. Titanium and nickel-enriched will be formed at the surface. The coexisting NiTi archwire will invite galvanic corrosion. The enriched matrix will serve as a reservoir for potential nickel release, especially when the surface is damaged, for example by mechanical bend of the archwire [9].

Orthodontic appliances unlike to medical uses of nickel alloys because they are not implanted; rather, they are placed in the oral environment and exposed to oral fluid. Many researches have recognized the potential biological implications of nickel release, focusing on the rate of corrosive products of alloys used in orthodontic treatment. The increase in corrosion rate might be at risk for patient especially patient with nickel hypersensitivity [10, 17].

5. Conclusion

The results showed that similar corrosion resistance and nickel release were presented in the preform-curved, cold bending, and cold bending then DERHT NiTi archwires. The DERHT bending archwires showed the lowest breakdown potential and the highest nickel release. The cold bending then DERHT archwires showed the lowest passive current density and the greatest protective oxide layer. Therefore, it is recommended that the cold bending then DERHT technique be used to bend a superelastic orthodontic NiTi archwire.

Reference

[1] Liu JK, Lee TM, Liu IH 2011 *Am J Orthod Dentofacial Orthop*. 140: p 166
[2] Iijima M, Ohno H, Kawashima I, Endo K, Mizoguchi I 2002 *Dent Mater*. 18: p 88

IOP Conf. Series: Materials Science and Engineering **191** (2017) 012038 doi:10.1088/1757-899X/191/1/012038

[3] Meling TR, Odegaard J 2001 *Am J Orthod Dentofacial Orthop*. 119: p 263
[4] Firstov GS, Vitchev RG, Kumar H, Blanpain B, Van Humbeeck J 2002 *Biomaterials*. 23: p 4863
[5] Kiattiwongse W KA, Santiwong P 2008 *Advanced Materials Research*. 55-57: p 249
[6] Hunt NP, Cunningham SJ, Golden CG, Sheriff M 1999 *Angle Orthod*. 69: p 433
[7] Widu F, Drescher D, Junker R, Bourauel C 1999 *J Mater Sci Mater Med*. 10: p 275
[8] Krishnan M, Seema S, Kumar AV, Varthini NP, Sukumaran K, Pawar VR, Arora V 2014 *Angle Orthod*. 84: p 358
[9] Shabalovskaya S, Anderegg J, Van Humbeeck J 2008 *Acta Biomater*. 4: p 447
[10] Locci P, Lilli C, Marinucci L, Calvitti M, Belcastro S, Bellocchio S, Staffolani N, Guerra M, Becchetti E 2000 *J Biomed Mater Res*. 53: p 560
[11] Eliades T, Athanasiou AE 2002 *Angle Orthod*. 72: p 222
[12] Iijima M, Endo K, Ohno H, Yonekura Y, Mizoguchi I 2001 *Dent Mater J*. 20: p 103
[13] Cisse O, Savadogo O, Wu M, Yahia L 2002 *J Biomed Mater Res*. 61: p 339
[14] Gu YW, Tay BY, Lim CS, Yong MS 2005 *Applied Surface Science*. 252: p 2038
[15] Hofer AM, Mori G, Fian A, Winkler J, Mitterer C 2016 *Thin Solid Films*. 599: p 1
[16] Hwang CJ, Shin JS, Cha JY 2001 *Am J Orthod Dentofacial Orthop*. 120: p 383
[17] House K, Sernetz F, Dymock D, Sandy JR, Ireland AJ 2008 *Am J Orthod Dentofacial Orthop*. 133: p 584

IOP Conf. Series: Materials Science and Engineering **191** (2017) 012039 doi:10.1088/1757-899X/191/1/012039

A study on the oxidation behavior of nickel alloys at elevated temperatures

Jung-Yeul Yun[1], Dahee Park[1], and Jei-Pil Wang[2]*

[1]Powder Technology Department, Korea Institute of Materials Science (KIMS), Changwon, Korea
[2]Department of Metallurgical Engineering, Pukyong National University, Busan, Korea

*E-mail: jpwang@pknu.ac.kr

Abstract. This study was conducted with the aim to investigate the high temperature oxidation behavior of Inconel 750 alloy, which has excellent high temperature corrosion resistance among Ni-based super alloys. In the study, the temperature was set at 700°C, 900°C and 1100°C in the atmospheric environment; the heat treatment holding time was increased to 1 hour, 6 hours and 12 hours at each temperature to investigate the weigh increase behavior at each condition; and the shape of the oxide layer formed on the surface and the distribution of the elements were analyzed by scanning electron microscopy (SEM) and energy dispersive X-ray spectroscopy (EDS).

1. Introduction

Ni-based superalloys have excellent corrosion resistance, oxidation resistance, strength and ductility and weldability at high temperature, so they are widely used to manufacture the parts requiring high temperature corrosion resistance and high temperature strength [1]. With the industrial development, materials are used in the gas atmosphere which is highly corrosive at high temperature such as coal gasification, petrochemical industry, or gas turbine. In this harsh atmosphere, Inconel 750 alloy among Ni-based super alloys is used [2].

Inconel is a heat-resistant Ni-based alloy containing 14-17% of chromium, 5-9% of iron, 2.5% of titanium, and less than 1% of aluminum, manganese, silicon and niobium. The Ni-based superalloy containing high Cr and Al forms a protective oxide film such as Cr2O3 and Al2O3 to maintain corrosion resistance even at high temperature [3, 4]. It also has excellent heat resistance, has oxidation resistance in high temperature oxidation stream, and is not immersed in an atmosphere containing sulfur. Its tensile strength, yield point and other properties do not change significantly until the temperature reaches about 600°C, indicating excellent mechanical properties. In addition, it is also resistant to organic and salt solutions. The inconel alloys are largely classified into four types: martensitic, ferritic, austenitic and precipitation strengthening alloy depending on the structure [5].

The Inconel 750 alloy, the subject of this study, is an age-hardening alloy with excellent corrosion resistance and oxidation resistance of Ni + Co> 70%, Cr 14 ~ 17% and Fe 5 ~ 9%. With high tensile strength and returning ability at high temperature, it is widely used for making springs, bolts and parts for turbine generator, plate spring, pressure tube, gas turbine, etc [5].

Although studies on the change of physical properties due to high temperature oxidation of Ni superalloy Inconel have conducted continuously, there have been few studies conducted on the recovery and recycling of the oxide scale generated by the oxidation of Inconel [6, 7, 8, 9, 10, 11]. The

Inconel is very high value metal among the metals used in the industry. Therefore, if the scraps from the inconel products can be recycled, it will yield very high returns. The recycling of Ni-based alloys is conducted mostly through wet process and there are very few studies [12].

This study, a base study on the possibility of dry recycling of the inconel among Ni-based alloys, investigated the high temperature oxidation behavior of Inconel 750 alloy in the atmosphere. When the Inconel 750 alloy was used at high temperature of 700°C, 900°C and 1100°C, the formation and deterioration process of the coating protecting the alloy base were observed according to the reaction with oxygen depending on the holding time. Scanning electron microscopy (SEM) and energy-dispersive X-ray spectroscopy (EDX) were used to observe the oxide scale formation behavior and the composition and structure of the oxide layer were investigated, was examined.

2. Materials

Table 1 below shows the chemical composition of general Inconel 750 and the material used in this test was a commercially-available Inconel 750 alloy (Figure 1). Inconel 750 alloy was cut to a size of $20 \times 30 \times 0.5$ (mm^3) and the cross-section in the thickness direction was observed using SEM.

Table 1. Chemical composition of Inconel 750 alloy

Inconel	Ni	Cr	Fe	Nb	Co	Mn	Cu	Al	Ti	Si	C	S
X-750	70.0	14.0-17.0	5.0-9.0	0.7-1.2	1.0	1.0	0.5	0.4-1.0	2.25-2.75	0.5	0.08	0.01

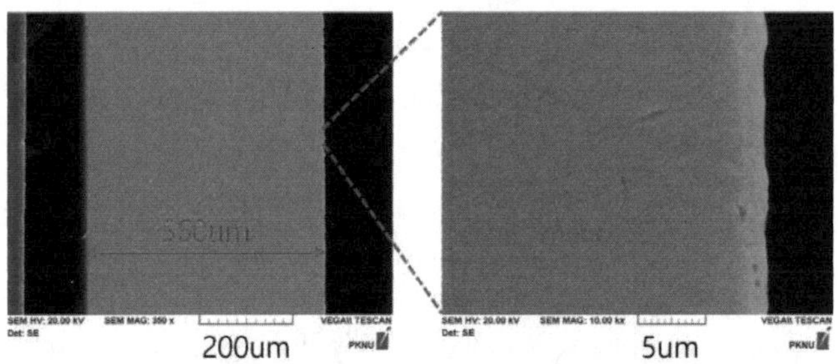

Figure 1. Cross section of Inconel 750 alloy.

Figure 2 is a schematic diagram of the electric furnace used for testing atmospheric high temperature oxidation of Inconel 750 alloy. The electric furnace was designed not to be blocked from the outside to maintain an air atmosphere of 1 atm. An alumina crucible was placed inside the electric furnace and Inconel 750 alloy was put into the crucible for oxidation test. The heater rod was connected to the outside of the furnace to maintain high temperature, and the test conditions of 700°C, 900°C and 1100°C were maintained by adjusting the voltage using a controller.

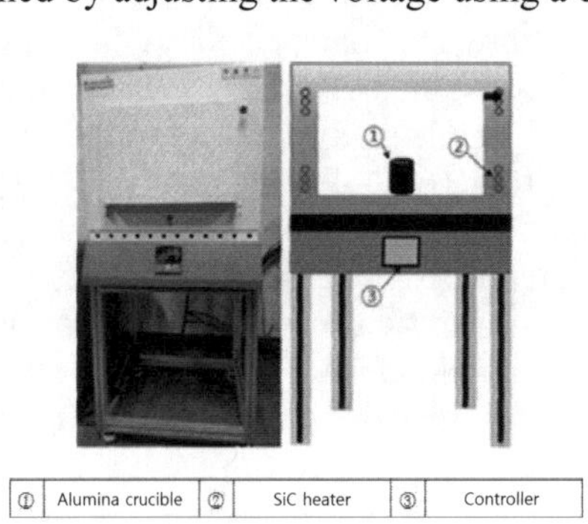

Figure 2. Schematic diagram of electric furnace used in this study.

Inconel 750 alloy was put into the electric furnace maintained at each test temperature, and the oxidation was conducted for 1 hour, 6 hours, and 12 hours. The weight increase due to oxidation was measured and the behavior of the formed oxide was observed and analyzed using scanning electron microscopy (SEM / EDS). In order to observe the thickness direction during scanning electron microscope observation, the sample was observed after fine mounting with a 0.1 μm diamond paste after cold mounting.

3. Results and discussion

Figure 3 shows the heat treatment temperature and the weight increase with time of Inconel 750 alloy. The temperature conditions were 700°C, 900°C and 1100°C, and the weight change at 1, 6 and 12 hours of holding time at each temperature were expressed. As the temperature and holding time increased, the weight increase became significant due to oxidation. When the holding time was 12 hours at 700°C, the weight increase was greater than that measured when the holding time was 6 hours at 1100°C, which indicates the oxidation behavior is greatly dependent on the holding time from 700°C to 1100°C rather than the temperature.

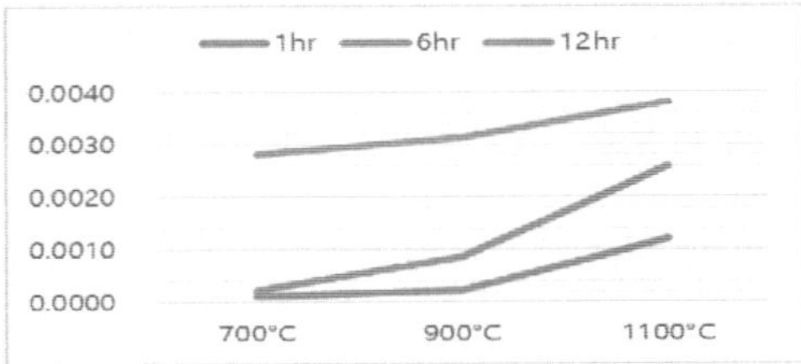

Figure 3. Changes in the weight increase of Inconel 750 alloy according to temperature and holding time.

Figure 4 shows the oxide layer shape and compositional change of the Inconel 750 alloy oxidized at 700°C for 12 hours through SEM / EDS. On the surface of cross-section, it was observed that the oxide layer was formed to a thickness of about 9 μm. Looking at the line profile and the EDS results of the oxide layer, the first oxide layer having a high Cr content was formed from the non-oxidized Ni base and then the second oxide layer having a high Ti content was formed above it. It was confirmed that Cr oxide and Ti oxide were formed on the surface in order to prevent oxidation at a high temperature of 700°C where the inconel alloy is mainly used.

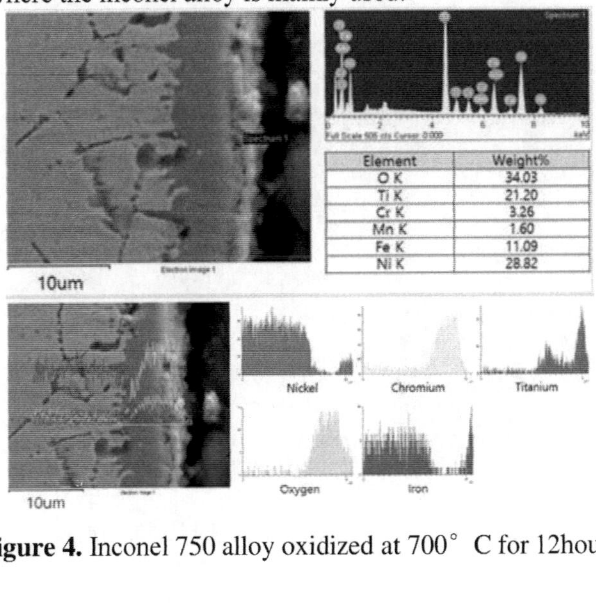

Figure 4. Inconel 750 alloy oxidized at 700° C for 12hours.

Figure 5 shows the oxide layer shape and compositional change of the Inconel 750 alloy oxidized at 900°C for 12 hours through SEM / EDS. It was confirmed that the oxide layer was formed with a thickness of about 5 μm on the surface of cross-section. Looking at the line profile and the EDS results of oxide layer, the Cr oxide layer exists from the Ni base but the Ti oxide layer is not remarkable, differently from the test results at 700°C. Ti oxide layer was peeled off and removed at high temperature, so it seems further researches are needed.

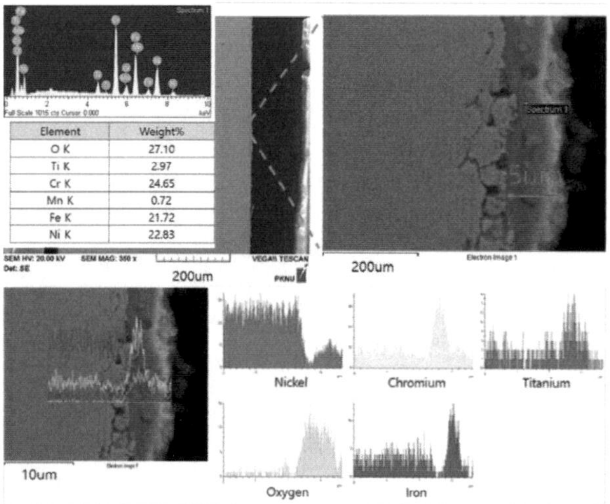

Figure 5. Inconel 750 alloy oxidized at 900° C for 12hours.

Figure 6 shows the oxide layer shape and compositional change of the Inconel 750 alloy oxidized at 1100°C for 12 hours through SEM / EDS. On the surface of the cross-section, oxide layer is formed to a thickness of about 14 μm. The line profile and the EDS results of the oxide layer are also similar to the test results at 900°C, showing that the Cr oxide layer exists but the Ti oxide layer almost does not exists. At a higher temperature of 900°C and above, the diffusion of Cr from the inside of the alloy to the surface gradually occurs and the Cr oxide layer which became thick has brittleness. This indicates that the oxide layer can be removed from the internal base.

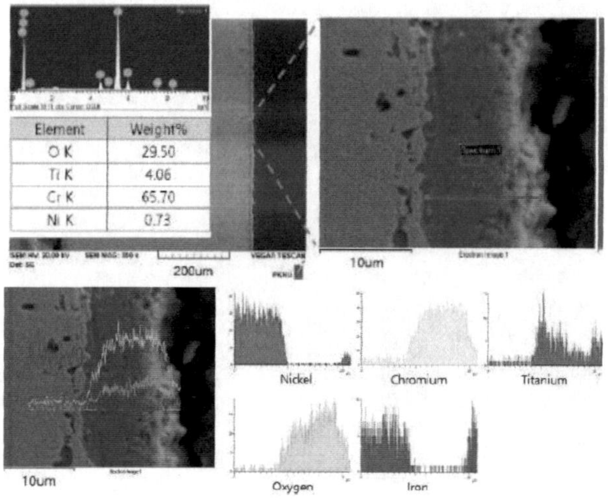

Figure 6. Inconel 750 alloy oxidized at 1100°C for 12hours.

4. Conclusions

In this study, heat treatment was conducted at 700°C, 900°C and 1100°C for 1 hour, 6 hours and 12 hours, respectively, to investigate the high temperature oxidation behavior of Inconel 750 alloy.

1. As oxidation temperature and holding time increased, the weight increased significantly. It was confirmed that the oxidation behavior was more remarkable according to the change of the holding time than of the temperature.
2. The high-temperature oxidation behavior of Inconel 750 alloy was observed through SEM / EDS. It was confirmed that Cr oxide layer and Ti oxide layer were formed from the Ni base to resist oxidation.
3. The Ti oxide layer was not formed at 900°C and 1100°C, and the thickness of the oxide layer was found to increase with temperature rise.
4. At a high temperature of 900°C and above, the Cr diffusion was gradually increased to the surface and the possibility of removing the Cr oxide layer with high brittleness formed on the surface was confirmed.

Acknowledgments

The work was supported by a grant from the Fundamental R&D Program for Core Technology of Materials funded by the Ministry of Trade, Industry & Energy, Republic of Korea.

References

[1] Lv P, Xiao S, Jie C, Zhang C, Liu X, and Guan Q 2017 Microstructure and high temperature oxidation resistance of nickel based alloy GH4169 irradiated by high current pulsed electron beam *Surf. Coatings Techno.* 309 401-409.
[2] Berthod P and Conrath E 2016 Creep and oxidation kinetics at 1100 °C of nickel-base alloys reinforced by hafnium carbides *Materials & Design* 104 27-36.
[3] Sitek R, Bolek T, Dobosz R, Plocinski T, and Mizera J 2016 Microstructure and oxidation resistance of aluminide layer produced on Inconel 100 nickel alloy by CVD method *Surf. Coatings Techno.* 304 584-591.
[4] Yang T, Huo Y, Liu Y, Rui Z, and Ji H 2017 Efficient formaldehyde oxidation over nickel hydroxide promoted Pt/γ-Al2O3 with a low Pt content *Applied Catalysis B: Environ.* 200 543-551.
[5] Xinhua C, Lei F, Huayi Y, Li L, Kaifa D, and Dihua W 2016 High-temperature oxidation behavior of Ni-11Fe-10Cu alloy: Growth of a protective oxide scale *Corro. Sci.* 112 54-62.
[6] Lee W S and Kim G M 1995 *J. Corr. Sci. of Kor.* 24(2) 124.
[7] Gullbransen E A and Wysong W S 1948 *TAIME* 175 628.
[8] Kim J C, Park Y K, Shin E K, Kim G M, and Choi S K 1997 *J. Corr. Sci. Soc. Korea* 26 245.
[9] Foley R T and Guare C J 1959 *J. Electrochem. Soc.* 106 (11) 936.
[10] Allen G C 1988 The oxidation of Inconel-690 alloy at 600 K in air *Appl. Surf. Sci.* 31(2) 220.
[11] Park S Y and Lee D B 2014 Oxidation of inconel 690 alloys at 800-1000°C in air *Appl. Mechan. Mater.* 607 7.
[12] Barbadillo J J 1983 *Phys. Metall. Mater. Sci.* 14(2) 329.

IOP Conf. Series: Materials Science and Engineering **191** (2017) 012040 doi:10.1088/1757-899X/191/1/012040

Improvement of the cavitation erosion resistance for Cr_3Si film on stainless steel by double cathode glow discharge

Hongqin Ding and Yujiang Qiu

School of Mechanical Engineering, Southeast University, Nanjing 211189, China

E-mail:dinghongqin2005@126.com

Abstract. In this study, sputter-deposited Cr_3Si film was prepared by double cathode glow discharge (DCGD) technique onto 304 stainless steel. The phase constituents, surface microstructure and chemical compositions of the film were examined by using X-ray diffraction (XRD), scanning electron microscopy (SEM) and transmission electron microscopy (TEM). After the DCGD process, the hardness of Cr_3Si film was 26 GPa, about 10 times of the stainless steel, 2.5 GPa. The cavitation erosion resistance of Cr_3Si film and stainless steel were investigated by using an ultrasonic vibration cavitation erosion system. After 30 hours of cavitation tests, the cumulative mass loss of Cr_3Si film was only 60% of the stainless steel. Compared with the untreated stainless steel, the cavitation erosion resistance of Cr_3Si film was improved. The cavitation mechanism of Cr_3Si film is due to the delamination and spalling of local surface layer derived from its inherent brittleness.

1. Introduction

Austenitic stainless steels are widely applied in many fields, such as fluid machinery, marine, hydraulic machines, machine tool spindle, and as well as in nuclear industries because of their excellent properties containing good mechanical properties, corrosion resistance and weldability, etc. In addition, austenitic stainless steels are also chosen as candidate materials in water lubricated bearings and seals [1-2]. However, the damage of cavitation erosion existed in these environments under the combined action of corrosion and mechanical wear, inevitably, which leads to their premature failure [3-6].

In the past decades, in order to enhance the cavitation erosion behavior of austenitic stainless steels, several surface modification techniques, including thermal spraying, laser surface modification, physical and chemical vapour deposition have been employed [7-10]. However, it is found that several problems still need to be solved though some active effects have been brought through the surface technologies. For example, TiN and CrN coatings are easier to generate microcracks and flake away from the tops of micro folds of the stainless steel matrix although they have good corrosion resistance and wear performance under cavitation erosion damage.

Among the transition metal silicides, Cr_3Si has excellent combination of wear, corrosion and oxidation resistance. In order to improve the cavitation erosion resistance of stainless steel, the Cr_3Si coating was prepared onto 304 stainless steel by DCGD technique. In view of this, the present work aims at evaluating the cavitation erosion resistance of Cr_3Si film, and attempts to reveal the cavitation mechanism of Cr_3Si film.

2. Experiment details

2.1. Test apparatus

Figure.1 (a) shows an ultrasonic vibration system to estimate the cavitation erosion behavior of specimens, Figure.1 (b) and Figure.1 (c) presents the schematic of this ultrasonic vibration system. During cavitation erosion tests, the beaker was filled with tap water, and the ultrasonic vibration cavitation erosion system has the amplitude of 6 μm and was ensured the frequency of 20 kHz with an axial vibration by the vibration horn to generate the cavitation bubbles at the tip surface. In addition, fresh water would replace the used water in the beaker in every half hour, and the temperature of the medium was kept at 20±5 ℃.

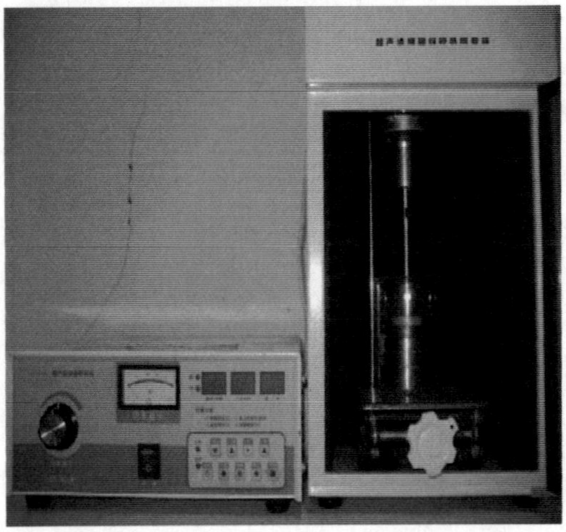

(a)

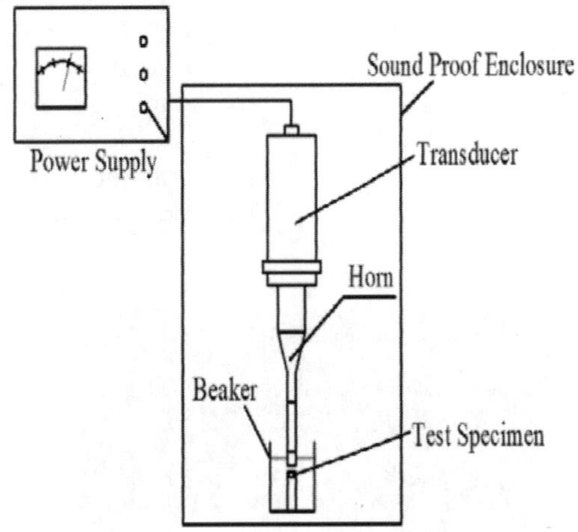

(b)

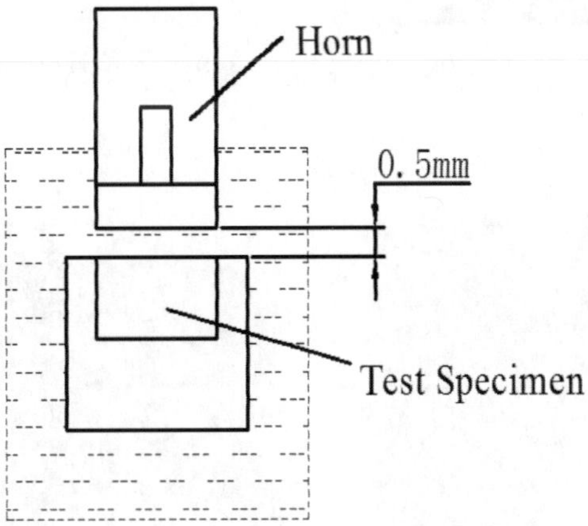

(c)

Figure 1. The ultrasound vibration system

2.2. Test samples

Austenitic stainless steel 304 was used as the substrate. The specimens (10×7×5mm) were polished with #5000 emery paper, and the surface roughness Ra < 0.2μm. Then, the Cr_3Si film was prepared onto 304 stainless steel by DCGD technique. Table 1 was the deposition parameters of Cr_3Si film employed in this work. Figure.2 shows the cross-sectional micrograph of Cr_3Si film with the thickness of 5 μm. Figure.3 shows a distinctive XRD pattern of the films by using a Smartlab (3) X-ray diffractometer (40kv and 30mA, Cu Kα radiation) with a step size 0.02° in the range of 20-80° and Cr_3Si phase can be seen from XRD pattern of the Cr_3Si film through a series of broad peak. Figure.4 (a) shows the TEM brighttield micrograph of the Cr_3Si film. Figure. 4 (b) shows HRTEM image of Cr_3Si film.

Table 1. Deposition parameters of Cr_3Si film.

Parameters	Cr_3Si
substrate temperature (℃)	800
treatment time (h)	3
target electrode bias voltage (V)	-900
substrate bias voltage (V)	-300
working pressure (Pa)	35
Ar flow rate (sccm)	60
parallel distance between the source electrode and the substrate (mm)	15

Figure 2. Cross-section SEM morphology of the as-deposited Cr_3Si film

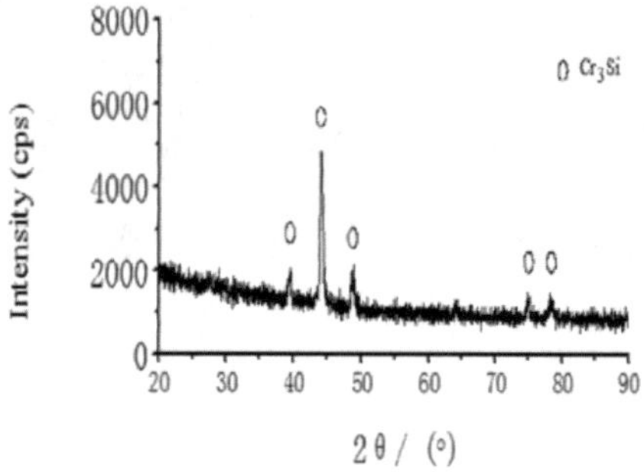

Figure 3. XRD pattern of the sputter-deposited Cr_3Si film

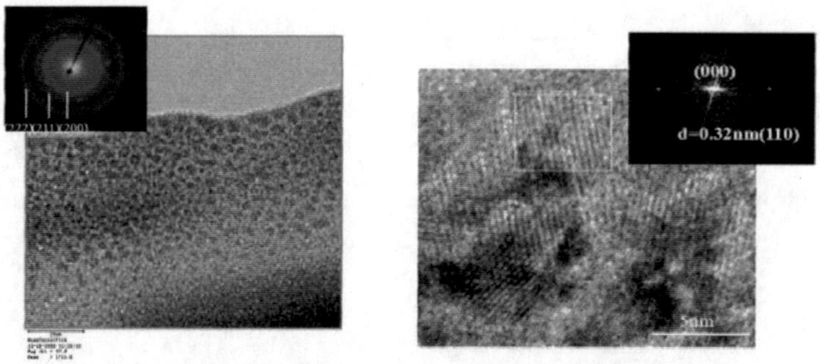

(a) (b)

Figure 4. TEM brighttield and HRTEM image

The nanohardness tester was used to measure the elastic modulus and hardness of the stainless steel and Cr_3Si film by using the method of Oliver and Pharr. Table 2 was the elastic modulus and nanohardness of stainless steel and Cr_3Si film.

Table 2. Properties of Cr_3Si film and substrate.

Properties	304 steel	Cr_3Si
film thickness (μm)	-	5
roughness (μm)	0.2	0.2
hardness (GPa)	2.5	26
elastic modulus (GPa)	193	300

2.3. Experiment procedures

Cavitation experiments were carried out with 30 hours in an ultrasonic vibration system. An ultrasonic cleaner (KQ118) was used to clean the cavitation samples with acetone, cavitation samples were weighed every two hours and dried in drying basin. The digital analysis balance was used in the mass changes of the samples at least three times and the precision of balance is 0.01 mg. The erosion test of each specimen is repeated at least three times to verify the result. The optical photographs were shown in Figure 5 for the specimens of 304 steel, Cr_3Si film after 30 hours experiment.

(a) (b)

Figure 5. Specimens after 30 hours of cavitation test, (a) 304 steel, (b) Cr_3Si film

3. Results and discussions

3.1. Mass loss

Figure 6(a) shows the mass loss of the stainless steel and Cr_3Si film specimens with the test time. Figure 6(b) shows the erosion rate of the specimens by curve fitting and derivation method of the mean mass loss. As shown in Figure 6, it is clear that distinct stages with four different erosion rates for the tested samples can be observed. As the erosion process continuing, weight loss of samples increases, accompanied by an increase in the number of erosion pits. The cumulative mass loss of the Cr_3Si film sample was 3.26 mg, which is only 60% of the stainless steel sample after 30 h of cavitation erosion experiment.

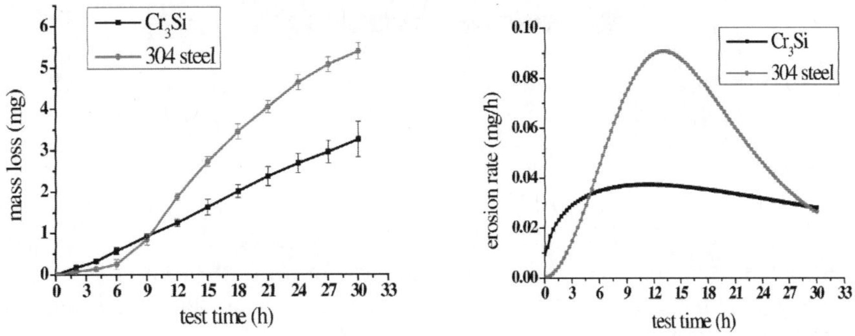

(a) (b)

Figure 6. The test curves of Cr_3Si film and 304 steel. (a) mass loss, and (b) erosion rate

3.2. SEM discussion of the erosion surfaces

Figure 7 shows the SEM images of the stainless steel and Figure 8 presents the Cr_3Si film after 30 h of cavitation erosion experiment. After the cavitation erosion experiment, a lot of erosion pits could be seen on the surface of the 304 specimen ranging from several microns to tens of microns in size in Fig.7 (a-c). At the beginning, plastic deformation of the surface occurs, causing to yield hardening, stress and fatigue cracks that result in the fatigue fracture of the stainless steel. Lots of slip band agglomerated and some cracks formed on the surface after their cavitation test as in the Ref. [11, 12]. As shown in Fig.8 (a-c), a few erosion pits with the diameters of tens of microns are found on the damaged area of the Cr_3Si film after 30 h test. The result indicates that the Cr_3Si film exhibit slight surface damage in the form of material loss and increased particles removal at grain regions of the film compared to the 304 stainless steel because of its higher hardness, wear, corrosion and chemical composition which effectively improves the cavitation erosion resistance. Hence, the resistance of the Cr_3Si coating is higher than the stainless steel. The cavitation erosion mechanism of Cr_3Si film is due to the silicon compound brittleness and the poor adhesion to the substrate.

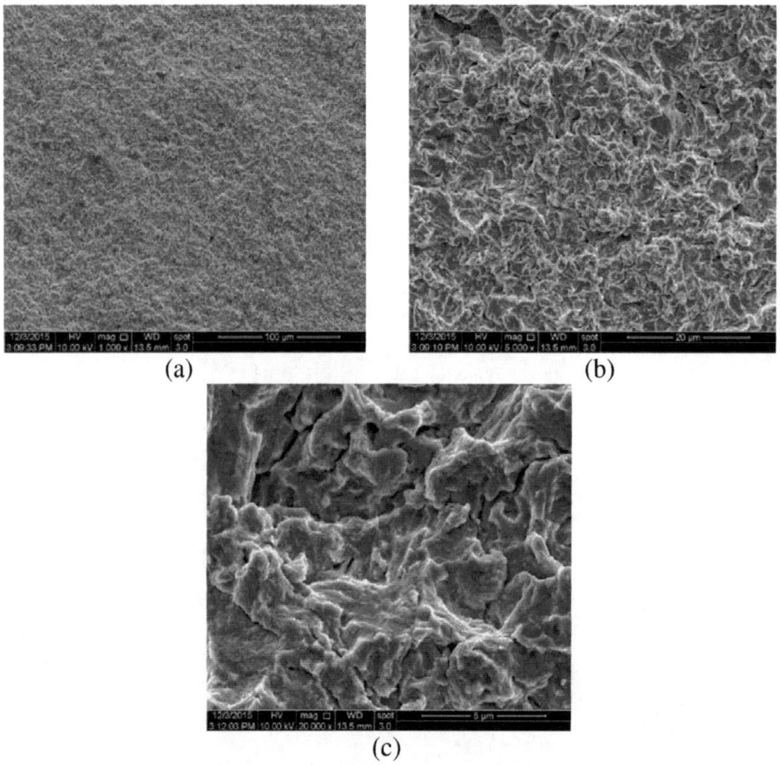

(a) (b) (c)

Figure 7. The SEM images of the specimen of 304 after 30 h test. (a) the damaged area (×1000), (b) the damaged area (×5000), (c) the damaged area (×20000).

4. Conclusions

In this work, the cumulative mass loss of the Cr_3Si film sample was only 60% of the stainless steel after 30 h of cavitation erosion experiment and Cr_3Si film has higher cavitation erosion résistance. The better adhesion force can improve cavitation erosion resistance of Cr_3Si film. The cavitation erosion mechanism of Cr_3Si coating could be attributed to not only the silicon compound brittleness but also the surface grain spalling with poor adhesion.

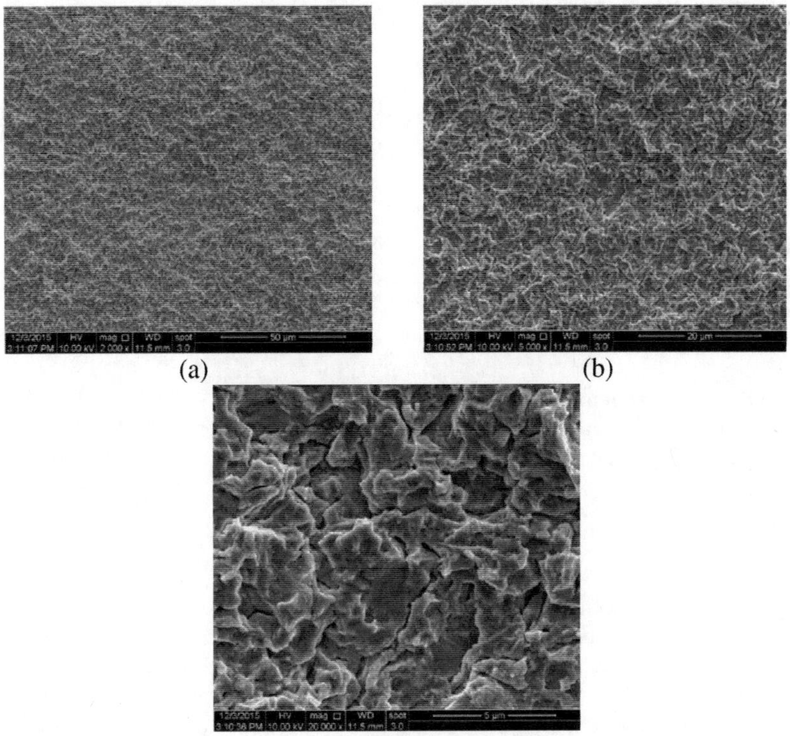

Figure 8. The SEM images of the specimen of Cr_3Si film after 30 h test. (a) the damaged area (×1000), (b) the damaged area (×5000), (c) the damaged area (×20000).

References

[1] Yoshimoto S, Oshima S, Danbara S and Shitara 2002 *ASME J.Tribol.*124 p 398-405.
[2] Yoshimoto S, Anno Y, Tamura M, Kakiuchi Y and Kimura K 1996 *ASME J.Tribol.*118 p 893-899.
[3] Abouel-Kasem A, El-Deen A E, Emara K M and Ahmed S M 2009 *ASME J.Tribol.*131 p 031605-1-031605-7.
[4] Abouel-Kasem A, Ahmed S M 2008 *ASME J. Tribol.*130 p 031601-1-031601-6.
[5] Abouel-Kasem A, Ahmed S M 2012 *ASME J. Tribol.*134 p 021702-1-021702-9.
[6] Karrab S A, Doheim M A, Aboraia M S and Ahmed S M 2012 *ASME J.Tribol.*134 p 011602-1-011602-6.
[7] A Krella, A Czyzniewski 2007 *Wear.* 263 p 395-401.
[8] A Krella, A Czyzniewski 2008 *Wear.* 265 p 963-970.
[9] A Krella, A Czyzniewski 2008 *Wear.* 265 p 72-80.
[10] A Krella, A Czyzniewski 2006 *Wear.* 260 p 1324-1332.
[11] Cheng F, Jiang S 2014 *Appl. Surf. Sci.* 292 p 16-26.
[12] Al-Hashem A, Caceres P G, Abdullah A and Shalaby H M 1998 *Corrosion.* 53 p 103-113.

Enhanced tensile, hardness and wear behaviors of pure aluminum matrix reinforced with steel chips via powder metallurgy technique

Mohamed Mahmoud Emara
Materials and Manufacturing Engineering Specialization, Manufacturing Engineering Department, Canadian International College (CIC), New Cairo, P.O. Box 59, 11835, Cairo, Egypt

mohamed_emara@cic-cairo.com, mmemara@aucegypt.edu

Abstract. The mechanical properties and wear behavior of aluminum matrix reinforced with steel machining chips was investigated. Pure aluminum was reinforced with 5, 7.5, and 10 wt% steel chips with an average size of 100 µm using powder metallurgy technique. Aluminum reinforced with 5 and 10 Wt.% SiC particles were manufactured for comparison. The investigation showed clear evidence that the addition of steel machining chips resulted in significantly low porosity levels in the aluminum matrix composites compared with the use of SiC as reinforcement. The mechanical properties (tensile and hardness) as well as the wear resistance were also observed to improve with the use of the steel machining chips as reinforcement. The results demonstrate the capability of steel machining chips to act as efficient reinforcing material and a reliable cost effective candidate in the development of aluminum matrix composites.

1. Introduction

Although aluminum reinforced with ceramic particulates enhances the tensile strength, still suffer from inadequate ductility [1, 2]. In order to achieve a higher strength and retain ductility of the composite, nano-sized ceramic particles are often used resulting in a metal matrix nano-composite [3, 4]. Many applications require components to be made with materials possessing light weight, high tensile and wear resistance as well as maintaining microstructure stability and high temperature resistance. For such applications requiring these highlighted complicated blend of properties, aluminum matrix composites have been found to be the most promising engineering material for selection. The development of aluminum matrix composites has relied on the use of ceramic reinforcements with silicon carbide and alumina the most commonly utilized due to their high hardness and wear resistances, refractory nature, relative availability and adequate cost

The choice of steel machining chips as reinforcement in the current research [5-7] was based on economic, environmental and metallurgical considerations. Steel machining chips is an industrial machine shops waste with very low cost implications. From a Mechanics point of view, the high deformation strains underlying chip formation causes chip microstructures to form an ultrafine grained microstructure which imparts significantly high strength to the chips with several higher magnitude than that of the bulk material [8-10]. Thus, it has the potential of improving the strength of a relatively softer matrix through load transfer from the matrix to the dispersed chips. At present, the use of steel machining chips as reinforcements in metal matrix composites has attracted very little interest. The present works evaluates the mechanical properties and wear behavior of aluminum matrix composites reinforced with steel machining chips via powder metallurgy technique.

2. Materials processing

The composite powders were fabricated from pure aluminum 99.7% with particle size average of 75μm. The steel chips used were discontinuous chippings from milling of medium carbon steel. Sieving of the chips was performed and chip size range of 100 μm was obtained to manufacture the composites. SiC powder with an average particle size of 30 μm was used to manufacture the control composites for good dispersion and mixing with the matrix material.

The powders were fabricated through mixing under argon atmosphere using a TURBULA Shaker for 1 hour at 96 rpm. This was followed by ball milling of the mixed powders under argon atmosphere in stainless steel jars containing 5 mm and 10 mm stainless steel hardened balls using the RETSCH 400MA high energy ball mill. The milling speed was maintained at 300 RPM. The ball-to-powder ratio (BPR) was 15:1, while the milling time was 3 hours. Consolidation of the milled powders were green compacted in a die under pressure of 475MPa for 1 hr at room temperature followed by sintering at 500°C for 1hr and hot extrusion at 4:1 extrusion ratio as shown in figure 1. The produced rods were air-cooled down to room temperature. The extruded rods were 10mm in diameter and approximately 100 mm long, as shown in figure 2.

Density was measured using Mettler Toledo densitometer. The hardness of the composites produced was evaluated on a Vickers hardness testing machine using 50 Kgf over dwell time of 15 seconds. Hardness of the pure aluminum and composites was measured on sections cut perpendicular to the extrusion direction. Hardness value measurements were carried out at room temperature. Five hardness indents were made on each specimen. The tensile properties of the composites were evaluated at room temperature using an Instron universal testing machine. The test was conducted at a strain rate of 0.1 mm/min. The testing procedure and basis for determination of the tensile properties were in accordance with the specifications of ASTM E8M-04 standard [11], as shown in 'figure 2'.

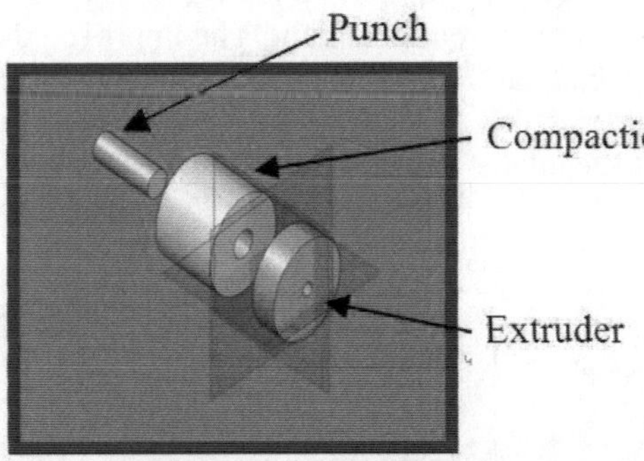

Figure 1. Manufacturing Setup.

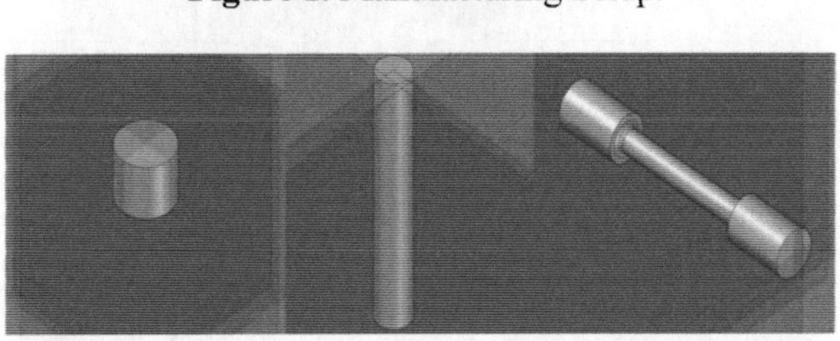

Figure 2. Billet, Extrudate and Tensile Test Specimen.

3. Density measurements

Table I, shows the theoretical and experimental densities for the manufactured composites as well as the percent porosity. It is observed that the composite densities are generally higher than that of the unreinforced aluminum, and that density increases with increase in the weight percent of the steel machining chips. This is expected as the density of steel (7.81 g/cm3) is higher than that of aluminum (2.71 g/cm3). It is also clear that the addition of steel machining chips in aluminum minimizes the porosity level in the aluminum matrix composites compared with the use of SiC as reinforcement. This can be attributed to the ductile behavior of the steel chips compared to SiC (brittle ceramic material) that incorporates high levels of deformation during the powder compaction, consolidation and hot extrusion which increases the grain refinement levels for both the aluminum and steel particulates and consequently reducing the porosity percentage.

Table 1. Density values and porosity percentage for the unreinforced aluminum and composites.

Sample composition	Theoretical density (g/cm^3)	Experimental density (g/cm^3)	Porosity (%)
Al-0wt.% steel chips	2.71	2.67	1.47
Al-5wt% steel chips	2.73	2.7	1.1
Al-7.5wt% steel chips	2.74	2.71	1.09
Al-10wt% steel chips	2.74	2.72	0.73
Al-10wt% SiC	2.66	2.63	1.12
Al-5wt% SiC	2.64	2.61	1.13

4. Mechanical properties

Table 2, represents the tensile properties of the composites derived from the stress–strain curves. It is observed that the 10 wt.% steel machining chips reinforced aluminum composite has the highest yield tensile strength with adequate percent elongation compared to the control samples. The increase in tensile properties accounts for the good bond between the aluminum matrix and the steel machining chips, which allows for the transfer and distribution of load from the matrix to the stronger (higher yield and ultimate strength than aluminum) reinforcement. The improved ductility is due to the higher plastic strain capacity which is significantly enhanced by the good matrix/reinforcement interface bonding and the retained ductility of the steel chips reinforcements is also a clear phenomenon accompanying powder metallurgy technique.

Table 2. Tensile properties of the unreinforced aluminum and composites.

Sample composition	Yield Strength (MPa)	Elongation (%)
Al-0wt.% Steel Chips	110	28
Al-5wt% Steel Chips	176	25
Al-7.5wt% Steel Chips	220	20
Al-10wt% Steel Chips	260	17
Al-10wt% SiC	161	10
Al-5wt% SiC	134	13

Figure 3 shows the hardness of the different composites produced. The hardness of the pure aluminum matrix composites is observed to increase with increasing the weight percent of steel machining chips. The hardness of the pure aluminum reinforced with 5 and 10 wt.% SiC particles serving as the control sample had hardness records slightly lower than that of the steel chips reinforced composite. This can be due to the relatively low percent porosity of the steel machining chips composites compared with the SiC reinforced aluminum composites.

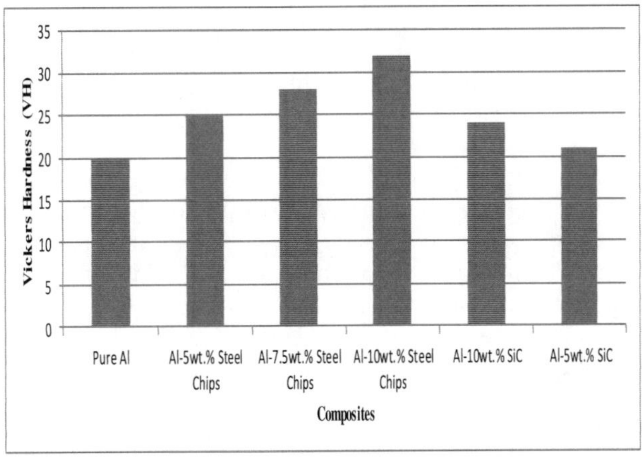

Figure 3. Vickers hardness results of the unreinforced aluminum and composites.

5. Wear behavior

The wear behavior of the composites was evaluated using a rotary platform Taber abrasion machine. The samples were placed on the abrasion machine rotary platform and gripped with a constant pressure. The rubbing action between the sample and the abrasive wheel during the rotating motion of the machine, results in the generation of loose composite debris from the sample surface. The wear index is calculated by measuring the loss in weight in milligrams per thousands cycles of abrasion. The lower the wear index, the better the abrasion resistance [12].

$$\text{Wear Index} = \frac{(\text{Initial sample weight} - \text{Final sample weight}) \times 1000}{\text{Test cycle time}}$$

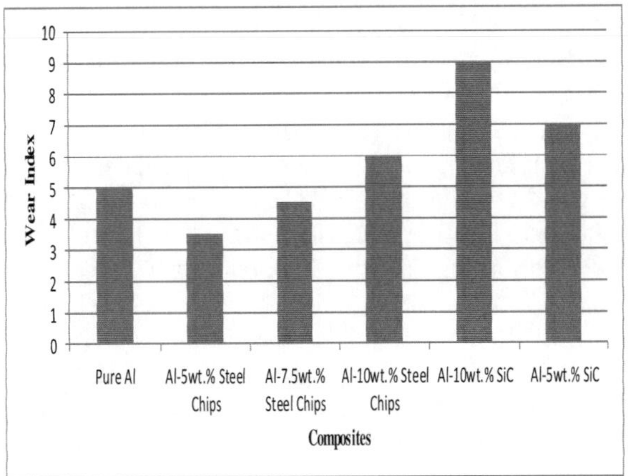

Figure 4. Wear index of the unreinforced pure aluminum and composites produced.

Figure 4 shows that the wear index increases with the increase in the weight percent of the steel machining chips but highest with the SiC reinforced aluminum matrix composite. Iglesias et al. [10] reported similar wear pattern in zinc aluminum based alloys reinforced with steel machining chips; where a high integrity interface observed between the metallic matrix and the steel chips serving as barriers to chip pullout and damage during wear. The high wear rate observed with the SiC reinforced aluminum matrix composites can be attributed to the relatively poor matrix/SiC particulate interface bonding which may facilitate the abrasive wear of the composite. Also SiC is a very brittle (ceramic) material that may suffer from fragmentation due to the rotational abrasive force applied.

6. Conclusions

1) The current work can underline the feasibility of incorporating steel machining chips as reinforcement in aluminum matrix composites as well as holding promises to improved mechanical properties.

2) The tensile, hardness and wear behavior of pure aluminum matrix composites reinforced with steel machining chips have been investigated in comparison with that reinforced with 5 and 10 wt.% SiC particulates.

3) The addition of steel machining chips in aluminum was found to minimize the porosity level in the aluminum matrix composites compared with the use of SiC as reinforcement.

4) The pure aluminum matrix strength was observed to improve with increaseing the steel machining chips wt.% as reinforcement with a retained ductility levels compared to that with the SiC reinforcements.

5) The hardness values were also observed to improve with the use of the steel machining chips as reinforcement.

6) The wear rate increases with the increase in the weight percent of the steel machining chips but highest with the SiC reinforced aluminum matrix composite.

References

[1] Hong S-J, Kim H-M, Huh D, Suryanarayana C, Chun BS, 2003 "Effect of clustering on the mechanical properties of SiC particulate-reinforced aluminum alloy 2024 metal matrix composites," *Materials Science and Engineering: A.*, 347(1–2), 198-204.

[2] Llorca J, Martínez JL, Elices M 1997 "Reinforcement fracture and tensile ductility in sphere-reinforced metal-matrix composites," *Fatigue & Fracture of Engineering Materials & Structures,* 20 (5), 689-702.

[3] Rajmohan T, Palanikumar K, Arumugam S 2014 "Synthesis and characterization of sintered hybrid aluminium matrix composites reinforced with nanocopper oxide particles and microsilicon carbide particles," *Composites Part B: Engineering,* 59 (0), 43-9.

[4] Ahamed H, Senthilkumar V 2010 "Role of nano-size reinforcement and milling on the synthesis of nano-crystalline aluminum alloy composites by mechanical alloying," *Journal of Alloys and Compounds,* 505 (2), 772-782.

[5] A Molina, A Torres-Islas, S Serna, M Acosta-Flores, R A Rodriguez-Diaz, J Colin 2015 "Corrosion, electrical and mechanical performance of copper matrix composites produced by mechanical alloying and consolidation," *Int. J. Electrochem. Sci.,* (10), p 1728–1741.

[6] R Zitoune, M El Mansori, K Vijayan 2013 "Tribo-functional design of double cone drill implications in tool wear during drilling of copper mesh/CFRP/woven ply," *Wear,* 302 (1–2), p 1560–1567.

[7] S Mallik, N Ekere, C Best, R Bhatti 2011 "Investigation of thermal management materials for automotive electronic control units," *Appl. Therm. Eng.,* (31), p 355–362.

[8] C Saldana, S Swaminathan, T L Brown, W Moscoso, J B Mann, W D Compton, S Chandrasekar 2010 "Unusual applications of machining: controlled nanostructuring of materials and surfaces," *ASME J. Manuf. Sci. Eng.,* 132 (3).

[9] T L Brown, S Swaminathan, S Chandrasekar, W D Compton, A H King, K P Trumble 2002 "Low-cost manufacturing process for nanostructured metals and alloys," *J. Mater. Res.,* 17 (10), p 2484–2488.

[10] P Iglesias, A E Jiménez, M D Bermúdez, B C Rao, S Chandrasekar 2013 "Steel machining chips as reinforcement to improve sliding wear resistance of metal alloys: study of a model Zn-based alloy system," *Tribol. Int.,* (65), p 215–227.

[11] ASTM E8M-04 2004 "Standard Test Methods for Tension Testing of Metallic Materials

[Metric], *ASTM International,* West Conshohocken.
[12] *http://www.taberindustries.com/taber-rotary-abraser*,2011-2016.

IOP Conf. Series: Materials Science and Engineering **191** (2017) 012042 doi:10.1088/1757-899X/191/1/012042

Structural comparison of Gilsonite and Trinidad Lake Asphalt using ^{13}C-NMR technique

Nader Nciri and Namjun Cho

School of Energy, Materials, and Chemical Engineering, Korea University of Technology and Education, 1600 Chungjeol-ro, Byeongcheon-myeon, Dongnam-gu, Cheonan-City, Chungnam-Province 330-708, Republic of Korea

E-mails: nader.nciri@koreatech.ac.kr, njuncho@koreatech.ac.kr

Abstract. The recent increased importance of natural asphalt as an alternative binder for sustainable road pavement has dictated that more knowledge should be acquired about its structure and properties. Earlier, Carbon-13 NMR spectroscopy has been applied to very few natural bituminous materials. In this work, two types of raw binders namely Gilsonite and Trinidad Lake asphalt (TLA) have been subjected to an extensive investigation by using ^{13}C-NMR technique. Results have shown that valuable chemical data can be readily withdrawn on aromatic ring structures and ring substituents in natural asphalts derived from different sources. The chemical significance of these findings will be discussed.

1. Introduction

Asphalt chemistry is a complex field and poses a considerable challenge to the chemist. Asphalts are known to be comprised of an exceedingly large number of organic molecules, ranging in size, mainly composed of carbon and hydrogen atoms. A better understanding of their compositions and properties is of utmost importance in the asphalt industries in order to apply adequate processing facilities.

The petroleum asphalt (PA) has been primarily used in road construction where it is employed as a glue or binder mixed with aggregate particles to create asphalt concrete. Its other main uses are for bituminous waterproofing products, including production of roofing felt and for sealing flat roofs [1]. It has been readily available, easy to handle, offers excellent performance, and has been the most economical. Today's environment however, has the asphalt industry searching for alternative asphalt sources. Concerns of energy independence, the exorbitant increase in oil prices and growing environmental concerns have invigorated the interest for alternative energy sources. The asphalt industries and the highway agencies have escalated research in the area of alternative PAs in order to meet the challenges presented today.

The use of alternative PA is not a new concept. Early in the 19th century, researchers investigated various forms of natural asphalts throughout the development of new building products. Gilsonite and Trinidad Lake asphalt (TLA) are among the fossil oils being evaluated as promising natural materials for the construction of flexible pavements. They are the black or dark brown bitumens found in seeps, veins, or permeating limestones or dolomites as a part of them. The Gilsonite of Utah is free from impurities, while the asphalt in the Trinidad Pitch Lake contains sand, clay, and shells up to 30 wt. % of the mass [2].

Gilsonite, originally known as "Uintaite", is found in commercial quantities in the Uinta Basin of northeastern Utah, in Uintah and Duchesne counties. It is a hard, brittle, native bitumen of variable but high softening point. Uintaite is uniform in purity, extremely low in sulfur, has a low specific gravity, a high content of volatile matter, a lustrous fracture varying from brilliant conchoidal to dull hackly or pencillated form, a typical brown streak and characteristic odor [3, 4]. It is composed mainly of aromatic asphaltite [5]. The low melting point, high solubility in oils, and near-perfect insulator qualities of Gilsonite historically has lent it to a wide variety of uses. Gilsonite has been used in manufacturing storage batteries, foundry forms, paints, varnishes, inks, and lacquers, and as a binder for road pavement [6].

TLA was first reported for paving in the United States about 1874. Trinidad asphalt exists in a number of deposits on the island of Trinidad, near the northeast coast of Venezuela. This asphalt has been used for pavement constructions and other applications throughout the world. TLA has a relatively uniform composition of 29 wt. % water and gas, 39 wt. % bitumen soluble in carbon disulfide, 27 wt. % mineral matter on ignition, and 5 wt. % bitumen that remains adsorbed on the mineral matter. The refined product averages 36 wt. % mineral ash with a penetration at 25 °C of about 2 (0.2 mm), a softening point of 99 °C, a flash point of 254 °C, and a sulfur content of 3.3 wt. %. The mineral matter typically contains particle sizes from 40 wt. % finer than 10 μm to 10 wt. % from the 30-75 μm range. Because of the influence of the fine mineral particles, many of them colloidal in size, the viscosity of TLA is higher than that of normal bitumens [7-9].

The use of Gilsonite and TLA as raw materials for diverse fields of application requires an immediate knowledge of their compositions and the existing molecular interactions. The molecular composition affects directly the physical properties of asphalt, such as softening point and viscosity, and in turn the performance of the final product. Therefore, in order to develop a rational approach to the conceptual understanding and practical resolution of the structural ambiguity, one must start with the study of natural asphalt itself. The molecular structure of this material has not yet been completely understood and is still the subject of some conjecture. Apart from its pertinence to road pavement technology, knowledge of the molecular structure of natural asphalt is the key to its possible future commercial utilization, to the elucidation of its origin and diagenetic history, and may conceivably shed some light on the history of the natural asphalt formations.

In our recent work on the "chemical characterization of Gilsonite bitumen" [10] published in the Journal of Petroleum & Environmental Biotechnology, substantial efforts have been devoted to elucidate the composition, structure, and properties of Gilsonite and TLA by using elemental analysis (EA), thin layer chromatography-flame ionization detection (TLC-FID), Fourier transform-infrared (FT-IR) and proton-nuclear magnetic resonance (^{1}H-NMR) spectroscopies. The chemical study led sucesuffully to the unravelling of the mystery of the source and origin of the asphalt formations in Utah and Trinidad Island, the relation between the natural asphalt deposits and their interrelation with conventional oil accumulations of the region. To provide further insights into the poorly defined structures of Gilsonite and TLA, the ^{13}C-NMR spectroscopy was carried out accordingly.

2. Materials and methods

2.1. Natural asphalts

Gilsonite was collected from Uintah Basin in northeastern portion of the state of Utah, the United States. It is made up of predominantly of pure asphalt, mineral matter, and other minor constituents. Gilsonite is a shiny, black substance similar in appearance to the mineral obsidian. It is brittle and can be easily crushed into a dark brown powder (Figure 1 A). In order to enable a more comprehensive discussion of the results, reference was made to other natural asphalt from the Pitch Lake at La Brea in the Island of Trinidad known as Trinidad Lake asphalt (TLA). TLA is a dark-brown colored cementitious substance. It is brittle and broke into small cellular glossy fragments (Figure 1 B). This asphalt is approximately 63 wt. % bitumen and the remainder being mineral and organic material. All asphaltic specimens were analyzed by the same ^{13}C-NMR technique to provide directly comparable chemical information.

IOP Conf. Series: Materials Science and Engineering **191** (2017) 012042 doi:10.1088/1757-899X/191/1/012042

Figure 1. Gilsonite (A) and Trinidad Lake asphalt (B) specimens.

2.2. ^{13}C-NMR spectroscopy

^{13}C-NMR spectra of natural asphalts were recorded on a VNMRS-400 spectrometer (Agilent Technologies Korea Ltd., Seoul, South Korea). NMR experiments were performed using 20~50 mg of bitumen dissolved in 0.6 mL of deuterated chloroform containing Tetramethylsilane as a chemical shift reference. The NMR spectra were collected with relaxation delay of 10.0 s, acquisition time of 1.3 s, and 5,000 repetitions.

3. Results and discussion

3.1. ^{13}C-NMR analysis

The ^{13}C solution NMR spectra of Gilsonite and TLA are displayed in Figure 2. The strong absorption between 76.5 and 78 ppm is due to solvent $CDCl_3$. The low field absorptions between 10 and 45 ppm are due to aliphatic sp^3 carbons, while the weak absorptions between 100 and 160 ppm are due to aromatics sp^2 carbons.

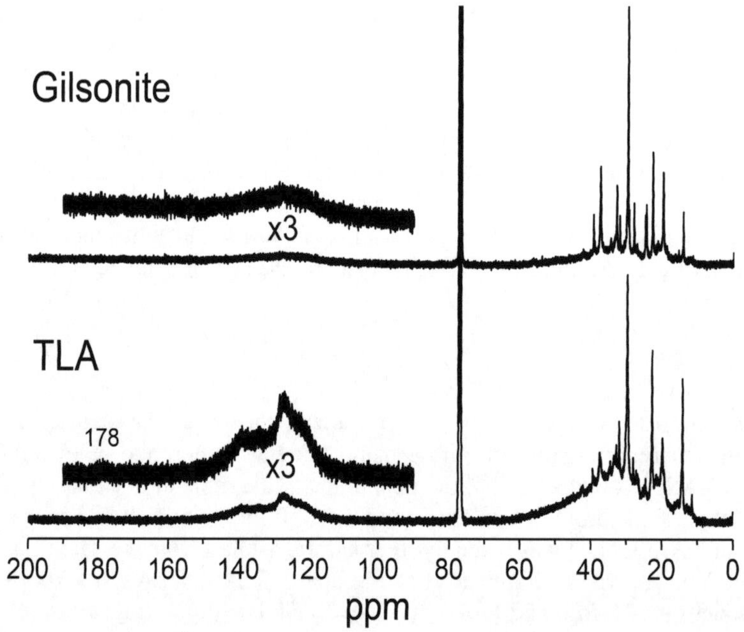

Figure 2. ^{13}C solution NMR spectra of Gilsonite and TLA.

In the ^{13}C-NMR spectrum of Gilsonite, the aromatic carbon band is very weak compared to that of TLA. This suggests that Gilsonite contains the major part of aromatic moieties of pyrrolic ring with limited mobility, while TLA has mainly the six-membered aromatic carbon rings with higher mobility.

This fact corroborates FT-IR spectra and TLC-FID analysis [10]. TLC-FID analysis showed that Gilsonite contains negligible amount of aromatics (~ 0 wt. %) and moderate quantity of resins (18.7 wt. %), while TLA contains 22.9 wt. % of aromatics and 37.4 wt. % of resins. The very small peak in ^{13}C spectrum of TLA appeared around 178 ppm is probably attributed to the C=O group of ester, carboxylic acid, or amide groups.

Figure 3 presents the expanded alkyl portion of the ^{13}C solution NMR spectra of Gilsonite and TLA, respectively. Interestingly, Gilsonite spectrum of the current study shows the major 10 peaks reported by Helms and his colleagues [5]. In this alkyl region, notable differences were detected between TLA and Gilsonite spectra, especially at peaks number 2, 4, 5, 8, 9, and 10, which were much weaker in intensity with TLA than Gilsonite. These differences could be used, of course, as marker bands to differentiate Gilsonite from TLA. Referring to the assignments of Helms et al. [5], it can be admitted that most of these peaks are ascribed to the carbons related to the side-chains, which illustrates that Gilsonite has more side-chains than TLA. This feature is fully supported by the H_γ/H_β ratio in ^1H-NMR spectra and by the higher saturates (4.1 wt. %) and resins (37.4 wt. %) contents in TLA [10]. Even though that Gilsonite appears poor in saturates (1.66 wt. %) and resins (18.7 wt. %) components [10], it comprises a relatively large amount of alkyl portion. This alkyl portion is essentially involved in the connection of pyrrole rings; basic constituents of asphaltene compound. This observation correlates favorably with the proposed model structure of Gilsonite asphaltene by Helms et al. [5].

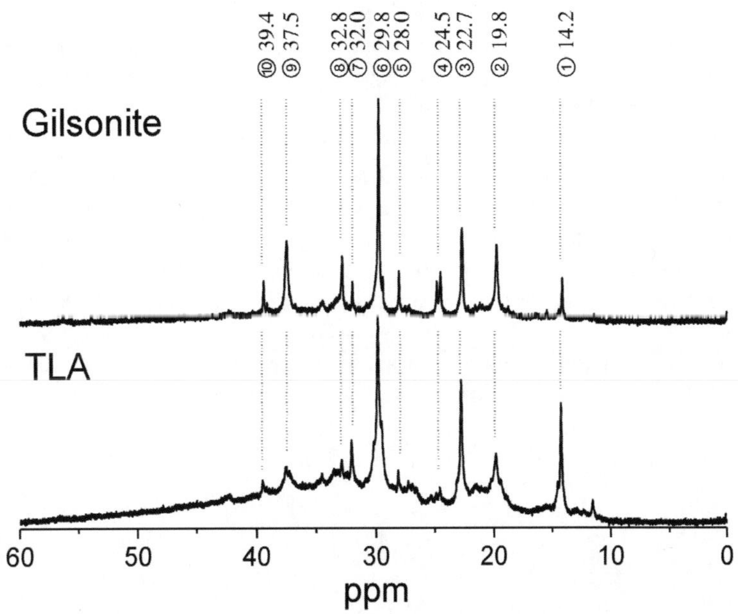

Figure 3. Alkyl portion of ^{13}C solution NMR spectra.

4. Conclusion

The present study enabled us to understand and compare in detail the chemical characteristics of Gilsonite and TLA. The relatively easy and convenient method of ^{13}C-NMR provided useful informations to identify the chemical structure of natural asphalts with the help of elemental analysis, FT-IR, TLC-FID, and ^1H-NMR data. Gilsonite has a significant amount of aromatic moieties with nonprotonated aromatic carbons as compared to TLA. The high content of asphaltenes (79.7 wt. %) found in Gilsonite contains pyrrole N–H groups and aromatic carbons connected with an important fraction of alkyl chains. TLA includes high concentration of oxygenated compounds and aromatic moieties with protonated carbons. The alkyl chains in TLA are more oxidized and sulfurized than Gilsonite; containing carbonyl, sulfide, and ether groups. Future work may involve the inclusion of additional techniques to characterize the natural asphalts such as X-ray diffraction (XRD), Raman spectroscopy (RS), thermogravimetric analysis (TGA), differential scanning calorimetry (DSC), etc.

Acknowledgments
This work was performed as a part of the project of the development of eco-friendly pavements to minimize greenhouse gas emission. We appreciate the financial support from the "Transportation Logistics research program (Grant 15TLRP-B079261-02)" of MOLIT (Ministry of Land, Infrastructure & Transport) and KAIA (Korea Agency for Infrastructure Technology Advancement) of the Republic of Korea.

References
[1] Sörensen A and Wichert B 2009 Asphalt and Bitumen *Ullmann's Encyclopedia of Industrial Chemistry* (USA: John Wiley and Sons, Inc.) DOI: 10.1002/14356007.a03_169.pub2.
[2] Woodruff E G 1935 Natural Asphalt and Its Relation to Oil Deposits *Tulsa Geological Society Digest.* 4 16-7.
[3] Gillson J L ed 1960 *Industrial minerals and rocks (non metallics other than fuels)* ed 2 (New York: The Amer. Inst. of Mining, Metall. and Petroleum Engineers) pp 1-934.
[4] Boden T and Tripp B T 2012 *Gilsonite Veins of the Uinta Basin, Utah* Report No 141 (Salt Lake: Utah Geological Survey) p 50.
[5] Helms J R, Kong X, Salmon E, Hatcher P G, Schmidt-Rohr K and Mao J 2012 Structural characterization of gilsonite bitumen by advanced nuclear magnetic resonance spectroscopy and ultrahigh resolution mass spectroscopy revealing pyrrolyic and aromatic rings substituted with aliphatic chains *Org. Geochem.* 44 21-36.
[6] Verbeek E R and Grout M A. 1992 Structural evolution of gilsonite dikes, eastern Uinta Basin, Utah *Hydrocarbon and mineral resources of the Uinta Basin, Utah and Colorado,* ed T D Fouch, V F Nuccio *et al* (Salt Lake: Utah Geological Association Guidebook 20) pp 237-55.
[7] Speight J G 2000 Asphalt *Kirk-Othmer Encyclopedia of Chemical Technology* (USA: John Wiley & Sons) pp 1–38 DOI: 10.1002/0471238961.0119160819160509.a01.
[8] Charles R and Grimaldi F 1996 Trinidad Lake Asphalt in Pavement Performance *W. Indian J. Eng.* 18 29-40.
[9] Widyatmoko I and Elliott R 2008 Characteristics of elastomeric and plastomeric binders in contact with natural asphalts *Constr. Build. Mater.* 22 239-49.
[10] Nciri N, Song S, Kim N and Cho N 2014 Chemical Characterization of Gilsonite Bitumen *J. Pet. Environ. Biotechnol.* 5:193 doi:10.4172/2157-7463.1000193 .

Experimental investigations of timber beams strengthened by CFRP and Rebars under bending

Zhou Nianqiang[1,2,a], **Shi Weixing**[1,b]

[1]College of Civil Engineering, Tongji University, Shanghai 200092, China
[2]College of Civil Engineering, Nanjing Forestry University, Nanjing 210037, China

E-mail: [a]xcznq@163.com,[b]swxtgk@126.com

Abstract. Wooden structure houses deteriorate over time due to environmental aging, fatigue, and other reasons. In order to solve this problem, composite timber beams strengthened by extra steel bars (rebar) and carbon fiber-reinforced plastic (CFRP) are studied experimentally in this paper. Specimens with various strenghting, ie., rebars only, CFRP only, and a combination of the two, were considered under four-point flexural tests. Failure, displacement and strain response and ductility capacity were evaluated for the present tested models. Dramatic enhancement of the capacity in addition to improved deflection and ductility were gained for the strengthened beams relative to the plain specimens, indicating the effectiveness of the reinforcement on the flexural strength of such composite beams.

1. Introduction

The Chinese have a long history of building wood houses, many excellent ancient architectures are wood structure. However, lots of wooden structure houses deteriorate over time due to environmental aging, fatigue, and other reasons. Traditional wood structure repair and reinforcement methods often require extensive repair site, easy to destroy the original appearance of the structure or bring other secondary effects.

Because of high geometrical plasticity, easy to cut and shape, high strength, light weight and easy construction, CFRP has become the common reinforcement material in the field of building engineering including wood structure [1-6]. In this paper, a new reinforcement method for timber beam by rebar and CFRP is introduced. The method is that the rebar is embedded and bonded in the timber beam by grooving at the bottom of the beam, and the groove is covered and by the CFRP cloth bonding. In other words, the timber beam was reinforced by two different materials, CFRP and steel bar. The synergy of timber, rebar and CFRP will determine reinforcement effects.

2. Experimental program

2.1. Test specimens and material properties

The timber used for the test beam is loblolly pine, an important tree species for afforestation and industrial timber species growing in southern China [7]. The main physical and mechanical properties of wood is shown in Table 1.

Carbon fiber cloth were made by Japanese carbon fiber silk, mainly for structural components of the tensile, shear and seismic strengthening. They had a net thickness of 0.111mm and the width of 200mm. Its mechanical properties were tested, and the results is shown in Table 1.

The rebar of Grade HRB335 is used in this test, and it is widespread used in building construction as reforcement, with good strength and plasticity. Because the tensile strength of steel is much higher than the timber, wood failure will not be pulled off the steel bar, or even yield, so the selection of a smaller diameter Φ10 screw steel, and the smaller diameter with a lower cost of steel. The steel bars used in the program were tested in the laboratory, giving average yield strengths of 380 MPa. The reinforcement is arranged in the tensile zone to be primarily subjected to tensile stresses. In this experiment, the steel bar and wood were bonded by anchorage glue. It is an important aspect of this study to observe the adhesive performance, and the researcher hope that the timber beams can work together with the rebar by the anchorage glue. This experiment uses XK-360 anchorage glue. XK-360 anchorage glue has the advantages of acid and alkali resistance, low temperature resistance, aging resistance, good heat resistance, no creep at room temperature, water logging resistance, long-term load stability in wet environment, resistance to welding, good flame resistance, seismic performance good and so on.

Table 1. properties of timber and strengthen materials

species	Compressive strength /Mpa	Flexural strength /Mpa	Flexural modulus of elasticit	Thickness /(mm)	tensile strength /MPa	Elastic Modulus /GPa	Bending strength /MPa	Elongation
loblolly pine	28.05	64.16	6.24					
CFRP				0.111	3426	265	-	1.51%
glue				-	50	-	85.5	2.5%
rebar					380			

2.2. Specimen design

Twenty-one timber beams were constructed for four-point bending. Table 2 summarizes the specimens program. The cross-sectional dimension of all specimens was 80 mm(width)×150 mm(depth)×2000 mm(length), as shown in fig.1(a). These beams were categorized into eight different groups. The variations from one group to another and beam designations are given in Table 1 and Figs.1. Seven beams in case 1 acted as control specimen. Two beams in case 2-3 were strengthened with 10-mm-diamter deformed bars. Two beams in case 4-5 were strengthened with steel bars and carbon wrap. Fig.1 shows the dimensions and reinforcement details. Two beams each in cases 6-8 were wrapped with different longitudinal or transverse configurations or varying depths of wrapping, using carbon fabrics.

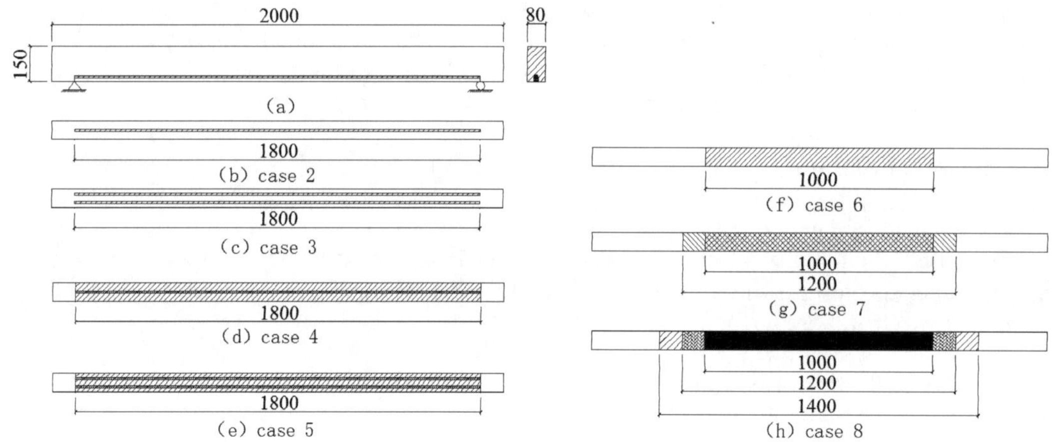

Figure 1. Strengthening scheme used in cases 2-8

Table 2. Beam designations

Group (1)	Case (2)	No. of beams tested (3)	Designation (4)
1	1	7	L-1,L-2...L-7(beams not strengthened)
2	2	2	1S-1,1S-2(The beam with one rebar)
	3	2	2S-1,2S-2(The beam with two rebars)
3	4	2	1S-1C-1,1S-1C-2 (The beam with a rebar and carbon wrap)
	5	2	2S-1C-1,2S-1C-2 (The beam with two rebars and carbon wrap)
4	6	2	1C-1,1C-2 (The beam with carbon wrap, width 80mm and length 1000mm)
	7	2	2C-1,2C-2 (The beam with carbon wrap, Two layers CFRP with width 80mm and length 1000mm, 1200mm)
	8	2	3C-1,3C-2 (The beam with carbon wrap, Three layers CFRP with width 80mm and length 1000mm, 1200mm, 1400mm)

2.3. Instrumentation, test set-up, and test procedure

The whole experiment was carried through on the electro-hydraulic servo controlled loading system in Nanjing Forestry University, as shown in fig.2.

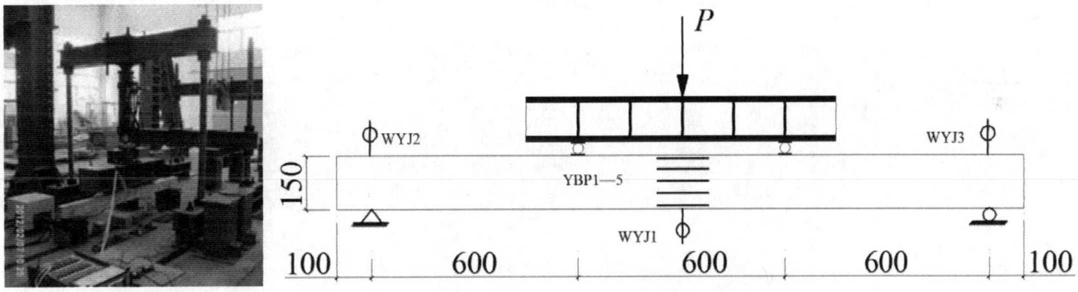

WYJ - Displacement meter; YBP - Strain gauges

Figure 2. Experimental set up and measurement

All the beams were tested under four-point bending with an effective span of 1.8m and a distance of 0.6m between the load points. Load was applied at a uniform rate (≈5kN/minute).and deflections and strains were recorded at regular intervals (1Hz).The specimens were preloaded before loading [8].

Three displacement meters were installed at the supporting points and the midspan of specimens. Five strain gauges were placed at the midspan of the beam. Two of them were used in the compression zone, three of others were used for the tensile zone and neutral axis. The distributions of meters and strain gauges are shown in Fig. 2.

3. Test results and discussions

3.1. Failure modes and analysis of observed mechanisms

The typical failure modes were identified as shown in fig. 3.

The failure in the plain timber specimens were quite abrupt with a sudden tensile rupture in the most time. The failure was initiated in different areas in various plain specimens, which is believed to be generally the result of the existence of knots .Most of the specimens split at the bottom or the scar near the middle of beam, however, a few of beams exhibition a large plastic deformation without fracture phenomenon, as shown in fig. 3a, 3b, 3c.

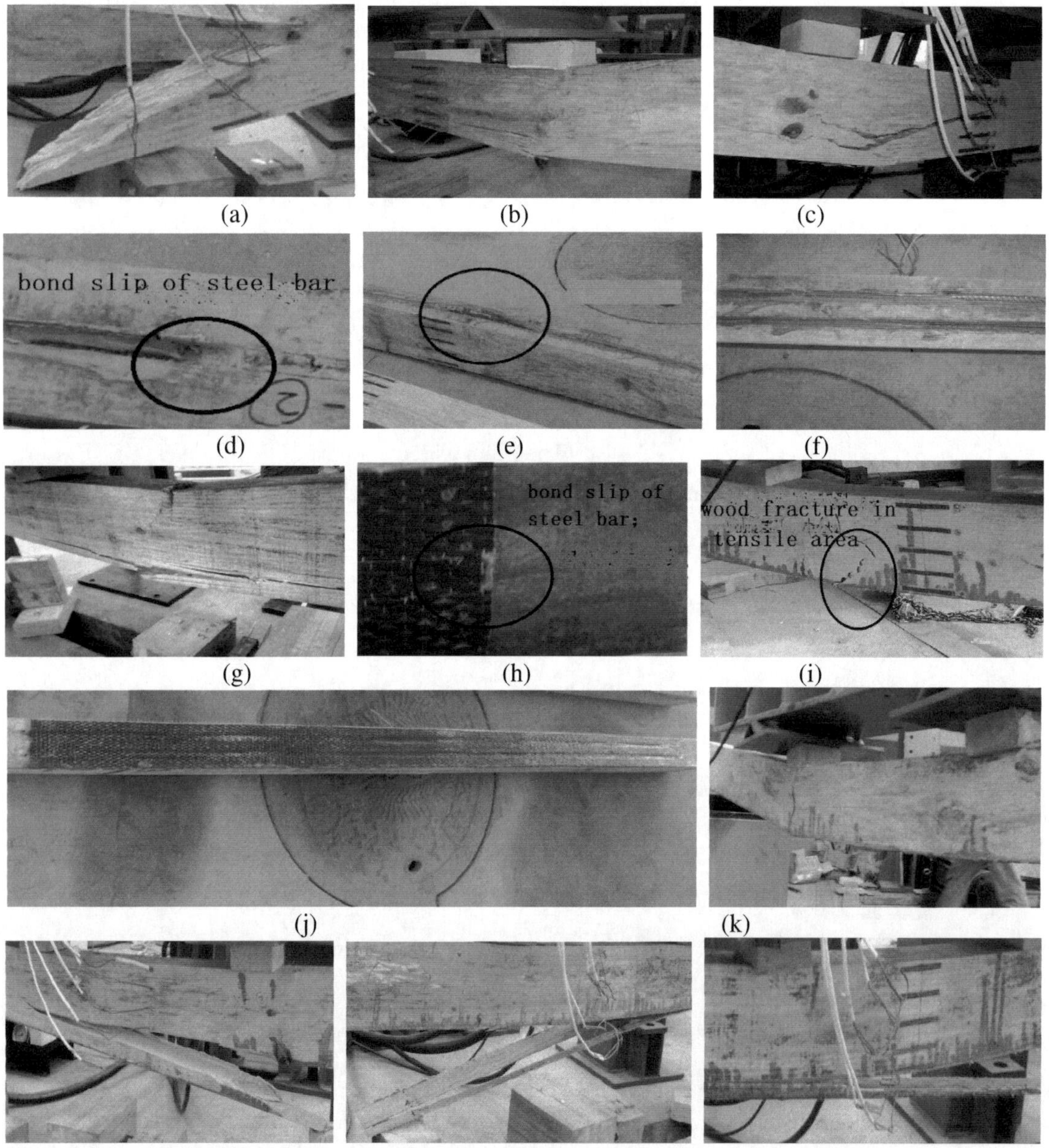

Figure 3. Typical failure modes of tested beams: (a)L-7; (b)L-3; (c)L-2; (d)1S-2; (e)1S-2(midspan); (f)2S-2; (g)2S-2; (h)1S-1C-2; (i)1S-1C-2; (j)2S-1C-1; (k)2S-1C-2; (l)1C-2; (m)2C-1; (n)3C-2

In case 2, the failure mode of the strengthening beam is brittle bending failure. Several cracks are found at the bottom of the beam with the load increasing, and one was developed into a main fracture. The wooden beam could restore original state again soon after load released, however, the bending deformation of rebar can not be restored into a straight line. A little steel slip observed in the beam groove at the end of the bar, and the central steel was protruding, as shown in fig.3d, fig.3e.

The reinforcement beam (In case 3) were damaged at shear bending region. Large deformation occured during the later period of the test. In the end, beams were broken and Compression deformation is very obvious in the top region. The rebars, however, there is no obvious deformation, and the side slip at both end is smaller, as shown in fig3f, fig 3g.

In case 4, the failure phenomena of two beams was differentiated. One of the specimens was cracked from the scar in the middle of the beam, and the carbon cloth broke suddenly and debonding from the bottom. The crack of the other specimen beam is generated in the tension zone of the timber beam, and the crack continues to develop with load increasing. When the beam damaged, the steel bar at the bottom of the slip occurred, and the carbon cloth along the length of the groove cut into three stripe, as shown in Fig. 3h and Fig. 3i.

In case 5, the failure mode of the strengthening beam is the compression deformation of the wood in the compression area. The whole beam has a large bending deformation and no obvious cracks in the tension end. The carbon fiber cloth is basically intact, as shown in Fig.3j and Fig. 3k.

In case 6-8, the failure mode of the beams wrapped with carbon fabric were characterized by the stripping of CFRP and the fracture at the scarring of beams, as shown in Fig.3l and Fig.3m.

3.2. Ultimat load and elastic modulus

Table 3 and 4 presents the values of ultimate load and elastic modulus obtained in the test. As compared with plain beams, most of all the strengthened specimens were significantly enhanced in flexural strength. The ultimat load of beams in bars-strengthened increased by 28%. The combination-strengthened (in case 4,5) possessed flexural strength 26%-32% larger than plain beams(in case 1).

Table 3. Ultimate load and elastic modulus of timber beam

Species code	**Ultimate load** /kN	**elastic modulus** /Mpa	Average ultimate load	Average **elastic modulus**
L-1	25.35	6109		
L-2	32.55	7275		
L-3	24.86	6545		
L4	30.66	5531	27.84	6107
L-5	22.87	6035		
L-6	34.71	6426		
L-7	23.90	4826		
Standard deviation	4.70	778		
Coefficient of variation	0.17	0.13		

3.3. Displacement vs load relationship

Fig. 4 shows the load versus deflection curves of all specimens. It can be seen from Fig.4a that the strength and ductility of 7 contrast beams are quite different, which indicates that the mechanical properties of the material itself are more discrete. Compared with the plain beams, the strength and rigidity of the strengthened beams are generally higher, as shown in Fig. 4b and Fig. 4c, they exhibit large plastic deformation at the later stage of loading, and the beams in the combination reinforcement program show a stable high strength and a higher stiffness and large plastic deformation of the mechanical characteristics, and the ductility phase strength remained unchanged, or even increased slightly.

Table 4. Ultimate load and elastic modulus of reinforced timber beam

Species code	Ultimate load /kN			elastic modulus /kN		
	Test Data	Average	Enhancing ratio (%)	Test Data	Average	Enhancing ratio (%)
1S-1	30.40	27.67	-0.6	6162	5478	-10
1S-2	24.94			4794		
2S-1	40.58	35.79	28	9945	9315	53
2S-2	31.01			8686		
1C-1	32.68	34.98	25.6	5974	6986	14
1C-2	37.28			7997		
2C-1	35.85	37.56	34.9	6186	7675	26
2C-2	39.27			9146		
3C-1	28.43	30.08	8.04	4839	4825	21
3C-2	31.73			4812		
1S-1C-1	37.50	34.97	26	6627	6943	14
1S-1C-2	32.44			7259		
2S-1C-1	35.44	36.91	32	7870	9679	58
2S-1C-2	38.39			11488		

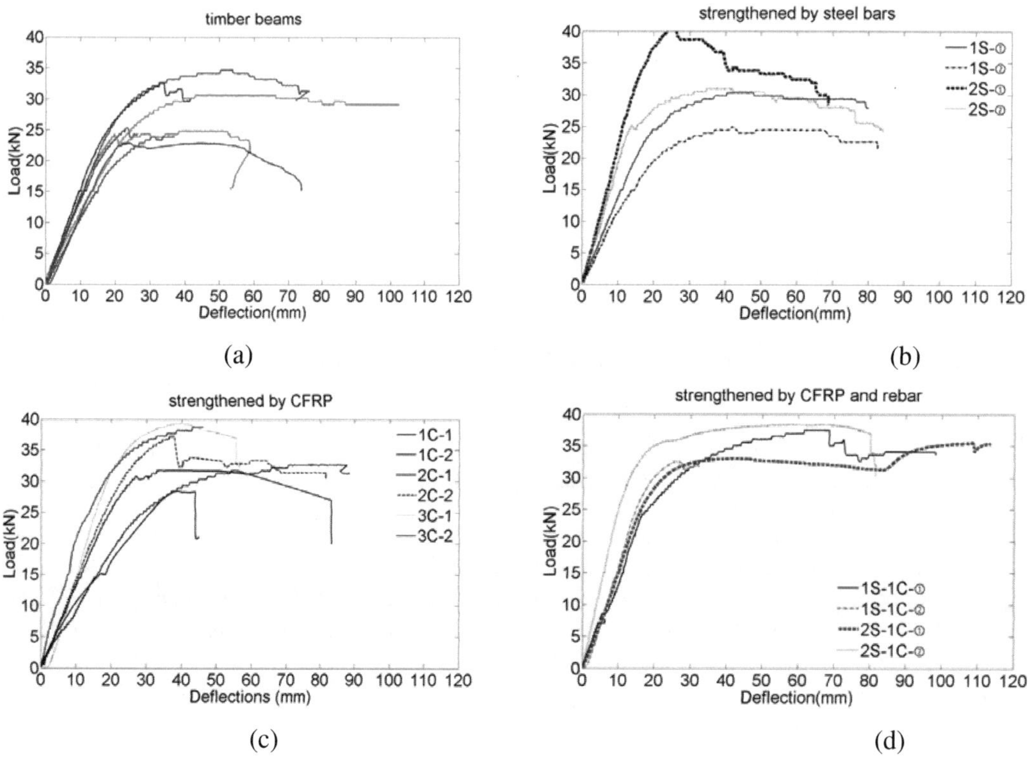

(a) (b) (c) (d)

Figure 4. Load versus deflection curve in four-point bending test

3.4. Analysis of strain response in beam span

The five typical strain values of the strain measuring points at the midspan cross section of various types of wood beams are shown in Fig. 5 which illustrates the strain distribution along the depth of the cross section. Figs. 5 show the following conclusions.

During the linear loading process, the longitudinal strain at the midspan cross section showed a linear distribution along the cross-sectional depth, indicating that the strain at the midspan cross section approximately agreed with the plane section assumption. As the load increased, the strain curve was not a straight line. The reason was that the glued connection, as one of the structural connections, was damaged and/or the material was damage, accompanied by structural rigidity changes.

The beam with rebar has an initial neutral axis of 15 mm below the center of the beam section, which indicates that the steel bar starts to exert its effect to bear the large load from the initial stage of the loading, which plays the role of reinforcement, as shown in fig. 5(b).

Both wraped beams and unwraped beams has its initial central axis in the center of the beam section, which indicated that the bottom of the carbon fiber cloth can not significantly improve the strength of the bottom of the beam. The tensile strain is relatively small at the later stage of loading, and the strain of the strengthened carbon beam is smaller than that of the carbon cloth, which indicates that the carbon cloth is not as good as the reinforcing bar. The initial neutral axis of the beam strengthened by two-layer steel bar and carbon fiber sheet is about 18mm below the center of the section and slightly decreased at the later stage, which indicates that the two type of strengthen material play important role during loading . The strain response at all levels of load is small, indicating that the overall stiffness was significantly improved.

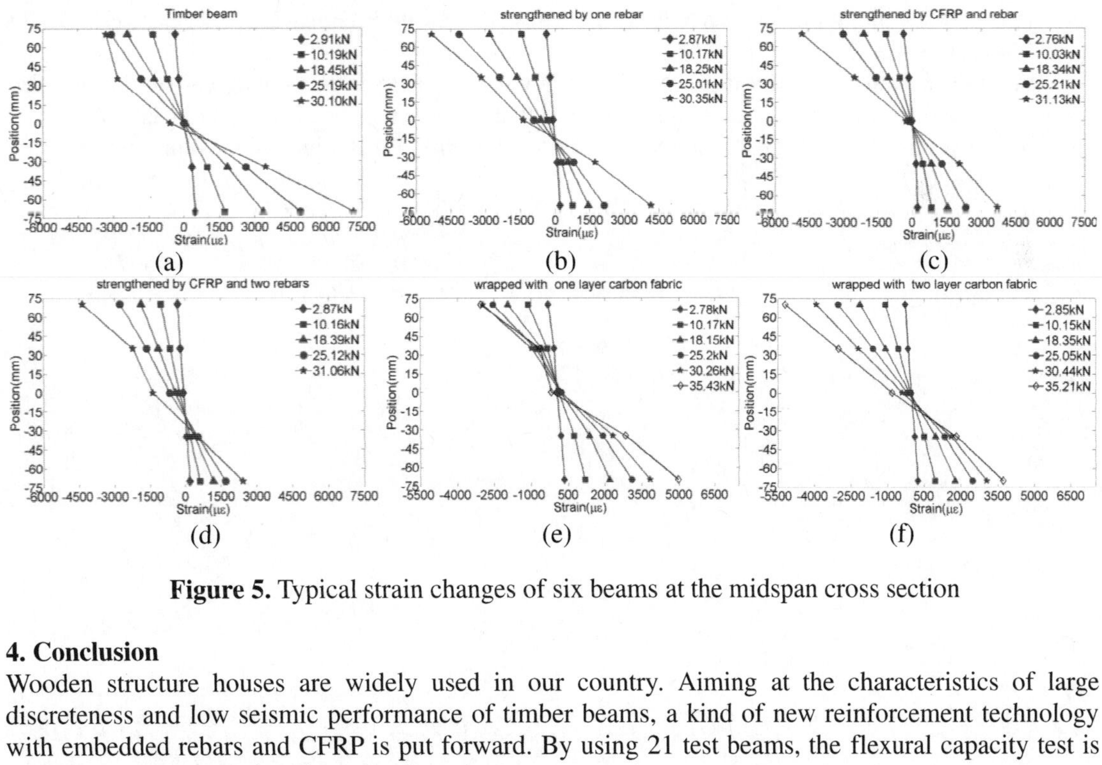

Figure 5. Typical strain changes of six beams at the midspan cross section

4. Conclusion

Wooden structure houses are widely used in our country. Aiming at the characteristics of large discreteness and low seismic performance of timber beams, a kind of new reinforcement technology with embedded rebars and CFRP is put forward. By using 21 test beams, the flexural capacity test is carried out, the experimental results show that.

(1) The timber beams with carbon or rebars are significantly better than the contrast beam. The rebar and carbon fiber sheets can improve the flexural rigidity and flexural capacity of the timber beams. The reinforced beams have certain signs before the failure. With the bending of the steel bars, the whole member can still provide a certain bending capacity. The two rebars and a layer of carbon

cloth can significantly improve the rigidity and strength of wood, and ductility better.

(2) In the linear loading process, the strain distribution all of beams were in accordance with the assumption of plane section. Rebars can improve the stiffness of the bottom of the beam, causing the initial neutral axis decreases, and improve rigidity greatly. However, the carbon sheet can do less.

(3) The results showed the interracial bonding was a decisive cause to influence the mechanical properties. For the gule is not dedicated glue for rebar and CFRP, some CFRP cloths was separated from timber beam, in the later stage of test, so as steel bar. So the effect of timber beam reinforcement should be further enhanced through improving the bonding performance.

References

[1] Xie Qifang, Zhao Hongtie, Xue Jianyang, Experimental Study on Flexural Behavior of Wood-Beams Strengthened with Carbon Fiber Sheet [J]. Industrial Construction, 2007, 37 (7): 104-107.

[2] Ma Jianxun, Jiang Xiangmin, Hu Ping, etc., carbon fiber reinforced wood beam bending test study [J]. Industrial Construction, 2005, 35 (8): 35-39.

[3] Plevris N, Triantafillou T C. FRP-Reinforced Wood As a Structural Material[J].Journal of Material in Civil Engineering ,1992,4 (3):300-317.

[4] Johns K C, Lacroix S1.Composite Reinforcement of Timber in Bending[J].Canadian Journal of Civil Engineering, 2000, 27 (5) :899-906.

[5] Antonio Borri, Marco Corradi, Andrea Grazini. A Method for Flexural Reinforcement of old Wood Beams with CFRP Materials [J]. Composites: Part B, 2005 , 36. 143-153.

[6] CHEN Dong-jian, ZHOU Zheng-wei, PAN Li-qun. Experimental study on mechanical properties of reinforced concrete beams strengthened with embedded steel bars [J]. Forest Engineering, 2011, 27 (4): 62-65.

[7] TONG Fang-ping, WU Ji-you, LONG Ying-zhong, et al. Effects of thinning on wood properties of loblolly pine [J]. Journal of Central South Forestry University, 2004, 24 (2): 23-27.

[8] GB / T50329-2012. Wood structure test method standard [S]. Beijing: China Building Industry Press, 2012.

IOP Conf. Series: Materials Science and Engineering **191** (2017) 012044 doi:10.1088/1757-899X/191/1/012044

Experimental and theoretical analysis of CFRP reinforced concrete beam anchored by CF anchors

Ning ZHUANG[1,a], Ye-ming MA[2,b], Hong-han DONG[3,c] and Da CHEN[4,c,*]

[1,2,3,4] College of Harbor, Coastal and Offshore Engineering, Hohai University, Nanjing, China

E-mail: [a]zhuangning1977@163.com, [b]1205140267@qq.com, [c]1037158873 @qq.com, [d]chenda@hhu.edu.cn

Abstract. Carbon fiber-reinforced polymer (CFRP) sheets is used to strengthen RC members widely. However, in the actual projects, premature separation of the CFRP sheet from the concrete surface is observed frequently, key issue in CFRP application is the performance of the connection between the CFRP sheet and the peripheral RC members. This paper presented the results of an experimental study to investigate mechanical properties of carbon fiber (CF) anchor dowels, a new CFRP anchoring method, and based on the experimental results, obtained the bearing capacity calculation formula of concrete beam strengthened with CFRP anchoring by carbon fiber anchor dowels .

1. Introduction

The carbon fiber reinforced polymer (CFRP) has a high tensile strength, light weight, strong corrosion resistance, simple and easy to operate compared with other reinforcement methods, and the construction process does not affect the duty of the structure. There have been quite a lot of studies report that externally bonded CFRP could effectively improve carrying capacity and change failure mode of RC members [1-3]. In order to improve bearing capacity of RC beams, the bonding strengthen between CFRP and RC members must be strong enough to make sure the full use of tensile strength of CFRP [4-6]. A study of behaviour of reinforced concrete beams strengthened with FRP found that CFRP sheets debonded on average when suffering 50% of their tensile capacity [7]. So premature debonding failure due to separation of laminates caused by excessive stress concentration at the end of laminate had been agreed to be the main problem. The risk of Debonding at laminate ends could be minimized using appropriate end anchors. Therefore, in order to more fully utilize the tensile capacity of the FRP sheet, carbon fiber (CF) anchors which were originally developed by the Japanese Shimizu Corporation were used as a new way to anchor FRP sheets.

2. Experimental study

2.1. Test program

In this study, two RC beams had been tested to failure by four-point bending method. Specimens had cross-sections of 200 mm width by 270 mm height and length of 2,260 mm. One of beams strengthened without anchoring as a control beam named L0, and the other beam named L1 was strengthened by

CFRP sheet and anchored by CF anchors at two ends. Beams strengthened with two steel bars of 12 mm diameter with a clear cover of 35 mm. In the test program, all the beam specimens were tested in four-point bending with an effective span of 1,960 mm and a shear span of 880 mm, as showed in Figure 1. Load was applied in displacement rate of 1.5 mm/min. the beam specimens were instrumented with linear variable differential transformer (LVDT) and strain gauges installed to measure vertical strain at the midspan of beams.

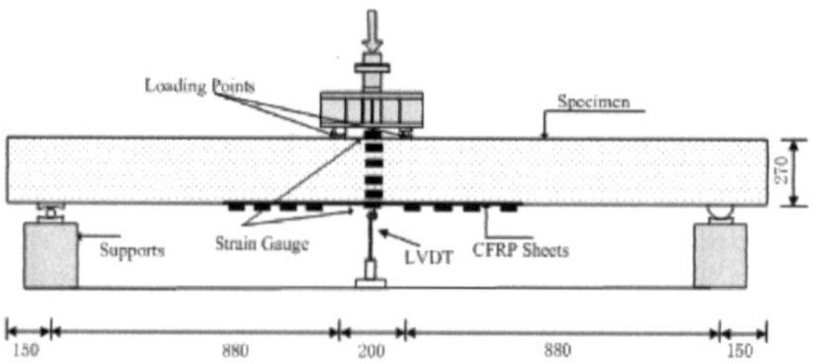

Figure 1. Load test setup

2.2. Settings of CF anchors

CF anchors were made as follow procedures: firstly cut a piece of CFRP sheets which as same as CFRP sheet material bonded in the beam span. Then CF anchors were saturated with epoxy and installed immediately into a predrilled and carefully cleaned holes in the beam face after CFRP sheet for strengthening was bonded. The ends of the CF anchors were fanned out over the CFRP sheet, as shown in Fig. 2.

(a) Drilling holes for anchors (b) Inserting CF anchors into predrilled hole

Figure 2. The mounting process of CF anchors

2.3. Experimental phenomena and results analysis

2.3.1. Cracks and failure mode. The cracks of control beam L0 was wider than strengthened beam L1 while the amount of cracks was less than L1's. Beam L0 shows a larger deflection than beam L1's when both of them was failed. It can be deduced that failure mode of beam L1 was the crush of concrete, which means beam L1 had a greater stiffness. The failure mode of beam that was not strengthened was flexural failure at the middle span, as shown in Fig. 3.

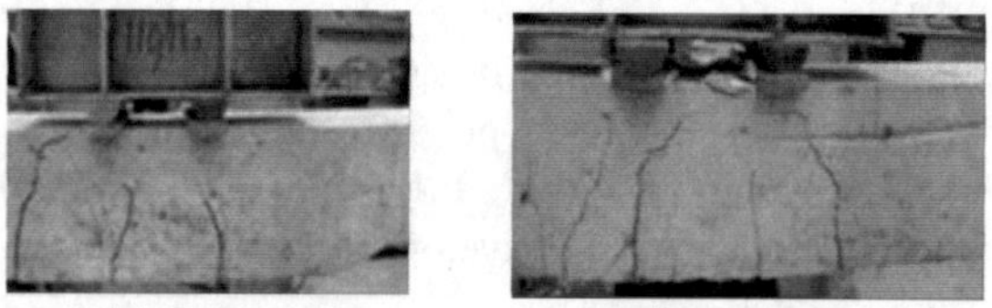

(a) Beam L0 (b) Beam L1

Figure 3. Fracture morphology of test beams

2.3.2. Distribution of strain. The readings of strain gauges mounted in the gauging points as shown in Fig.1 are shown in Fig. 4.

There was an obvious linear distribution of strain along the altitude-direction of the section in the middle span in the elastic stage, which basically agreeing with the plane section assumption. The location of neutral axis of beam L0 was close to the center of the specimen, while the location of neutral axis of beam L1 was lower than L0's because of the CFRP's high elasticity modulus.

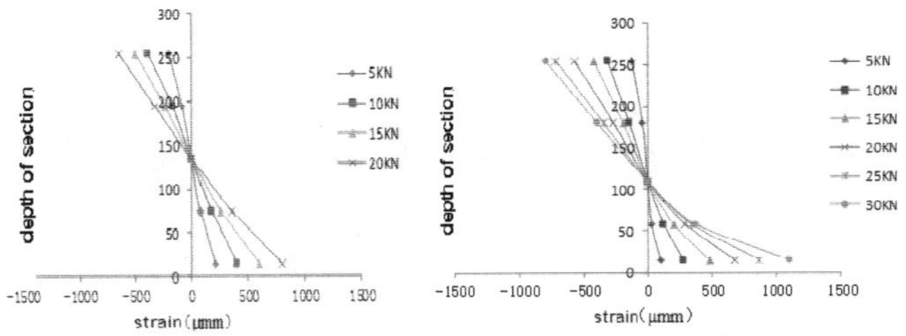

Figure 4. Strain distribution in the middle span of beams

3. Theoretical analysis

The calculation model of the ultimate bearing capacity of the beam strengthened with CFRP sheets was deduced based on failure mode which is concrete crushing. Ultimate compressive strength of concrete was used as a control parameter in the progress of deducing.

3.1. Basic assumptions

a) Hypothesis of plain section which had proved in the beam span strain analysis.

b) CFRP only bears axial tension and do not bear bending moments.

c) Take no account of the concrete resisting tensile effect.

d) Using elastic-all plastic type curve to describe the change of steel bar, regardless of the strengthening phase; Using linear elastic stress - strain curve to describe the change of carbon fiber, magnitude of stress is equal to the strain of CFRP multiplied by the modulus .

3.2. Ultimate bearing capacity under beams destroying

The sectional strain and stress distribution along the section height for CFRP strengthened beams anchored by CF anchors was given in Fig.5.

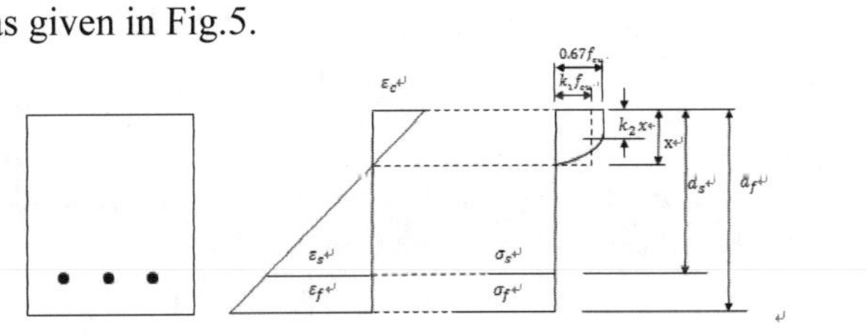

Figure 5. Sectional strain and stress distribution along the beam height

Considering compressive strain was positive, the stress and the steel strain of CFRP sheet was negative. According to the assumption of plane, the CFRP strain ε_f and the steel bars strain ε_s could be calculated as follows:

$$\varepsilon_f = \varepsilon_{cf}\frac{x - d_f}{x}, \qquad \varepsilon_s = \varepsilon_{cf}\frac{x - d_s}{x}. \tag{1}$$

Where x was the distance between concrete compression zone edge and the neutral axis, d_s was the distance between concrete compression zone edge and centroid of steel bars, d_f was the distance between concrete compression zone edge and CFRP's centroid. According to The Chinese Concrete Design Specification (GB50010-2012), the concrete strain could be calculated:

$$0 \le \varepsilon_c \le \varepsilon_0, \sigma_c = f_c\left[1 - \left(1 - \frac{\varepsilon_c}{\varepsilon_0}\right)^n\right], \qquad n = 2 - \frac{1}{60}(f_{cu,k} - 50)\cdot \tag{2}$$

$$\varepsilon_0 \le \varepsilon_c \le \varepsilon_{cu}, \sigma_c = f_c, \ \varepsilon_0 = 0.002 + 0.5(f_{cu,k} - 50)\times 10^{-5}, \ \varepsilon_{cu} = 0.003 - 0.5(f_{cu,k} - 50)\times 10^{-5}. \tag{3}$$

Where σ_c was concrete compressive stress, f_c was the design value of concrete compressive strength, ε_0 was the concrete compressive strain when the concrete compressive stress value was equal to f_c, and when the calculated value of ε_c was less than 0.0002, take it as 0.0002. $f_{cu,k}$ was the compressive strength standard values of concrete cube. n was a given coefficient, when the calculating value of n was greater than 2.0, take it as 2.0. In this paper, the concrete strength level was 30 MPa, so writer set the value of ε_{cu} as 0.0033, n as 2.0.Concrete resultant compressive stress C was calculated as follows:

$$C = k_1 f_c bx, \quad k_1 = \frac{\int_0^{\varepsilon_{cf}} \sigma_c d\varepsilon_c}{f_c \varepsilon_{cf}}. \tag{4}$$

$$0 \le \varepsilon_c \le \varepsilon_0, k_1 = \left(\frac{\varepsilon_{cf}}{\varepsilon_0} - \frac{\varepsilon_{cf}^2}{3\varepsilon_0^2}\right), \qquad \varepsilon_0 \le \varepsilon_c \le 0.0033, k_1 = 1 - \frac{\varepsilon_0}{3\varepsilon_{cf}}. \tag{5}$$

Where b was the width of the beam, k_1 was the average stress coefficient. The static equilibrium equation is:

$$k_1 f_c bx + \sum_{i=1}^{n} \sigma_{si} A_{si} + \sigma_f A_f = 0. \tag{6}$$

Where σ_{si} was the stress of steel bars, σ_f is the stress of CFRP, A_{si}, A_f is a total area of the ith layer and the sectional area of CFRP respectively. n is the total number of steel bars layers. The σ_{si} and σ_f were calculated:

$$|\varepsilon_s| < \frac{f_y}{\gamma_s E_s}, \sigma_{si} = E_S \varepsilon_{si}, \quad |s_s| \ge \frac{f_y}{\gamma_s E_s}, \sigma_{si} = \frac{\varepsilon_{si}}{|\varepsilon_{si}|} \frac{f_y}{\gamma_s}, \quad \sigma_f = E_f \varepsilon_f \ge -\frac{f_f}{\gamma_f}. \tag{7}$$

Where E_s, E_f was the elasticity modulus of steel bar and CFRP respectively. f_f was the tensile strength of CFRP, while f_f was the yield strength of steel bar. γ_s, γ_f was the material subentry coefficient of steel bar and CFRP, and both positive.

If D was the distance between the resultant compressive stress and the edge of the concrete compressive zone, and x was the height of compression zone, then:

$$D = k_2 x, \qquad k_2 = 1 - \frac{\int_0^{\varepsilon_{cf}} \varepsilon_c \sigma_c d\varepsilon_c}{\varepsilon_{cf} \int_0^{\varepsilon_{cf}} \sigma_c d\varepsilon_c}. \tag{8}$$

CFRP achieving ultimate tensile strain or concrete compressive strain reaching to 0.0033(the ultimate compressive strain) were failure signs of RC beam. When the failure mode was the concrete crush (ε_{cu} = 0.0033), from (1) CFRP sheet's stress distribution ε_f could be got, and according to (5), the average stress coefficient k_1 could be calculated.

Based on (1) to (8), the neutral axis depth x could be caudated:

$$|\varepsilon_s| < \frac{f_y}{\gamma_s E_s}, x = -\left(\sum_{i=1}^{n} E_s \varepsilon_{si} A_{si} + E_f \varepsilon_f A_f\right) \Big/ \left(1 - \frac{\varepsilon_0}{0.0099}\right) f_c b. \tag{9}$$

$$|\varepsilon_s| \ge \frac{f_y}{\gamma_s E_s}, x = -\left(\sum_{i=1}^{n} \frac{\varepsilon_{si}}{|\varepsilon_{si}|} \frac{f_y}{f_y} A_{si} + E_f \varepsilon_f A_f\right) \Big/ \left(1 - \frac{\varepsilon_0}{0.0099}\right) f_c b. \tag{10}$$

Then the ultimate bending moment of CFRP reinforced concrete beams was calculated:

$$M_u = k_1 f_c bx\left(\frac{h}{2} - k_2 x\right) + \sum_{i=1}^{n} \sigma_{si} A_{si}\left(\frac{h}{2} - d_s\right) + \sigma_f A_f\left(\frac{h}{2} - d_f\right). \tag{11}$$

Where h was the height of the concrete beam. The right three parts of formula were the contribution of concrete, steel and CFRP to the ultimate capacity.

4. The comparison between the theoretical and the experimental value

The theoretical and the experimental ultimate bearing capacity of experiment beams was shew in Fig.6.

The ultimate bearing capacity of the beam's theoretical value was 45.8kN, which was only 3.4% higher compared to experimental value of 44.3kN. The error may be caused by the bond between concrete face and CFRP sheet. The two kinds of values were very close this indicated that the theoretical formula was corrected according to the hypothesis of the test.

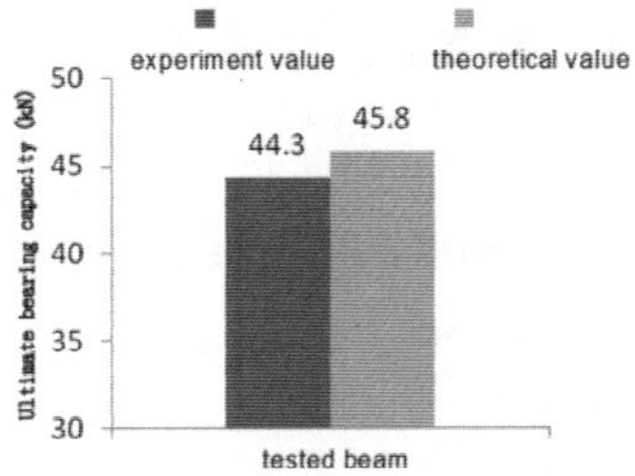

Figure 6. the ultimate bearing capacity value of experiment beams

5. Conclusion

According to the experiment, the failure model of CFRP strengthened beam that was anchored by CF anchors was crash of concrete, while failure model of beam that was not strengthened was flexural failure at the middle of beam. In the process of loading, the measured beam strain along the section height was the linear distribution and basically accord with flat section assumption. Based on the test results, the formula of the bearing capacity of the beam anchored by CF anchors was established in this paper. Compared the experiment and the theoretical analysis, we concluded that the theoretical calculated value shew a good agreement with experimental data, which indicated that the theoretical formula was relatively reasonable.

Acknowledgements

This project is funded by the National Natural Science Foundation of China 51679080 and 51379073.

References

[1] Gangarao H V, Vijay P: Bending behavior of concrete beams wrapped with carbon fabric, J. Journal of Structural Engineering, 1998,124 (1): 3-10

[2] Zhuang, Ning, Zhou, Yujue, Sun, Haodong: Effects of steel reinforcement corrosion on carbon-fibre-reinforced polymer repaired slabs. Proceedings of The Institution of Civil Engineers-Structures and Buildings, 2016, 169(1):46-53

[3] Nguyen D M, Chan T K, Cheong H K: Brittle failure and bond development length of CFRP-concrete beams, Journal of Composites for Construction, 2001, 5(1): 12-17

[4] Deng Z C: Bending mechanical properties research for CFRP strengthen reinforced concrete beam, J. China Journal of Highway and Transport, 2001, 14(2): 45-51

[5] Deng Z C, Li J H: Experimental and theorical study of bending performance of hybrid fiber reinforced concrete beam, Journal of Engineering Mechanics, 2009, 26(2): 115-123

[6] Imran A., Bukhari., Robert Vollum: Saeed Ahmad., Shear Strengthening of Short Span Reinforced Concrete Beams with CFRP Sheets, Arab J Sci Eng., 2013, 38:523-536

[7] Bonacci J F, Maalej F: Behavioral trends of RC beams strengthened with externally bonded FRP, Journal of Composites for Construction, 2001, 5(2): 102-113

IOP Conf. Series: Materials Science and Engineering **191** (2017) 012045 doi:10.1088/1757-899X/191/1/012045

Improve isolation materials and recommendation of residential buildings in Kazakhstan

Yuliya Kim[1 a], Cheng Sun[2,b] Zhmagul Nuguzhinov[3 c]

[1,2] School of Architecture, Harbin Institute of Technology, Heilongjiang Cold Region Architectural Science Key Laboratory, Xidazhi Street 66, Harbin, Heilongjiang Province, China
[3]Kazakhstan Multi-Profile Institute of Reconstruction and Development (KazMIRD) at Karaganda State Technical University, Karaganda, Kazakhstan

E-mail: [a] yuliyakim8610@yahoo.com, [b]suncheng@hit.edu.cn, [c] kazmirr@mail.ru

Abstract. The article shows the problem of high volumes of energy use in Kazakhstan and introduction of energy-efficient technologies, production and causes of the problem. To improve the thermal comfort, needed primarily to improve the thermal insulation properties of materials. This article examines the characteristics of thermal insulation materials and their application in the construction process in Kazakhstan. The material properties are calculated and shown as formulas, which can use in the future in the design of residential buildings. The construction studied of passive technology in the residential area. By the design process and the combination of design variables selected option.

1. Introduction
Technologies of energy [1] saving in cold regions for improvement of the inhabited thermal environment throughout a long time were the primary objective of a research. The leading technologies of energy saving consist use of the solar system, structure of thermal insulation.

A climatic feature of Kazakhstan characterized by a cold climate, which demands during the winter period of heating burning coal that bears serious problems with environmental pollution. At the introduction of energy saving technologies and improvement of thermal characteristics by the critical moment are the protecting designs. A practical design in the green building is the use of resource-saving way out the efficient use of the building during construction. The main principles in the development and practice include the use of green materials, sustainable site development, improvement of environmental quality in the room and improve the efficiency of energy and water consumption. Reducing the impact on the environment, reduce the use of water and energy, as well as contribute to a healthier workspace, which increases the level of comfort and performance of its occupants is the main direction.

Green Building Rating System is the assessment, which is used to measure the environmental performance of the building throughout the building's life cycle in order to quantify the quality of the performance of buildings. This article attempts to consider in Kazakhstan the basic parameters of green building technology for the next generation of guidelines and recommendations. Developers and

[1] This work was supported by National Natural Science Foundation of China (51578172)

designers can establish a platform for efficient energy and can guide the builders defend the stability of the value by integrating construction quality strategies.

For this study, it is the area of Astana is the capital. Climate classification map is according to Koppen-Geiger climate change in Kazakhstan, due to its vast territory and relevant to 4 different zones. For Astana and the surrounding area, which is in the DFB region, characterized by snow, full of moisture and extremely continental weather (cold and freezing winters and hot summers) [1].

Over the past few decades, Kazakhstan has experienced an upward trend in energy consumption, mainly due to rapid urbanization and industrialization. In recent decades, energy consumption in the residential and commercial sectors in Kazakhstan has increased significantly, according to the statistics of Kazakhstan. In figure 1 indicates that the demand for energy is steadily growing [2].

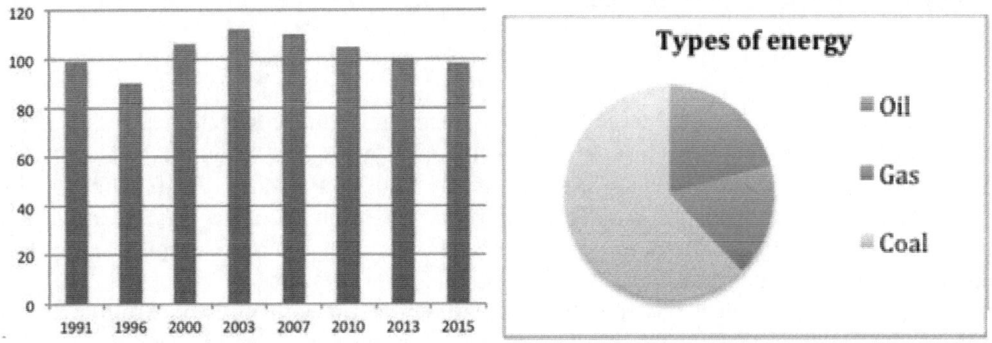

Figure 1. Electricity supply about the previous year in Kazakhstan

2. Improve passive technologies

The purpose of the optimization and implementation of passive technologies improved thermal comfort, a set of strategies to reduce energy consumption. For correct selection find the right balance, which depends on the economic and environmental benefits of energy.

Improving the living conditions of thermal comfort and environmental influences new green building materials. [3]

2.1. General evaluation

The assessment of Energy and Atmosphere category show with the commissioning of the building energy systems and energy efficiency of buildings [4]. Initial construction of the energy system requires an integrated delivery system in operation, which provided input to the authority granted by the schematic design phase to the maintenance of the building [5]. This is intended to ensure that the building energy systems are functioning well and as expected. Since this practice is not yet a part of the construction practice and requires commissioning professionally, it needs to put much effort, work and costs for the implementation of this entry tasks into operation. The energy efficiency requires the building to develop by the standard 2.04-21-2004. The construction of the building has requirements a very high level regarding insulation and energy-efficient features. However, it is not part of the building design and construction practices in Kazakhstan. Today in Kazakhstan is used when designing new software projects for energy modelling is another point for implementation. When using an approach based on the international experience result, identify some characteristics to be included in the base of the building to determine the energy budget.

2.2. External walls

Exterior walls play a major role in maintaining the heat inside the building. From the architectural side of the wall does not bear the aesthetic component. Thermal insulation as schematically represented in Figure 2. To identifies the optimal insulation of external walls admission energy-efficient properties of materials one of the external walls.

Calculation of thermal resistance coefficient attic floor should include: coloring concrete twice (not involved in the calculation); screed of cement sand mortar M150, reinforced mesh ∅5V-carbon Solid， surface waterproofing, reinforced concrete slab.

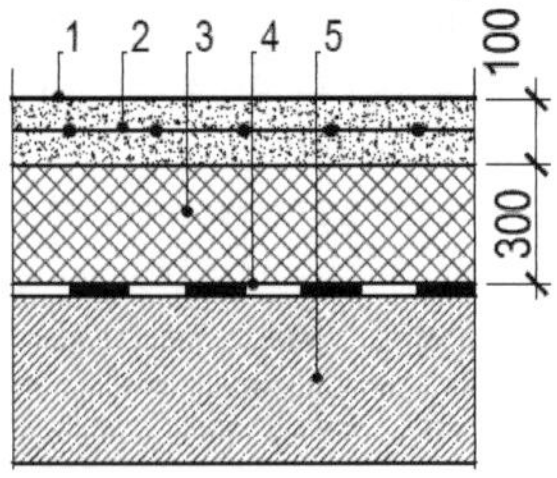

1.Coloring Concrete twice (not involved in the calculation);2. Screed of cement sand mortar M150, reinforced mesh ∅5V-1, cell 200×200 mm: δ =60 mm, p_0=1800 kg/m^3, λ_p=0,76 W/m·℃ 3.Carbon Solid 1000: δ =300 mm, p_0=38 kg/m^3, λ_p=0,032 W/m·℃ ;4. Surface waterproofing δ=4 mm, p_0=1400 kg/m^3, λ_p=0,26 W/m·℃ ;5. Reinforced concrete slab: δ=250 mm. p_0=2500 kg/m^3, λ_p=1,92 W/m·℃ .

Figure 2. Thermal resistance coefficient attic floor

As for heaters, Bimetallic sectional radiators of the brand, as well as connectors, recessed into the floor with forced convection. Doors piping and wiring horizontal pipe technical and commercial facilities provide polyethylene reinforced pipes, laid in the floor construction. All pipelines, vertical risers to cover Ø40 mm pipe insulation made of foam rubber, 9 mm thick and more Ø40mm - tubular foam insulation 13mm thick. Distribution pipes (for each apartment heating systems), set up to the floor construction, and insulated tubular foam insulation thickness of 9 mm.

In residential areas, the project provides for exhaust ventilation with mechanical prompting the installation of internal axial fans. In the public areas of the project includes the supply and exhaust ventilation with the mechanical drive. Air supplied to the loft insulated mats URSA, insulation thickness - 50 mm, covered with foil, the cover layer of fiberglass. Mathematical analysis can be performed using the following formulas [6], [7].

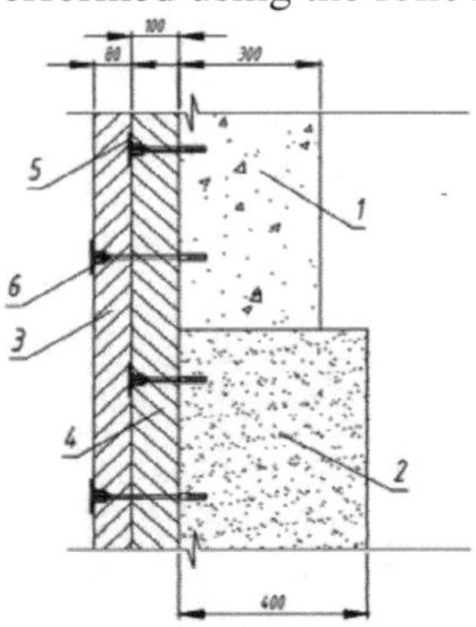

1.Masonry of aerated concrete block: δ =300 mm, p_0=600 kg/m^3, λ_p=0,026 W/m·℃ .
2.Reinforced concrete (not involved in the calculation).
3. Mineral Plate Ecover Vent-Facade 80 δ =80 mm, p_0=80 kg/m^3, λ_p=0,039 W/m·℃
4. Mineral Plate Ecover Vent-Facade 60 δ =100 mm, p_0=60 kg/m^3, λ_p=0,041 W/m·℃;
5. Umbrella anchor the upper layer (not involved in the calculation

Figure 3. Calculation of the coefficient of resistance of heat exterior walls

Calculation of resistance to heat transfer coefficient on the ground floor (Elevator hall, staircase, 1st floor hallway) include: gres, cement-sand screed, monolithic foam concrete, concrete class, compacted soil backfill, reinforced concrete slab.

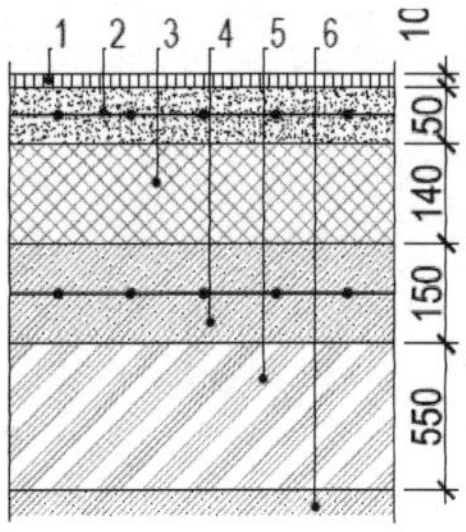

1. Gres: δ =15 мм; p_0=2800 kg/m^3, λ_p=3,49 W/m·℃; 2. Cement-sand screed δ =50 mm; p_0=1800 kg/m^3, λ_p=0,76 W/m·℃; 3. Monolithic foam concrete: δ =140 mm, p_0=600 kg/m^3, λ_p=0,22 W/m·℃ ; 4. Concrete class B25: δ =150 mm, p_0=1800 kg/m^3, λ_p=0,76 W/m·℃ ; 5 Compacted soil backfill: δ =550 mm, p_0=1600 kg/m^3, λ_p=0,47 W/m·℃ ; 6. Reinforced concrete slab: δ=200 mm. p_0=2500 kg/m^3, λ_p=1,92 W/m·℃.

Figure 4. Reinforced concrete slab

IOP Conf. Series: Materials Science and Engineering **191** (2017) 012045 doi:10.1088/1757-899X/191/1/012045

The coefficient of heat transfer resistance uniform transparent building envelope is calculated as follows(1)[8]:

$$R_{\Sigma} = \frac{1}{\alpha_{\text{int}}} + \sum_{i=1}^{n} R_i + \frac{1}{\alpha_{ext}} = \frac{1}{\alpha_{\text{int}}} + \sum_{i=1}^{n} \frac{\delta_i}{\lambda_i} + \frac{1}{\alpha_{ext}},\ m^2C^\circ / W \tag{1}$$

α_{int}, α_{ext} — heat transfer coefficients of the inner and outer surfaces of the building envelope;

R_i — thermal resistance of each layer of the structure;

λ_i — the thermal conductivity of each material in the layer structure of settlement conditions.

The above non-uniform thermal resistance to heat opaque building envelope is determined by the formula(2):

$$R_{\Sigma 0} = \frac{1}{\alpha_{\text{int}}} + \sum_{j=1}^{j} \frac{R_i * F_i}{F_{\Sigma}} + \frac{1}{\alpha_{ext}} \tag{2}$$

R_j — thermal resistance of the homogeneous zone;

F_j — the area of each thermally homogeneous area;

$F\Sigma$— area of the building envelope

The calculated checked value of the thermal conductivity of the walls of resistance to limit the temperature drop using the following relationship (3):

$$\Delta t_0 = \frac{n(t_{\text{int}} - t_{ext})}{R_0 \alpha_{\text{int}}} \tag{3}$$

n - coefficient reflecting the dependence of the position of the outer surface of the partition structure in relation to the outside air;

t_{int} - the calculated average temperature of the internal air of the building;

t_{ext} - the estimated temperature in the cold season;

α_{in} - heat transfer coefficient of the inner surface of the building envelope.

The heat transfer coefficient of resistance through the exterior building envelope is calculated as follows(4):

$$K_m^{tr} = \frac{\frac{A_w}{R_w} + \frac{A_F}{R_F} + \frac{A_{ed}}{\text{R}_{ed}} + \frac{A_c}{R_w^r} + n\frac{A_{cl}}{R_{cl}} + n\frac{A_f}{R_f^r} + \frac{A_{f1}}{R_{f1}^r}}{A_e^{sum}} \tag{4}$$

A_w, R^r_w— area and reduced resistance to heat exterior walls;

A_F,R^r_F— fillings of light apertures (windows, stained-glass windows, lamps);

A_{ed},R^r_{ed}— inside and outside doors;

A_c,R^r_C— combined coatings (including over the bay windows);

A_{c1},R^r_{cl}— attic floors;

A_f,R^r_f— basement floors;

A_{f1},R^r_{f1}— ceilings over passages and under the bay windows;

A^{sum}_e— the total area of the internal surfaces of external walling.

$$Q_h^y = (Q_h - (Q_{\Delta nt} + Q_s) \cdot \upsilon \cdot \xi) \cdot \beta_h \tag{5}$$

v — reduce heat gain coefficient due to the thermal inertia of walling, are taken into account $v = 0{,}8$;

ζ— effectiveness ratio of auto heat supply in heating systems, are taken into account $\zeta = 0{,}95$;

β_h — coefficient taking into account the additional heat consumption of the heating system, associated with discrete nominal heat flow heaters, are taken into account $\beta_h = 1{,}11$.

The construction process is necessary the use of local materials. At the initial stage of the project help the developer to choose the right materials. The advantages of energy-efficient technologies are especially popular in residential buildings. To improve the energy efficiency potential are encouraged to increase the use of renewables. The primary objective of reducing the environmental and economic impacts associated with energy consumption. The use of passive technology allows buildings to achieve the lowest energy requirements by adjusting the heat gain and heat loss.

3. Conclusion

In conclusion, the green technology for energy efficiency in Kazakhstan (Astana) is possible only in the case of applications with regulatory standards in Kazakhstan. Further research on the breakdown of energy-saving and energy costs has indicated that dominated by the performance of the air conditioning system, heat-insulating system, but the difference in the tariff systems between cities. Additional savings achieved by energy efficiency measures were attractive.

The authors suggest a green building in Kazakhstan to develop a green building, which bases on the context in Kazakhstan. The development of passive technology provided for region. However, the criteria for the pre-conditions and credit should be based on local design and building construction methods. The most significant result of discussions, among all stakeholders, and to discuss it architects, engineers, ecologists, governments, communities and others.

References

[1] A reference IVPH, 2014, World map of the Koppen-Geiger climate classification updated. Adopted from http://koeppen-geiger.vu-wien.ac.at/present.htm

[2] Ministry of National economy of the Republic of Kazakhstan committee on Statistic available http://www.stat.gov.kz/faces/homePage;jsessionid=yMn2Y00cJJNDpKWKGL3pLGmFZ6nv6ZRcmQFgsDHX95rTn7p29HXS!-1918550083!NONE?_afrLoop=1321418147775328#%40%3F_afrLoop%3D13214181477753288%26_adf.ctrl-state%3Ddns1fc6t0_4

[3] Wang Z, Han F. New Green Building Materials [J]. Applied Mechanics & Materials, 2016, 851:639-642.

[4] Chen Y, Ng S T. Integrate an Embodied GHG Emissions Assessment Model into Building Environmental Assessment Tools [J]. Procedia Engineering, (2015), 118:318-325.5.

[5] Barbosa R, Vicente R, Rui F D S. Comfort and buildings: climate change vulnerability and strategies [J]. International Journal of Climate Change Strategies & Management, (2016)

[6] Kazakhstan construction standard 2.04-21-2004

[7] Suzer O. A comparative review of environmental concern prioritization: LEED vs other major certification systems. Journal of Environmental Management. 2015; 154: 266-283.

[8] LEED v4 for BUILDING DESIGN AND CONSTRUCTION, Updated April 5 (2016) Published by U.S. Green Building Council.

IOP Conf. Series: Materials Science and Engineering **191** (2017) 012046 doi:10.1088/1757-899X/191/1/012046

Analysis of defects of overhead facade systems and other light thin-walled structures

L.ENDZHIEVSKIY[1,a], A.FROLOVSKAIA[1,b], Y.PETROVA [1,c]

[1]School of Engineering and Construction, Siberian Federal University, 82 Svobodny, Krasnoyarsk, 660041 Russia

E-mail: [a]end-lev@yandex.ru, [b]tereshkova81@mail.ru, [c]yevtifeva@mail.ru

Abstract. This paper analyzes the defects and the causes of contemporary design solutions with an example of overhead facade systems with ventilated air gaps and light steel thin-walled structures on the basis of field experiments. The analysis is performed at all stages of work: design, manufacture, including quality, construction, and operation. Practical examples are given. The main causes of accidents and the accident rate prediction are looked upon and discussed.

1. Introduction

Currently, the building market at all levels is overfilled with materials, technologies and design solutions. The flow of research and developments in various areas is so huge that it is not always easy even for a specialist to understand what is better to use for the object construction in this or that case.

Modern materials and design solutions must meet requirements of a customer such as longevity, durability and strength, high thermal performance, environmental friendliness and cost effectiveness, simplicity and speed of assembly, flexibility of structures, etc. All progressive methods and technologies practically satisfy these indicators. But in fact it turns out that some structures ask for more detailed, qualitative study and approach [1].

The authors of the monograph [2] have shown that accidents of buildings and structures occur due to either human activities (violation of production technology, lack of discipline, inattention, etc.) or natural hazards (earthquakes, hurricanes, floods, etc.). At that statistics shows that the accident is the result of a number of complex reasons: poor funding of measures to ensure safe operation and timely repairs, insufficient study of the causes of the occurred tragedies that often lead to repetitive errors during design, construction and installation.

A study of the causes of accidents and disasters will enable us to better understand the patterns in the structures, buildings and installations, as well as to attract the attention of scientists, planners and builders to disadvantages, shortages, and drawbacks of materials, technologies and design solutions.

In this paper, the authors analyze two types of modern design solutions, which are used in practice. Overhead facade systems (OFS) and buildings from light steel thin-walled structures (LSTS) are taken as objects of research. These constructive solutions are characterized by the fact that, according to manufacturers, they are easy and simple to be assembled. In advertising they are compared with the children's toy design kit.

2. Overhead facade systems

Facade systems of modern buildings are as popular [3,4], as complex in design, installation and operation. Any technical mistake associated with their installation or operation becomes immediately obvious, in contrast to defects of internal structures.

In the OFS design there appear the following defects, leading to their partial collapse: the lack of a unified approach and the calculation method of the bearing and fastening elements, reliable data about the real behavior of the structure, calculations of thermal effects and other factors.

In the analysis of defects, depending on the quality of materials and products one can mention: the use of substandard materials or their replacement by materials of a lower quality, uncertified without the consent of the manufacturer of facade systems.

For example, the replacement of clasps from stainless steel by low carbon steel has led to the fact that some operation years later most of the façade elements have come out of its design position (Fig. 1), which has resulted in unpresentable appearance of the building.

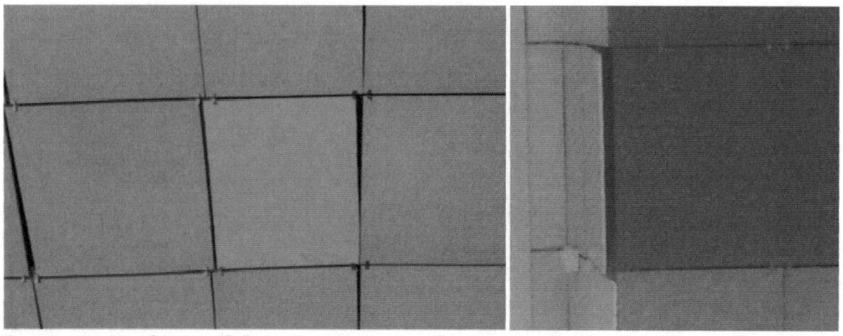

Figure 1. Defects of façade systems: facing walls design inconformity

Replacement of composite cladding panels by a lower quality class of flammability led to the fire of the facade of a multistory apartment building (Fig. 2) in September 2014 in Krasnoyarsk (Russia).

Figure 2. A residential building is on fire in Krasnoyarsk (Russia)

Inflammation of overhead façade systems is quite common fact.

In the researchers point of view the main causes of the façade system fires are the failure of fire prevention measures and savings in all phases of construction. Unscrupulous construction participants artificially create low cost structures design by eliminating unnecessary, from their point of view, trials, studies, calculations, installation of additional fire nicks, etc.

One of the things that contribute to the fire danger is combustible materials, the spread of flaming droplets during melting of materials and other factors [5]. One of the most flammable materials in the design of facade systems is a hydro windproof film (membrane), which performs the function of protecting the insulation layer against weathering. As a result of its burning, the fire spreads throughout the structure of the facade (Fig. 3). Noncombustible membrane application leads to an increase in the cost of the entire system.

Figure 3. Membrane burning in the facade system design

To eliminate fires the builders have recently started to refuse its use. The result is that the insulation is not protected from atmospheric influences, and after a while it begins to disintegrate into fibers, and this leads to the fact that the fibers clog the air gap between the insulation and the system facing material. Therefore, there is a violation of the functioning of the whole structure of the overhead facade system with a ventilated air gap (Fig. 4). But the gap is required for effective removal of moisture and vapors migrating from the premises through the outer wall to the street [6]. Its application is clearly stipulated in the design of a façade system, and the size is calculated depending on the main characteristics of the thermal protection enclosing structures (resistance to heat transfer) [7].

Figure 4. Result of hydro windproof membrane refusal

On the basis of the above facts, we can conclude that while trying to solve one problem, we find a new one. That is like going in circles. And so it will be as long as someone from the experts does not suggest a new solution or a new material that will satisfy all the interested parties.

In construction one can mention such disadvantages as insufficient qualification of workers and engineers, low degree or absence of control, violation in the sequence of operations ("chase" for terms).

Fig. 5 shows defects in overhead wall systems, resulting from poor quality of the installation. The destruction of the cladding material occurs as a result of the fact that in the erection of the vertical profiles temperature joints have not been provided, moreover, they were concreted in the pavement around the building. As a consequence, thermal deformation of the profiles occurred, qualitative changes in the design of calculating schemes of bearing elements and the values of the loads acting on them took place.

Figure 5. Destruction of the cladding material

Fig. 6 shows the defects in the installation of overhead facade system caused by the negligence of the builders.

Figure 6. Defects in the process of overhead façade system installation

Having analyzing the defects, we come to the conclusion that the use of low-quality overhead façade l systems does not meet requirements as for energy savings and presentable appearance of the building as a whole. Overhead facade systems in the construction market have been used for more than a dozen years, but for their reliable and safe use not only monitoring to identify and study errors for their further elimination are require, but more accurate and strict control at all stages of work with them.

3. Light steel thin-walled structures

One of the main problems in the LSTS construction is that the building is destroyed almost immediately (brittle fracture) and fully, and the defective items cannot be applied again (for the second time).

When designing this constructive solution it is possible to distinguish such aspects as the lack of regulatory framework applicable on the territory of the Russian Federation, the overestimation of mechanical characteristics, etc.. Manufacturers point out fire resistance, eco-friendliness and energy saving as advantages. These indicators do not directly depend on the LSTS structures application and, according to the authors, on the contrary do not show the use of thin-walled profiles. These indicators depend on the quality of the insulating layer and the type of facing material for covering walls or ceilings and surfaces [8]. Consequently, these structures must be viewed as a single system, and manufacturers need not only to calculate load-bearing structures, but also consider other materials. For example, builders use polystyrene as an insulation material in façade systems for the sake of economy, but when it burns it emits not only harmful black smoke, but the styrene, causing respiratory organs damage in humans [7].

Fig. 7 shows the alarm status of a corrugated frameless shell of the arched building with a span of 21 m in the period of installation. The deficiency of the design was the result of its delayed grouting (fixing) in the foundation. Furthermore, experts proved overestimation of mechanical properties of the steel in the certificate after the tests [9].

Figure 7. Loss of local stability of a corrugated shell

The factors influencing defectiveness in the manufacture of structures are: low-quality production and original manufacturing defects, failure to comply with the technology application of protective coatings.

Stage of construction of buildings is partially characterized by the violation of rules of installation and lack of proper control. For example, the collapse of LSTS structures (Fig. 8) resulted from changing the cover disk and the loss of overall stability due to the neglect of the assembly process [9].

Figure 8. LSTS destruction

Examples of accidents of LSTC buildings testify to the need for a comprehensive approach using structural measures in the development of this technology. Failure analysis of the considered structures allows to establish the main causes of accidents:

- Defects and low quality construction works,
- Project non-compliance,
- Use of the materials and structures of insufficient strength,
- Replacement of structures or parts without the approval of design organizations,
- Overload of the bearing structures in the process of operation,
- Lack of reliable means and methods of corrosion protection

Insufficient knowledge of the work of some structures under loads is also looked upon as one of the causes of failure.

Predictions of accidents:

- Tendencies of reduction of accidents of buildings and structures are not being observed.
- - Significant influence of human factor on the accident rate and the severity of the consequences of the accident are not to be excluded.
- - The severity of the consequences of accidents can be foreseen and the total number is reduced by elimination of repetition of similar errors and the causes of the accident, the "barbaric" operation and illiterate operation of maintenance work.

4. Conclusion

In conclusion we can say that each technology has its pros and cons. Having chosen "cheap, shoddy" technology we can economize, but make a huge mistake at that, which will not be corrected. It is important to have an adequate assessment of the real level of possible danger and severity of

IOP Conf. Series: Materials Science and Engineering **191** (2017) 012046 doi:10.1088/1757-899X/191/1/012046

consequences of potential accidents, and an understanding that ensuring the security of buildings and structures is an important and significant component.

References

[1] S. Ivorra, J. Garcна-Barba, M. Mateo, C. Perez-Carramicana, A. Macia. Partial collapse of a ventilated stone facade. Diagnosis and analysis of the anchorage system. Engineering Failure Analysis: Vol. 31 (2013), pp. 290–301
[2] L.V. Endzhievskiy, A.V .Tereshkova, The history of failures and catastrophes. Krasnoyarsk: Sib. Feder. University (2013), p. 440
[3] M.A. Shameri, M.A. Alghoul, K. Sopian, M. Fauzi M. Zain, Omkalthum Elayeb. Perspectives of double skin facade systems in buildings and energy saving. Renewable and Sustainable Energy Reviews: Vol. 15 (2011), pp. 1468–1475
[4] Curtain Wall Systems: A Primer, Committee on Curtain Wall Systems. Edited by Ali M. Memari,Ph.D., P.E. M.ASCE, American Society of Civil Engineers (2013)
[5] L. Junmei, X. Xuefei, H. Cheng, L. Yanfeng, Y. Chenchen, L. Shanshan. Numerical studies on effects of cavity width on smoke spread in doubleskin facade. 2012 International Symposium on Safety Science and Technology. Procedia Engineering: Vol. 45 (2012), pp. 695 – 699
[6] Recommendations on design of overhead façade systems with a ventilated air gap for new construction and reconstruction. Administration of Perspective Design and Standards of Moscow Committee of Architecture #20 (2002)
[7] Information on http://fasadec.ru
[8] Y. Cheng, in Recent Trends in Cold-Formed Steel Construction, 1st Edition, edited by Woodhead Publishing, Denton, Texas (2016)
[9] Information on http://forum.dwg.ru

IOP Conf. Series: Materials Science and Engineering **191** (2017) 012047 doi:10.1088/1757-899X/191/1/012047

Investigation of using Polyelectrolytes as an Interlayer on Polymer Solar Cells

Wei-Chih Chen1 , Yung-Cheng Hsiao1 , Yi-Chiang Huang1 , Hsu-Feng Lee1 and Wen-Yao Huang1,a

1 Department of Photonics, National Sun Yat-sen University, 80424 Kaohsiung, Taiwan, R.O.C.

wayne49914020@gmail.com

Abstract. A new approach to improve hole extraction anode interfacial layer by introducing polyelectrolytes in polymer solar cells (PSCs). The polyelectrolytes interfacial layer is prepared simply spin-coating on the ITO substrate. Remarkable improvement in the open-circuit voltage(Voc) and short-circuit current density (Jsc) of the PSCs could be achieved upon the introduction of polyelectrolytes anode interfacial layer.To study the effect of polyelectrolytes anode interfacial layer on the device efficiency. The polyelectrolytes are analyzed, exhibited good thermal stability and high transmittance over 85% in visible light region. According to our experiments and measurements, insertion of polyelectrolytes anode interfacial layer can decrease spatial barriers at the active layer/ITO interfaces, planarize the ITO substrate and modify surface of ITO.The PSCs under the optimized structure of ITO/SA8/P3HT:PCBM/LiF/Al exhibited open-circuit voltage of 0.62 V, short-circuit current density of 7.15 mA/cm^2, fill factor of 54.84%, and power conversion efficiency of 2.43% at AM 1.5G of 100 mW/cm^2

1. Introduction

The mixture of two polymer Poly(3,4-ethylenedioxythiophene) (PEDOT) : poly(styrene sulfonate) (PSS) are well-known materials as the hole extraction anode interfacial layer in PSCs[1]. Lithium fluoride that was usually plated on cathode was an inorganic compound. It can enhance electron injection which constitute a coupling layer to improves the polymer/metal contact[2]. The PEDOT :PSS were substituted by polyelectrolytes in our previous work. Two kinds of novel materials were developed to have high solubility in our study. In this work, we applied Two kinds of materials, SA8(Fig. 1a) and SP4c(Fig. 1b), to substitute hole transport layer and electron injection layer.

Kang et al. (2008) reported a buffer layer for a polymer solar cell. Insertion of the PTFE buffer layer at the anode of the ITO/organic interface can significantly improve the Jsc, Voc, and PCE of PSCs[3].

Xu et al. (2013) through a low temperature and inexpensive method of spin-coating a copper acetylacetonate ($Cu(acac)_2$) 1,2-dichlorobenzene solution onto ITO. The device with CuO_x anode buffer layer demonstrated improved PSCs performance compared with PEDOT:PSS anode buffer layer[4].

In our lab previous work, Zhou et al. used a molecular structure, 6F (Fig. 1c), as anode interfacial layer in PSCs[5]. The structure and characteristic have reported in his thesis. Zhu and Zhang et al. utilized a small molecular, 6F, that designed and synthesized two kinds of polymers, SA8 and SP4c, respectively. Both of their structure and characteristic have reported in their thesis , respectively[6,7].

IOP Conf. Series: Materials Science and Engineering **191** (2017) 012047 doi:10.1088/1757-899X/191/1/012047

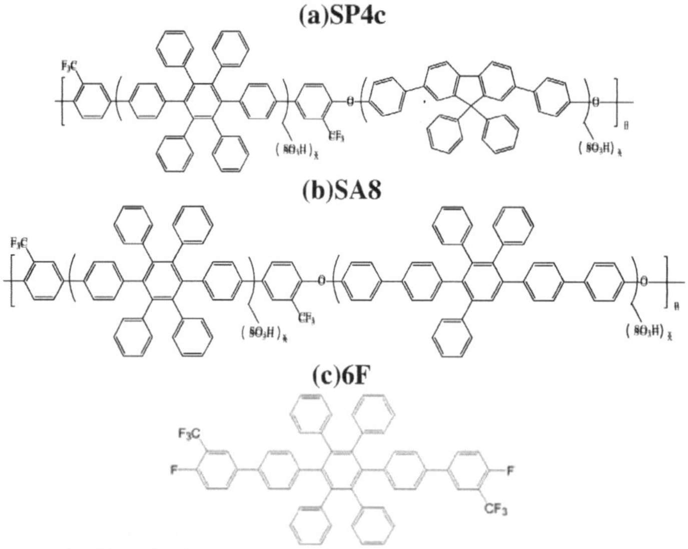

Figure 1. Chemical structures of compounds SP4c and SA8 and 6F.

2. Experimental

2.1. Materials and characterization

The temperature of thermal degradation (Td) was obtained at the point of 5% weight loss by thermal gravimetric analysis (TGA) which was conducted on a Perkin Elmer PYRIS 1. Under nitrogen condition, both of two polymers performed an excellent thermally stable ability. Under a heating rate of 10°C min^{-1}, Td5% was observed of 298°C and 284°C for SA8 and SP4c.

The poly(3-hexylthiophene) (P3HT) : phenyl-C61-butyric acid methyl ester (PCBM) was as an active layer in this study[8-11]. In order to investigate the effect of SA8 and SP4c as the buffer layer, polymer solar cells were fabricated with configuration of ITO/PEDOT :PSS/P3HT :PCBM/SA8 or SP4c or LiF or none/Al[4].

To achieve energy level matched, the work function of SA8 and SP4c were measured to be -5.04eV and -4.95eV, respectively. Highest occupied molecular orbital (HOMO) of PEDOT :PSS was observed at -5.0eV photo-electron spectroscopy(AC-2). Demonstrated energy level of polymers was suitable as anode interfacial layer. HOMO was conducted on a RIKEN KEIKI Surface Analyzer model(AC-2).

A 40 nm thick layer of PEDOT :PSS was spin-coated onto the glass substrates. The morphological stability and film-forming ability of the three samples was observed by atomic force microscopy. The atomic force microscopy images of all the samples was shown in Figure 4.

2.2. Device fabrication and testing

PSS indicates that the PEDOT has been doped with polystyrene sulfonic acid in order to enhance the conductivity of the film[12,13]. A PEDOT :PSS-covered ITO anode reduces the operating voltage and increases the device lifetime remarkably, which was attributed to a lower hole injection barrier between ITO and the hole transport layer as well as a smoother surface than bare ITO[14]. Therefore, there have been a lot of reports about improveed hole injection[15]. It was interesting that the PEDOT :PSS was substituted by polyelectrolytes. The structure used in this work were ITO/PEDOT :PSS or SA8 or SP4c or none/P3HT :PCBM/ LiF/Al. Then compared with PEDOT :PSS that has a similar work function(Table 1). All manipulations involving air-sensitive experiments were carried out in the nitrogen glove box.

The performances of the solar cells were obtained from J-V characteristics measured using a Keithley 2400 instrument under 100 mW cm^{-2} with AM1.5 G[16].

The EQE measurements were performed using solar simulator (Newport 91160A) obtained from National Renewable Energy Laboratory calibrated silicon cell photodiode used as a reference diode.

Table 1. Work function of the SP4c、SA8、PEDOT:PSS

	SP4c	SA8	PEDOT :PSS
Work function (eV)	4.95	5.04	5.0

3. Results and discussions

Since rigid of the polymer material for the backbone of benzene structure has a good thermal stability. According to the information[17,18], the first pyrolysis temperature was about 200 ~ 400 ℃ with sulfonate group. Then, the second pyrolysis was from about 400 ~ 600 ℃ by backbone pyrolysis (Fig. 2).

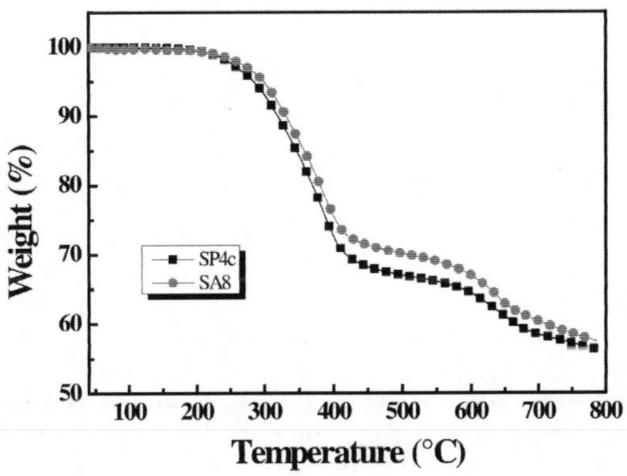

Figure 2. TGA curves of the polymers.

Two polymer electrolyte material SP4c, SA8 was dissolved in dimethyl sulfoxide (DMSO) to form a 0.05wt% solution. The device fabrication throughed spin coating a cathode interface layer. The devices structure were displayed in Table 2. The J-V curves characteristic of devices were displayed in Fig 3, related data are shown in Table 3.

Table 2. The structure of devices.

Device1 : ITO / PEDOT :PSS / P3HT :PCBM / Al

Device2 : ITO / PEDOT :PSS / P3HT :PCBM / SP4c / Al

Device3 : ITO / PEDOT :PSS / P3HT :PCBM / SA8 / Al

Device4 : ITO / PEDOT :PSS / P3HT :PCBM / LiF / Al

The short-circuit current (Jsc) of SP4c and SA8 improved from 6.15mA / cm2 to 7.26 mA / cm2 and 8.32 mA / cm2(Table 3) , respectively. Compared with lithium fluoride, the open circuit voltage (Voc) of SP4c and SA8 can not reach to 0.62V. It can also be seen the characteristics of SP4c and SA8 from the fill factor (FF). They were not improved between the active layer and cathode.

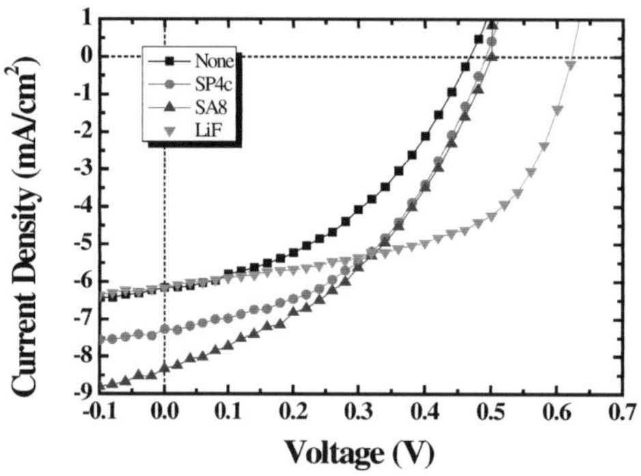

Figure 3. J-V curves of the different cathode interfacial layer

Table 3. Devices characteristic of PSCs at the different cathode interfacial layer

Cathode interfacial layer	V_{oc}(V)	J_{sc} (mA/cm^2)	FF(%)	PCE(%)
None	0.46	6.15	43.41	1.23
SP4c	0.50	7.26	45.77	1.66
SA8	0.50	8.32	40.84	1.70
LiF	0.62	6.15	55.48	2.12

In order to confirm that the polyelectrolytes can optimize interface roughness, we scanned surface by the atomic force microscopy. For the roughness of the anode interface layer, the ITO, ITO / PEDOT : PSS, ITO / SP4c, ITO / SA8 Root Meat Square(RMS) roughness value were obtained 4.30, 4.16, 3.83 and 3.13 nm (Table 4), respectively. The SA8 showing a relatively flat surface, 3.13nm, thereby enhanced holes capture, in favor of the active layer spin coating.

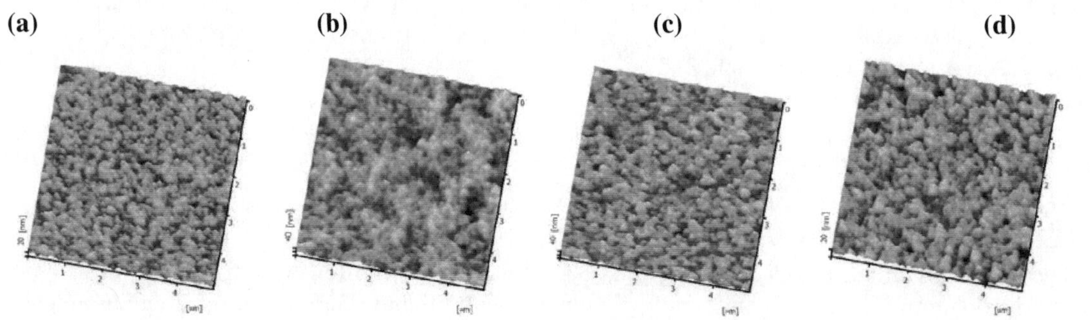

Figure 4. AFM images of dimension 5 × 5 μm. (a) ITO、(b) ITO/PEDOT :PSS、(c) ITO/SP4c、(d) ITO/SA8

Table 4. The roughness of the different anode interface layer

	ITO	ITO/PEDOT :PSS	ITO/SP4c	ITO/SA8
Ra (nm)	3.51	3.30	3.03	2.53
RMS (nm)	4.30	4.16	3.83	3.13

In this work, Both of SP4c and SA8 have energy level about 5.0eV. Control devices without the presence of the PEDOT :PSS hole transporting layer (Device 1) and devices based on PEDOT :PSS (Device 4) were also fabricated for comparison (Table 5).

Table 5 The structure of devices.
Device1 : ITO / P3HT :PCBM / LiF / Al
Device2 : ITO / SP4c / P3HT :PCBM / LiF / Al
Device3 : ITO / SA8 / P3HT :PCBM / LiF / Al
Device4 : ITO / PEDOT :PSS / P3HT :PCBM / LiF / Al

According to the results depicted in Fig 5. Polymer solar cells (PSCs) reached a maximum short-circuit current (Jsc) and power conversion efficiency of 2.43% and an external fill factor (FF) of 54.84% was also determined. Such results indicate that, introduced into the polyelectrolytes contribute to improve the efficiency of PSCs.

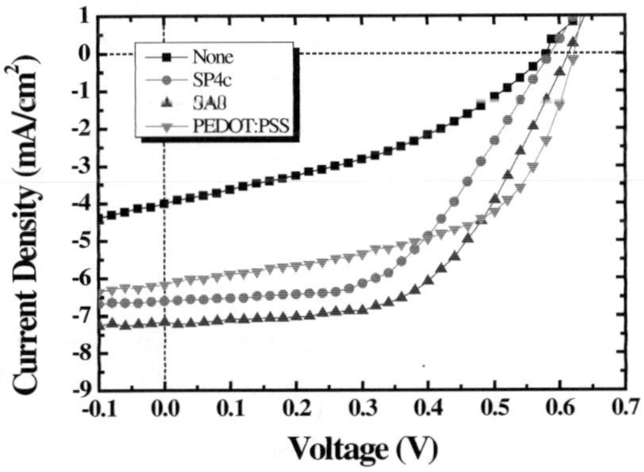

Figure 5. J-V curves of the different anode interfacial layer

The best performance was achieved in device 3 (Table 6), which showed the short-circuit current (Jsc) of 7.15mA/cm2, the open circuit voltage (Voc) of 0.62V (Fig. 5). Enhance the short-circuit current (Jsc) was attributed to a relatively flat surface, reducing the overall device resistance, resulting ITO surface properties change. Overall, the anode interface layer applied polyelectrolytes to fabricate PSCs device obtained good results.

Anode interfacial layer film transmittance was an important condition for the absorption of light. In order to increase the light absorption of the active layer, the anode interfacial layer material must have a low extinction coefficient in the range of visible light wavelength. The transmittance of the spin-coated thin film of polyelectrolytes was measured by the UV-Vis Spectrometer. PEDOT :PSS and polyelectrolytes was observed above 40nm and 10nm, respectively.

IOP Conf. Series: Materials Science and Engineering **191** (2017) 012047 doi:10.1088/1757-899X/191/1/012047

Table 6. Devices characteristic of PSCs at the different anode interfacial layer

Anode interfacial layer	V_{oc}(V)	J_{sc} (mA/cm^2)	FF(%)	PCE(%)
None	0.58	4.02	38.30	0.89
SP4c	0.58	6.60	52.29	2.00
SA8	0.62	7.15	54.84	2.43
PEDOT :PSS	0.62	6.15	55.48	2.12

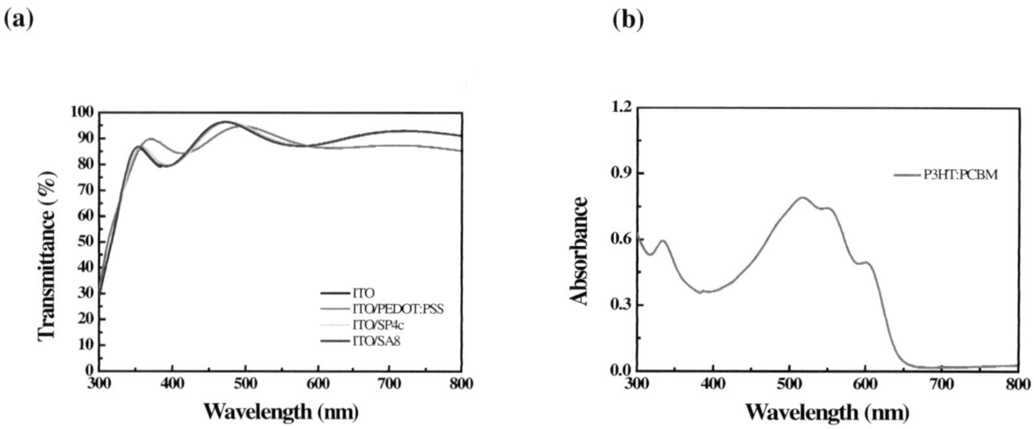

Figire 6. UV-vis spectrum of the (a) anode interfacial layer (b) active layer

The transmittance of all the samples was shown in Figure 6(a). It demonstrated polyelectrolytes was different from PEDOT :PSS advantage on transmittance. In addition, we measured the absorption spectra of the active layer was shown in Figure 6(b).

To further understand the impact of polymer on ITO, contact angle measurement was a useful system. Anode interface layer surface energy can be calculated by liquid contact angle, and explained hydrophilicity. In this work, deionized water and diiodomethane was dropped to make the contact angle measurements. For deionized water, the PEDOT :PSS, SP4c, SA8 exhibited contact angle of 40.9°, 28.1°, 20.8°, respectively. For diiodomethane, the PEDOT :PSS, SP4c, SA8 exhibited contact angle of 30.1°, 35.4°, 35.7°, respectively. The surface free energy of the PEDOT :PSS, SP4c, SA8 was 63.7, 68.6, 73.3 mJ/m^2 , respectively. This result was consistent with the function of the sulfonate group. It was attributed to the increase in chemical polarity and enhanced surface free energy. The difference was reflected in efficiency.

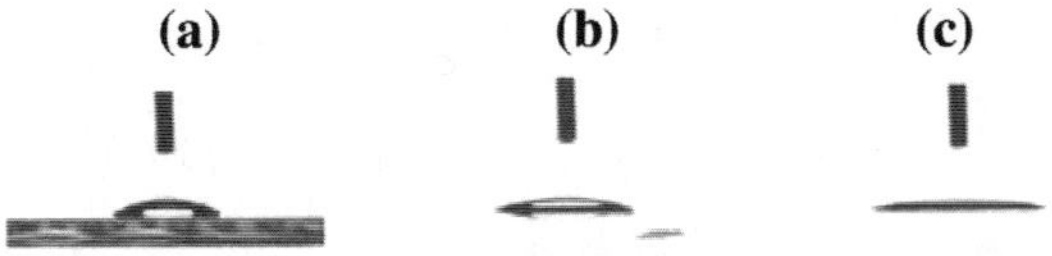

Figure 7. Deionized water of contact angles formed by sessile liquid drops on (a)PEDOT:PSS (b)SP4c (c)SA8

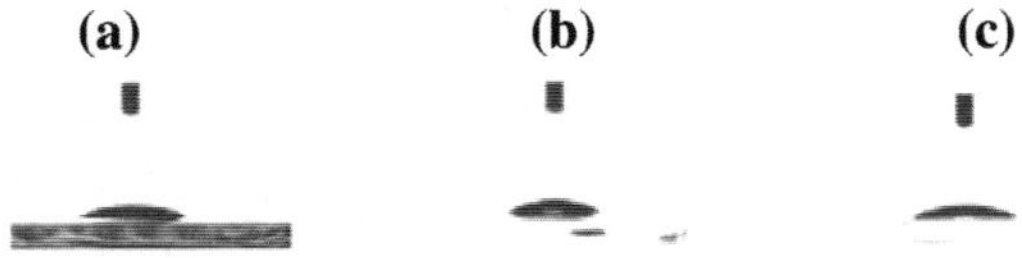

Figure 8. Diiodomethane of contact angles formed by sessile liquid drops on (a)PEDOT:PSS (b)SP4c (c)SA8

4. Conclusions

In this paper, SA8 has such excellent thermal stability and morphological stability. The polyelectrolytes material is formed an anode interface layer by spin coating. The polyelectrolytes, SA8, was dissolved in dimethyl sulfoxide (DMSO) production element and was made polymer solar cells device. The short-circuit current (Jsc) was obtained of 7.15mA/cm2 at AM 1.5G of 100 mW/cm^2, higher than devices based on PEDOT :PSS. The fill factor of 54.84%, and power conversion efficiency of 2.43% can be achieved.

References

[1] R. Po, C. Carbonera, A. Bernardi and N. Camaioni, *Energy Environ. Sci.*, 4, 285–310 (2011).
[2] S. E. Shaheen, G. E. Jabbour, M. M. Morrell, Y. Kawabe, B. Kippelen, N. Peyghambarian, M.-F. Nabor, R. Schlaf, *Appl.Phys.Lett.*, 84, 2324 (1998)
[3] Kang, Bonan, L. W. Tan, and S. R. P. Silva. *Appl. Phys. Lett.* 93, 133302, (2008).
[4] Q. Xu, F. Wang, Z. Tan, L. Li, S. Li, X. Hou, G. Sun, X. Tu, J. Hou and Y. Li, *ACS Appl. Mater.* Interfaces, 5, 10658–10664, (2013).
[5] Y. H. Zhou (2015). Investigation of using multi-fluoro monomers as buffer layer on organic solar cells. Unpublished master' s thesis, National Sun Yat-sen University, Kaohsiung.
[6] M. H. Zhu (2012). Synthesis and Application of Poly(arylene ether)s for Proton Exchange Membrane. Unpublished master' s thesis, National Sun Yat-sen University, Kaohsiung.
[7] J. X. Zhang (2013). The effect of substitute numbers of benzene ring on the proton conductivity of alternative sulfonated poly(arylene ether)s. Unpublished master' s thesis, National Sun Yat-sen University, Kaohsiung.
[8] G. Li, V. Shrotriya, Y. Yao and Y. Yang, *J. Appl. Phys.*, 98, 043704 (2005).
[9] G. Li, V. Shrotriya, J. Huang, Y. Yao, T. Moriarty, K. Emery and Y. Yang, *Nature Materials*, 4, 864–868 (2005).
[10] W. Ma, C. Yang, X. Gong, K. Lee and A. J. Heeger, *Adv. Funct. Mater.*, 15, 1617–1622 (2005)
[11] R. He, L. Yu, P. Cai, F. Peng, J. Xu, L. Ying, J. Chen, W. Yang and Y. Cao, *Macromolecules*, 47, 2921–2928 (2014).
[12] T. M. Brown, J. S. Kim, R. H. Friend, F. Cacialli, R. Daik and W. J. Feast, *Appl. Phys. Lett.*, 75, 1679–1681 (1999).
[13] Vosgueritchian, Michael, Darren J. Lipomi, and Zhenan Bao. *Adv. Funct. Mater.* 22 , 421 (2012).
[14] F. Zhang, A. Petr, H. Peisert, M. Knupfer and L. Dunsch, *J.Phys. Chem.* B, 108, 17301–17305 (2004).
[15] R. Kang, S. Oh and D. Kim, *ACS Appl. Mater. Interfaces*, 6, 6227–6236 (2014).
[16] D. G. Collins, W. G. Blattner, M. B. Wells and H. G. Horak, *Applied Optics*, 11, 2684–2696 (1972).

[17] H. Y. Jung, K.Y. Cho, K. A. Sung, W. K. Kim and J. K. Park, *J. Power Sources*, 163, 56–59 (2006).
[18] H. Y. Jung, K. Y. Cho, K. A. Sung, W. K. Kim, M. Kurkuri and J. K. Park, *Electrochimica Acta*, 52, 4916–4921 (2007).

IOP Conf. Series: Materials Science and Engineering **191** (2017) 012048 doi:10.1088/1757-899X/191/1/012048

Comparison of metallurgical coke and lignite coke for power generation in Thailand

Sudlop Ratanakuakangwan[1,*] and Somkiat Tangjitsitcharoen[1]

[1]Department of Industrial Engineering, Faculty of Engineering, Chulalongkorn University, Bangkok 10330, Thailand

[*]Corresponding author: Sudlop.Ra@student.chula.ac.th

Abstract. This paper presents and compares two alternatives of cokes in power generation which are the metallurgical coke with coke oven gas and the coke from lignite under the consideration of the energy and the environment. These alternatives not only consume less fuel due to their higher heat content than conventional coal but also has less SO_2 emission. The metallurgical coke and its by-product which is coke oven gas can be obtained from the carbonization process of coking coal. According to high grade coking coal, the result in the energy attitude is not profitable but its sulfur content that directly affects the emission of SO_2 is considered to be very low. On the other hand, the coke produced from lignite is known as it is the lowest grade from coal and it causes the high pollution. Regarding to energy profitability, the lignite coke is considered to be much more beneficial than the metallurgical coke in contrast to the environmental concerns. However, the metallurgical coke has the highest heating value. Therefore, a decision making between those choices must be referred to the surrounding circumstances based on energy and environment as well as economic consideration in the further research.

1. Introduction

According to Thailand energy situation in 2015, more than 60 percent of natural gas are used in the power generation which causes the dependency of one particular fuel. However, the Thailand Power Development Plan 2015-2036 (PDP2015) has formulated the policies focusing on fuel diversification by reducing the use of natural gas in power plant and increasing the clean coal and the alternatives energy [1].

Coke is a porous carbonaceous material that naturally has higher heating value than coal. There are two main types of coke which are the petroleum coke and the metallurgical coke.Petroleum coke or pet coke is a by-product of petroleum refining industry. It is usually used in power generation but it releases higher SO_2 in the air than other sources in contrast to metallurgical coke or met coke [2 - 3].

Met coke and coke oven gas (COG) are produced inside coke oven chamber from coking coal. COG is a source of energy and is generally used to preheat the blast furnace. Lately, it become more common that COG is employed in the gas turbine cogeneration for producing electricity which means that the production from coking coal yields both met coke and COG that both can be utilized in power generation [4].

On the other hand, the coke can also be made from long-flamed lignite coal. Making the lignite coke upgrades the quality of fuel which means that it will require less amount. Producing lignite coke might be another source of energy for the power generation in the near future [5].

IOP Conf. Series: Materials Science and Engineering **191** (2017) 012048 doi:10.1088/1757-899X/191/1/012048

The aim of this paper is to propose and compare the uses of different cokes in power generation from metallurgical coke with coke oven gas and lignite coke as a fuel. Those alternatives will be considered based on energy and environment aspects.

2. Classification of coals

Coal has four major compositions which are fixed carbon, ash, volatile matters and moisture. The fixed carbon content represents the heating value (HV). The higher proportion of fixed carbon, the higher HV it has. According to coal components, coal can be divided into four types which are anthracite, bituminous, sub-bituminous and lignite. The approximate components of these coals are shown in Table 1 [5 - 6].

Table 1. Types of coal by components

Components	Anthracite	Bituminous	Sub-Bituminous	Lignite
Fixed carbon (%)	85 - 98	45 - 85	35 - 45	20 - 35
Ash content (%)	10 - 20	3 - 12	≤ 10	10 - 50
Moisture (%)	< 15	2 - 15	10 - 45	30 - 60
Sulfur (%)	0.6 - 0.8	0.7 - 4.0	< 2	0.4 - 1.0

Besides classification of coal compositions, coal can also be identified by application into two types, thermal coal and coking coal. Normally, the thermal coal is used in power generation because of its low cost and abundant sources around the world. On the other hand, the amount of coking coal is less which leads to higher cost and it has the caking property that is required for producing the steel. Therefore, the coking coal is the material for making metallurgical coke which is used in blast furnace for steel production [6].

3. Met coke production

In the process of coal carbonization to met coke, coking coal is heated within the range of 900°C to 1200°C without the absence of oxygen inside the coke oven chamber [7]. During this process, the volatile matters leave the coal mass which is called coke oven gas [4]. The energy consumed for coking process is about 3.5 GJ/ton of coke [8]. The transfer ratio from coking coal to met coke and COG is shown in the Table 2 [4].

Table 2. Transfer ratio from coking coal to met coke and COG

Coking Coal (ton)	Met Coke (ton)	COG (m^3)
1	0.7	354.2

As a result of the met coke process, the percentage of fixed carbon inside coke will be increased due to the reduction of volatile matters and moisture. Referring to the use of high grade coking coal, the estimated components of coking coal and met coke are illustrated in Table 3 [3, 9 - 11].

Table 3. Components of coking coal and met coke

Components	Coking coal[3]	Met coke[4]
Fixed carbon (%)	55.35	87.65
Ash content (%)	8.99	8.85
Volatile matters (%)	27.83	0.75
Moisture (%)	7.83	2.75

[3] Average data from [3] and [9]
[4] Average data from [10] and [11]

4. Lignite coke production

The Lignite is pulverized and fed into the heating boilers using fluidized bed technology through drying, pyrolysis and partial gasification process. The volatile matters that vaporizes from the lignite can be obtained above the bed. It has rich thermal energy that will be burned to maintain the high temperature inside the boilers. It is estimated that 1 ton of lignite can generate the thermal energy of 6,276 MJ. The lignite coke or thermocoke which is the solid residue left under the bed is gathered. The amount of thermocoke obtained compared to the input of lignite is in the ratio of 4:1. Hence, both lignite coke and thermal energy are produced in the same boiler [5]. The components of lignite and lignite coke are expressed in Table 4 [5].

Table 4. Components of lignite and lignite coke

Components	**Lignite (Input)**	**Lignite coke (Output)**
Fixed carbon (%)	20 - 22	74.2
Ash content (%)	7 - 8	8.2
Volatile matters (%)	45 - 48	10
Moisture (%)	33	7.6

5. Comparison of pet coke, met coke and lignite coke

Pet coke is the by-product obtained from oil refinery. Due to its low cost and high heating value, it is usually used in power generation. Power generation of pet coke, met coke or lignite coke is done by sending one of them which is pulverized into fine particles and then sent into a furnace with air for a combustion. Thermal energy from the furnace is used to heat water into steam to rotate the turbine to generate electricity as shown in Figure 1 [12].

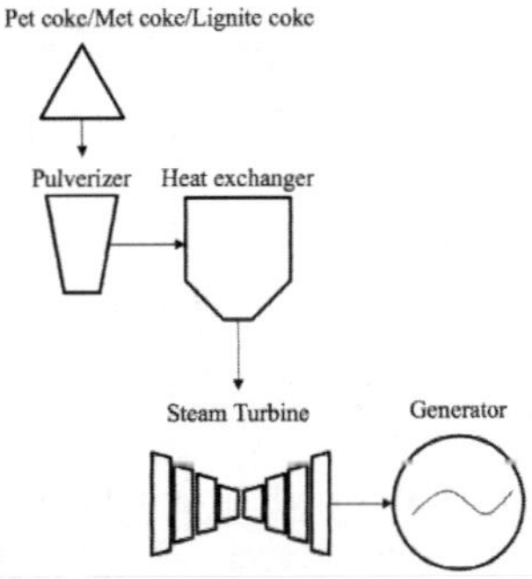

Figure 1. Simple power generation process from pet coke, met coke or lignite coke.

The characteristics which are determined the usability of coke in power generation are the components and the hardness. Hardgrove Grindability Index (HGI) is a measurement of the hardness of coke. The lower index means the higher grinding energy is required for that particular coke before used in power generation. Normally, pet coke's HGI lies between 35 and 45 [13].
HGI can be calculated using the following formula;

$$\text{HGI} = 105 - \text{R}\,(1.16 + 0.002\text{R})\,0.4\text{A} \tag{1}$$

Where R is the summation of the percentage of volatile matters and moisture. A is the percentage of ash content [14]. Using the components of met coke and lignite coke, the HGI of them is calculated in the formula (1). The characteristics of pet coke, met coke and lignite coke are shown in Table 5.

Table 5. Characteristics of pet coke, met coke and lignite coke

Components	Pet coke[5]	Met coke	Lignite coke
Ash content (%)	0.3 - 5.0	8.85	8.20
Volatile matters (%)	8 - 15	2.75	10.00
Moisture (%)	2 - 10	0.75	7.60
HGI	**35 - 45**	**91**	**36**

[5] From [15]

IOP Conf. Series: Materials Science and Engineering **191** (2017) 012048 doi:10.1088/1757-899X/191/1/012048

With the similarity of the components and HGI of pet coke, met coke and lignite coke shown above, the met coke and lignite coke both can be used as a fuel in power generation as well as the pet coke with the different heating contents.

6. Monitoring of energy and environment hazard

6.1 Energy profitability

Tables should be centred unless they occupy the full width of the text.

The Higher Heating Value (HHV) or Gross Calorific Value (GCV) of coal can be calculated by following formula;

$$HHV\ (MJ/kg) = 37.777 - 0.647M - 0.387A - 0.089VM \quad (2)$$

Where M, A, VM are the percentage of moisture, ash, and volatile matters inside the coal, respectively. As the components mentioned above, the HHV of coking coal, met coke, lignite, lignite coke and pet coke are calculated using formula (2) as shown in Table 6 [16].

Table 6. Higher Heating Value of coal and coke

Components	Coking coal	Met Coke	Lignite[6]	Lignite coke	Pet coke[7]
Ash content (%)	8.99	8.85	7.50	8.20	2.65
Volatile matters (%)	27.83	0.75	33.00	7.60	11.5
Moisture (%)	7.83	2.75	46.50	10.00	6.00
Higher Heating Value (kJ/kg)	**26,755**	**32,506**	**9,385**	**28,796**	**31,846**

[6] Average from Table 3.

[7] Average from Table 4.

The energy profitability is calculated by the differentiation of energy input required and energy output of both process. The energy input is consisting of coal and energy required for the carbonization processes. The energy output is including coke type, by-product and excess energy output. The energy profitability is shown in Table 7.

Table 7. Energy profitability between met coke and lignite coke

	Properties	**Met Coke**	**Lignite coke**
Energy Input		**Coal**	
	Coal Type	Coking coal	Long-flamed Lignite
	Coal HHV (kJ/kg)	26,755	9,385
		Energy	
	Energy Input[8] (kJ/kg coal)	2,975	1062.5
Energy Output		**Coke**	
	Coke Type	Met Coke	Lignite coke
	Coke HHV (kJ/kg)	32,506	28,796
	Amount (kg/kg coal)	0.7	0.25
		By-product	
	By-product	Coke oven gas	-
	By-product HHV (kJ/m3)	19,900	-
	Amount (m^3/kg coal)	0.3542	-
		Energy	
	Energy Output (kJ/kg coal)	-	6,276
	Net Energy Profitability (kJ/kg coal)	**73**	**3028**

[8] Estimated from [8]

The net energy profitability of the met coke is obtained about 73 kJ/kg of coal while the lignite coke's is 3028kJ/kg as shown in Table 7. Thus, the production of lignite coke is considered to be more profitable in the energy aspect.

6.2 Environmental hazard

Coal is considered to be highly-polluted fuel as the high proportion of ash and SO2 emission. SO2 which is determined by sulfur content in coal is one of the main environmental problems of coal fired power plant. For example, if all coal fired boilers use low-sulfur coal (less than 0.6 percent) instead of high sulfur coal (more than 3 percent), SO2 emission would be decreased by 1.5 Mt/year [17].

Regarding to the assumption that does not consider the SO2 treatment due to the complexity of its process, the sulfur content in coke directly affects the SO2 emission. Pet coke has the highest sulfur content within the range of 3% to 8% [15]. Met coke tends to have lower sulfur with 0.675% by average weight [3,9] than lignite coke that has the potential to have more sulfur due to the higher proportion of its source from lignite but still less than pet coke.

As mentioned above, it is understood that met coke has the highest heating value as shown in Table 6. Also, met coke emits less SO2 as compared to lignite coke and pet coke. The comparison of heating value and SO2 emission is concluded in Table 8. Thereby, met coke is the best alternative under the consideration of heating value and environmental hazard.

Table 8. Comparison of heating value and SO2 emission

Heating Value	**SO_2 emission**
Met coke > Pet coke > Lignite coke	Met coke < Lignite coke < Pet coke

7. Conclusions

This paper proposed and compared alternatives using conventional coal types in a new way under two considerations which are environmental hazard and energy aspect. The first option is to use coking coal in carbonization process in coke oven to produce metallurgical coke and coke oven gas in order to use them in power generation. Another one is to use lignite coke that created from the lowest grade of coal that is lignite. Referring to energy profitability, the production of lignite coke is considered to be more beneficial than met coke in contrast to environmental hazard. Therefore, a decision making between those two choices must consider the specific heating value. Both alternatives increase their heating value which means that they will be needed less number of coke to generate the same amount of power. Met coke has the highest heating value while the lignite coke's is the lowest. For environmental viewpoint, met coke is also inclined to emit the least SO2 emission. However, an economic evaluation will be considered for the final decision to select which alternatives in the further research.

Acknowledgments

This research has been performed and supported by the Scholarship of the Graduate School, Chulalongkorn University to commemorate the 72nd anniversary of his Majesty King Bhumibala Aduladeja.

References

[1] Energy Policy and Planning Office 2015 *Thailand Power Development Plan 2015-2036*
[2] Chen J and Lu X 2007 *Resources, Conservation and Recycling* 49 203
[3] Valia H S 1994 *Blast Furnace Ironmaking* 2 10
[4] Razzaq R, Li C, Zhang S 2013 *Fuel* 113 287
[5] Vorob'ev V P, Orlov P P, Lslamov S R, Stepanov S G 2015 *Steel in Translation* 45 207
[6] Bowen B H, Irwin M W 2008 *Coal Characteristics, CCTR Basic Facts File*
[7] Falcon R 2013 *Coal Geology, Types, Ranks and Grades, Coal, Coke and Carbon in the Metallurgical Industry*

IOP Conf. Series: Materials Science and Engineering **191** (2017) 012048 doi:10.1088/1757-899X/191/1/012048

[8] International Energy Agency 2007 *Tracking Industrial Energy Efficiency and* CO_2 *Emissions*
[9] Díez M A, Alvarez R, Barriocanal C 2002 *International Journal of Coal Geology* 50 389
[10] *Coal producing companies and product specification* 2013 Retrieved from http://www.resourcesandenergy.nsw.gov.au/investors/investment-opportunities/coal/nsw-coal-producing-companies-and-product-specifications#_anglo-_american-__2013-_australia
[11] SUEK AG 2015 *Coal specification* Retrieved from http://www.suekag.com/coalspecification/?specification=3
[12] Hirayama Y, Hichida M, Yamamoto Y, Makiura S, Arakawa Y, Okamoto A 2007 *Operation results of power station with petroleum coke firing boiler Mitsubishi Heavy Industries Ltd. Technical Review No. 4* 44
[13] PEC Consulting 2015 *Comparative Properties of Bituminous Coal and Petroleum Coke as Fuels in Cements Kilns* Retrieved from http://pecconsultinggroup.com/newsflash/comparative-properties-of-coal-and-petcoke
[14] Sinha N C, Moitra A K, Majumdar S 1962 *Journal of Mines, Metals & Fuels* 10 1
[15] Góral D, Wylenżek A 2012 *The efficient coal alternative. Petroleum coke-fired CFB boilers in Europe Foster Wheeler presented at Coal Gen Europe*
[16] Mesroghli Sh, Jorjani E, Chelgani S C 2009 *International Journal of Coal Geology* 79 49
[17] You C F, Xu X C 2010 *Coal combustion and its pollution control in China, Energy* 35 4467

IOP Conf. Series: Materials Science and Engineering **191** (2017) 012049 doi:10.1088/1757-899X/191/1/012049

Enhanced Electrochemical Performances of Si/Graphite Composite Anode Using Commercial Promising Waterborne Conductive Binder

Shu Huang[1], Jianguo Ren[2], Rong Liu[1], Min Yue[2], Youyuan Huang[2], Guohui Yuan[1,*]

[1] School of Chemistry and Chemical Engineering, Harbin Institute of Technology, Harbin, 150001, P. R. China

[2] Shenzhen BTR New Energy Materials Inc., Shenzhen, 518000, P. R. China

E-mail address: yghhit@163.com (Guohui Yuan).

Abstract. Silicon has attracted increasing attention as an anode material in Li-ion batteries (LIBs) owing to its high theoretical capacity. However, electrochemical properties of Si-based anodes are hindered by the electrode swelling during the charge/discharge process. In this paper, we developed a novel polymer binder with high electronic conductivity and self-healing function leading to stable cycling. These properties are attributed to the polymer structure which could provide more free Li+ leading to higher conductivity and capacity retention of the anodes. It is believed that this novel conductive binder could be a promising candidate for commercial application for the Si/graphite composite anodes in Li-ion batteries.

1. Introduction

Lithium ion batteries (LIBs) as an enabling energy storage technology present ever-increasing demands of high capacity and stable electrochemical performances [1, 2]. Silicon-based anode [3-8] has attracted increasing attention owing to its high theoretical capacity. Polymer binder [9], one of the major components of the electrode, is used to link active materials and conducting particles together onto the current collector and ensure the integrity of the electrodes during subsequent cycling. However, conventional poly (vinylidenefluoride) (PVDF) binder needs N-methyl-2-pyrrolidone (NMP) as the solvent which is toxic to the environment. In addition, the addition of polymer binders would restrict the electrochemical performances of electrodes for the poor conductivity. Thus, novel and effective waterborne binders with a low addition amount are urgently needed.

In recent years, waterborne polymer binders have particularly attracted more and more attention. Many researchers prepare conductive polymer as the binders for Si-based anodes to increase the conductivity of the electrodes leading to excellent electrochemical properties. Many studies focused on conductive polyamide imide (PAI) based polymer binder [10-12] presenting improved initial Coulombic efficiency and excellent retention for Si-based electrode for the high electronic conductivity upon cycling. Liu and his group prepared conductive polymer as the binders, such as PFFOMB [13] binders, PEFM [14, 15] conductive binder, conductive poly(9,9-dioctylfluorene-co-fluorenone-co-methylbenzoic ester) (PFM) [16, 17] binder, poly(1-pyrenemethyl methacrylate-co-methacrylic acid) (PPyMAA) [18] binder and aqueous poly(3,4-ethylenedioxythiophene):poly(styrenesulfonate) (PEDOT:PSS) [19], leading to enhanced cycle

stability and the specific capacity for Si-based anodes. Thus, the usage of conductive polymer binder could be an efficient way to enhance the performances.
In this paper, we report a novel conductive polymer binder, a kind of lithiated ionic polymer, for graphite anode materials through an in-situ hydrothermal method. The novel binder leads to the excellent cycling stability during the cycling and low impendence for Si/graphite composite anode which would be promising for further commercial application.

2. Experimental

2.1. Synthesis of the polymer binder

The aqueous binder was synthesized by styrene (Alfa Aesar), n-butyl acrylate (Alfa Aesar), methacrylate (Alfa Aesar) and lithium hydroxide (Alfa Aesar). For comparison purpose, the polyvinylidene fluoride (PVDF) solution of (NMP) was used as a control binder.

2.2. Preparation of electrodes and Li-ion cells assembling configuration

The prepared polymer binder was mixed with Si/graphite material (<100 nm, BTR Co., Ltd, China), and the weight ratio of the electrode was 96.5% Si/graphite, 1.5% CMC as the thicker and 2% binder for the anode. The composition of the PVDF electrodes was 96.5% Si/graphite and 3.5% PVDF (weight). The coin-type half cells were assembled with lithium metal as the counter electrode in Argon filled glove box. The electrolyte was consisted of 1 mol/L LiPF6 in a mixture of organic carbonates: 2:3:1 EC/DEC/EMC (volume).

2.3. Morphology and electrochemical characterization

Morphology test was used Hitachi S4800. Galvanostatic charge-discharge cycling of cells was tested between 0.01 and 1.5 V at 25oC using a Land battery tester (Wuhan, China) for 50 cycles at a current density of 0.1C. Electrochemical impedance spectroscopy (EIS) was tested on a CHI 660E electrochemical work station over the frequency range of 10-3-105 Hz with an amplitude modulation of 5 mV.

3. Results and discussion

3.1. Morphology test

The SEM image (Fig.1a) of the polymer emulsion particles shows that the emulsion particles are spherical and the average diameter of the particles is about 200nm. From Fig.1b and c, the appearance and interspace of the electrodes made by conductive binder and PVDF binder show no obvious difference which demonstrates that the conductive binder could prepare stable slurry with Si/graphite composite anode. In addition, Fig. 1b demonstrates that the conductive binder is dispersed uniformly in the slurry because black points which represent the conductive binder are equably distributed on the surface of the graphite particles. Thus, the conductive binder would be promising to improve the electrochemical performances.

Figure 1. SEM images of (a) binder emulsion particles, (b) electrode using conductive binder before cycling, (c) electrode using PVDF binder before cycling.

3.2. Cycling performance

The cycling performances show that the electrode with conductive binder presents more stable cycling performance with the retention of 92.3% after 50 cycles even with a high rate of 0.1C compared to PVDF electrode with retention of 40.2%. Fig. 2b shows the rate capability of the graphite electrodes with the conductive binder and PVDF binders. The specific capacity of the conductive binder electrode performs no obvious decrease during the first 20 cycles owning to the conductivity of Li+. Thus, the conductive LIP binder could effectively enhance the cycling performance and the rate capacity because the improved conductivity of the electrode and the conductive binder could provide more free Li+ which would be beneficial to the electrochemical performances.

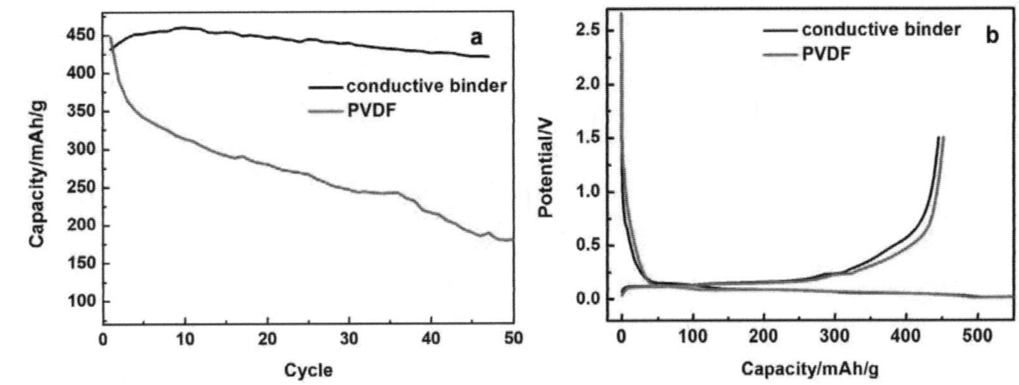

Figure 2. (a) Cycling performances of the Si/graphite electrodes with conductive binder and PVDF binder. (b) Potential profiles of the Si/graphite electrodes using conductive and PVDF binder.

3.3. EIS spectroscopy

The electrochemical impedance spectroscopy (EIS) of graphite electrodes with conductive binder and PVDF were tested before cycling. From Fig. 3, the semicircle diameter of the conductive binder electrode is much smaller than PVDF binder electrode which indicates a lower impendence of the conductive binder electrode. The results demonstrated that the conductive binder could enhance the conductivity of electrode and would be beneficial to decrease the impendence and increase the properties for the following cycling.

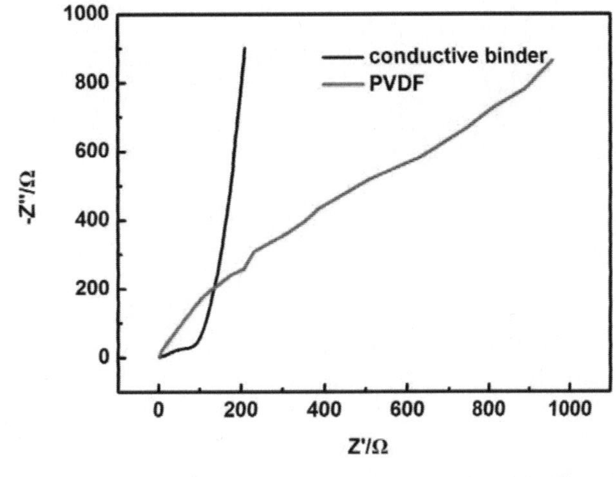

Figure 3. Nyquist plots for graphite electrode prepared with LIP and PVDF binder.

4. Conclusion

In summary, a conductive polymer binder has been prepared by using an in-situ thermal technique based on the low-cost for Si/graphite composite anodes. This conductive binder could effectively decrease impendence and enhance cycling stability owning to the increased conductivity and more free Li+ based on the polymer structure. In contrast, the cells using PVDF binder present fast capacity fading and PVDF binder needs NMP solvent which is toxic to the environment. Thus, this research demonstrates that waterborne polymer binder is not only environmental friendly but also beneficial to improve the electrochemical performances owning to the chemical structure of the polymer. The results of this study also provide a pattern for further researches and commercial application.

5. References

[1] S. W. Lee, S. W. Choi, S. M. Jo, B. D. Chin, D. Y. Kim, K. Y. Lee, Electrochemical properties and cycle performance of electrospun poly(vinylidene fluoride)-based fibrous membrane electrolytes for Li-ion polymer battery, J. Power Sources Vol. 163 (2006), P. 14-46

[2] N. Nitta, F.X. Wu, J.T. Lee, G. Yushin, Li-ion battery materials: present and future, Mater. Today Vol.18 (2015), P. 252-264

[3] M. Gu, Y. He, J.M. Zheng, C.M. Wang, Nanoscale silicon as anode for Li-ion batteries: The fundamentals, promises, and challenges, Nano Energy Vol. 17 (2015), P. 366-383.

[4] M. L. Terranova, S. Orlanducci, E. Tamburri, V. Guglielmotti, M. Rossi, Si/C hybrid nanostructures for Li-ion anodes: An overview, J. Power Sources Vol. 246 (2014) , P. 167-177.

[5] M. Ko, S. Chae, J. Cho, Challenges in Accommodating Volume Change of Si Anodes for Li-Ion Batteries, ChemElectroChem Vol. 2 (2015) , P. 1645-1651.

[6] M. Ashuri, Q.R. He, L.L. Shaw, Silicon as a potential anode material for Li-ion batteries: where size, geometry and structure matter, Nanoscale Vol. 8 (2016), P. 74-103.

[7] H. Wu, Y. Cui, Designing nanostructured Si anodes for high energy lithium ion batteries, Nano Today Vol. 7 (2012), P. 414-429.

[8] S. Goriparti, E. Miele, F. De Angelis, E. Di Fabrizio, R.P. Zaccaria, C. Capiglia, Review on recent progress of nanostructured anode materials for Li-ion batteries, J. Power Sources Vol. 257 (2014) , P. 421-443.

[9] S. L. Chou, Y. Pan, J. Z. Wang, H. K. Liu, S. X. Dou, Small things make a big difference: binder effects on the performance of Li and Na batteries, Phys. Chem. Chem. Phys. Vol. 16 (2014) , P. 20347-20359.

[10] N. S. Choi, K. H. Yew, W. U. Choi, S. S. Kim, Enhanced electrochemical properties of a Si-based anode using an electrochemically active polyamide imide binder, J. Power Sources Vol. 177 (2008) , P. 590-594.

[11] H. Wu, G. H. Yu, L. J. Pan, N. Liu, M. T. McDowell, Z. Bao, Y. Cui, Stable Li-ion battery anodes by in-situ polymerization of conducting hydrogel to conformally coat silicon nanoparticles, Nat. Comms. Vol. 4 (2013), P. 1943.

[12] T. Yim, S. J. Choi, Y. N. Jo, T. H. Kim, K. J. Kim, G. Jeong, Y. J. Kim, Effect of binder properties on electrochemical performancefor silicon-graphite anode: Method and applicationof binder screening, Electrochim. Acta Vol. 136 (2014), P. 112-120.

[13] G. Liu, S. D. Xun, N. Vukmirovic, X. Y. Song, P. O. Velasco, H. H. Zheng, V. S. Battaglia, L. W. Wang, W. L. Yang, Polymers with Tailored Electronic Structure for High Capacity Lithium Battery Electrodes, Adv. Mater. Vol. 23 (2011), P. 4679-4683.

[14] M. Y. Wu, X. C. Xiao, N. Vukmirovic, S. D. Xun, P. K. Das, X. Y. Song, P. O. Velasco, D. D. Wang, A. Z. Weber, L. W. Wang, V. S. Battaglia, W. L. Yang, G. Liu, Toward an Ideal Polymer Binder Design for High-Capacity Battery Anodes, J. Am. Chem. Soc. Vol. 135 (2013) , P. 12048-12056.

[15] M. Y. Wu, X. Y. Song, X. S. Liu, V. Battaglia, W. L. Yang, G. Liu, Manipulating the polarity of conductive polymer binders for Si-based anodes in lithium-ion batteries, J. Mater. Chem.

A Vol. 3 (2015), P. 3651-3658.

[16] H. Zhao, Z. H. Wang, P. Lu, M. Jiang, F. F. Shi, X. Y. Song, Z. Y. Zheng, X. Zhou, Y. B. Fu, G. Abdelbast, X. C. Xiao, Z. Liu, V. S. Battaglia, K. Zaghib, G. Liu, Toward Practical Application of Functional Conductive Polymer Binder for a High-Energy Lithium-Ion Battery Design, Nano Lett. Vol. 14 (2014), P. 6704-6710.

[17] H. Zhao, N. Yuca, Z. Y. Zheng, Y. B. Fu, V. S. Battaglia, G. Abdelbast, K. Zaghib, G. Liu, High Capacity and High Density Functional Conductive Polymer and SiO Anode for High-Energy Lithium-Ion Batteries, ACS Appl. Mater. Interfaces Vol. 7 (2015), P. 862-866.

[18] H. Zhao, Y. Wei, R. M. Qiao, C. H. Zhu, Z. Y. Zheng, M. Ling, Z. Jia, Y. Bai, Y. B. Fu, J. L. Lei, X. Y. Song, V. S. Battaglia, W. L. Yang, P. B. Messersmith, G. Liu, Conductive Polymer Binder for High-Tap-Density Nanosilicon Material for Lithium-Ion Battery Negative Electrode Application, Nano Lett. Vol. 15 (2015), P. 7927-7932.

[19] D. Shao, H. X. Zhong, L. Z. Zhang, Water-Soluble Conductive Composite Binder Containing PEDOT: PSS as Conduction Promoting Agent for Si Anode of Lithium-Ion Batteries, ChemElectroChem Vol. 1 (2014) , P. 1679-1687.

IOP Conf. Series: Materials Science and Engineering **191** (2017) 012050 doi:10.1088/1757-899X/191/1/012050

Evolution features of coal matrix porosity with the variation in temperature and stress

Jianlou Li[1,2], Buqing Li[3]

[1]School of Resources and Civil Engineering, Suzhou University, Anhui Suzhou 234000, China

[2]Coal Mine Exploration Engineering Technology Center of Anhui province, Anhui Suzhou 234000, China

[3]School of Information Engineering, Suzhou University, Anhui Suzhou 234000, China

E-mail: jianlouli1973@126.com

Abstract. To clarify the differentiated effect of temperature and stress on coal matrix pores, the coals from No. 8 coal bed in Taoyuan coal mine of Huaibei Coalfield, Anhui Province, were chosen as experimental samples. Cryogenic nitrogen adsorption methods were used in laboratories to test the matrix pore parameters of the dried coal samples treated under the temperature of 60℃, 70℃, 80℃, 90℃ and 100℃. Also efforts were made to test the parameters of matrix pore structure of dried coals after being treated under the stress of 10 MPa, 15 MPa, 20 MPa and 25 MPa respectively. The results show that: the specific surface area of coal matrix pore decreases, the average pore diameter and median pore diameter increases with the increase of temperature. As the stress goes up, the coal matrix pore first experiences increase in specific surface area, average pore diameter and median pore diameter, followed by a slight decrease after rising to a certain point. Therefore, the change in specific surface area of coal from the same coal bed can, to some extent, reflect the variation in temperature and ground stress.

1. Introduction

The Coal-mass pores can be regarded as a dual-pore structure system comprising fracture and coal matrix pores [1]. Also it can be seen as a tri-element structure system consisting of cleat, microscopic fracture and coal matrix pores [2]. Based on the pore genesis, coal pore is classified into gas pore, residual plant tissue pore, corrosion pore, inter-granular pore, etc [3]. There are three types of pore shape [4, 5], which is identified based on mercury injection curve characteristics in mercury injection experiment or the loop curve characteristics in liquid nitrogen adsorption experiment. The coal pore size varies from nanoscale to millimeter [6]. With the downward extension of coal mining, the exploited coal bed will experience constant growth in temperature and stress. What effect will temperature and stress have on coal matrix pores which are the major storage space for coal bed gas? Studies focusing on this topic are of great significance to clarifying the laws of storage and migration of coal bed gas.

According to research literatures concerning coal matrix pores [7-11], the specific pore volume and specific surface area of coal matrix, during the low and medium coal rank, will gradually decrease with progress in coal rank. For the same coal rank, the specific pore volume and specific surface area of coal matrix will rise with the growth in stress. The changing characteristics of microscopic pores from high-rank coal samples, with the variation in temperature and stress, were studied in literature

IOP Conf. Series: Materials Science and Engineering **191** (2017) 012050 doi:10.1088/1757-899X/191/1/012050

research [12]. However, the changing characteristics of microscopic pores from low-rank coal samples were not mentioned. Above all, previous research literatures failed to make in-depth analysis of the affecting mechanisms of ground stress and temperature on coal matrix pores.

In this study, coal samples were taken from No.8 coal bed in Taoyuan coal mine of Huaibei Coalfield. The coal samples were heated in labs under relatively low temperature (≦100℃) and underwent stress-elevation simulation experiment respectively. With the application of cryogenic nitrogen adsorption methods, structural features of coal matrix pores were ascertained. And meanwhile, analysis was made of the evolutionary characteristics of coal matrix pores during the increase in temperature and stress.

2. Equipment, samples and methods involved in the study

2.1. Experimental equipment

The Gemini VII 2390t full automatic and quick analyzer for specific surface area and pores, manufactured by U.S. Micromeritics, was used in the study. The equipment consists of refrigeration system, vacuum-pumping system, gas injection system, measuring system and data analysis system. Accessories include heating dehydration pre-treatment system, electronic scale, etc. The equipment is shown in Figure 1.

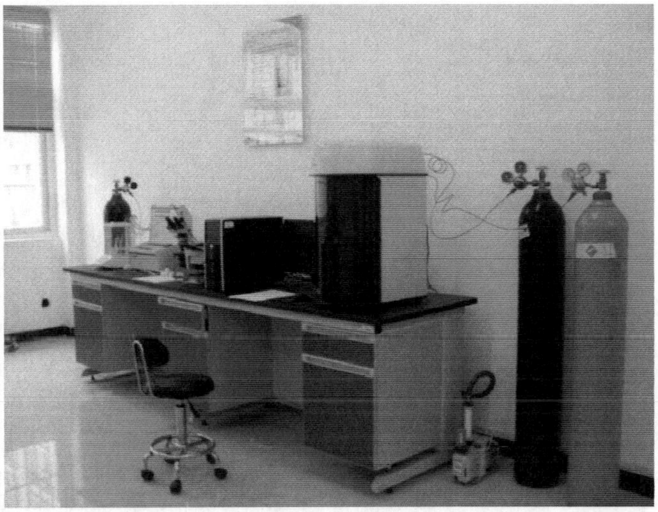

Figure 1. Experimental equipment of cryogenic nitrogen adsorption.

2.2. Experimental samples

The coal samples are taken from No. 82 coal bed in Taoyuan coal mine of Huaibei Coalfield, which is located in the southeast of Huaibei Coalfield. Taoyuan coal field features a mono-inclined structure, which strikes north-south and inclines toward the east. The position of the field in regional geological structure is shown in Figure 2.

The coal-bearing formation in Taoyuan coal field belongs to permo-carboniferous system. Among the major minable seams, 82 coal bed lies beneath the Shihezi formation and varies from 0 to 4.10 meters in thickness, with the average thickness of 2.01m. 82 coal bed has simple structure, sandstone roof and mudstone baseboard. Its rank is gas coal.

2.3. Experimental methods

(1) Pre-treatment of samples: Seal the newly-collected samples before taking them to the labs, where they are broken into small granules with a rubber hammer. Screen them with 60-mesh and 80-mesh sieves, followed by picking out around 50 grams of coal granules with 60-80 mesh in diameter as

back-up samples. Then, take 10 experiment-dedicated glass tubes and fill each with around 1.5 grams of experimental samples. Weigh and record the actual mass of each sample. Put the tubes filled with samples into the heating devices where the heating temperature is set to be 50℃. Meanwhile, isolate the samples with nitrogen from the external air to prevent coal oxidation during the heating process. Dry the coal samples until no water left. Reweigh the mass of each dried coal sample. Finally, store the 10 tubes filled with samples in room temperature for further treatment.

(2) Heating steps: Randomly take 5 tubes filled with coal samples and heat them for 2 hours under the temperature of 60℃, 70℃, 80℃, 90℃ and 100℃ respectively. Meanwhile, isolate the samples with nitrogen from the external air to prevent coal oxidation. Then, cool the samples to room temperature before cryogenic nitrogen adsorption tests one by one.

(3) Steps for stress-elevating experiment under room temperature: Take 5 tubes filled with coal samples. Elevate the stress on them with 769YP-15A type tablet presser to 10MPa, 15MPa, 20MPa and 25MPa respectively. Every stress-elevating stage last for two hours. Then, conduct cryogenic nitrogen adsorption tests one by one.

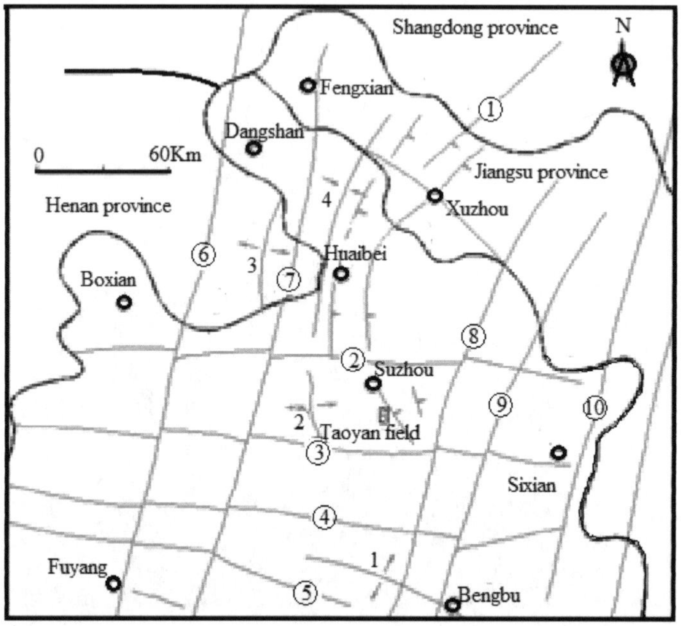

①Fengpei fault ②Subei fault ③Guangwu-Guzhen fault ④Taihe-Wuhe fault ⑤Liufu fault ⑥ Xiayi fault ⑦Fengxian-Kouziji fault ⑧Guzhen-Changfeng fault ⑨Lingbi-Wudian fault ⑩Tanlu fault

Figure 2. Geological structure background of sampling area [13].

3. Experimental results and analysis

3.1. Results and analysis of the heating experiment

Distribution features of specific surface area of 5 coal samples in Figure 3.

According to Figure 3, under the temperature of 60℃, the maximal peak of specific surface area comes where the pore width ranges from 3.0nm to 2.7nm. Under the temperature of 70℃, the maximal peak of specific surface area also comes where the pore width ranges from 3.0nm to 2.7nm. Under the temperature of 80℃, the maximal peak of specific surface area comes where the pore width ranges from 2.7nm to 2.4nm. When the temperature rises to 90℃, shrink in the specific surface area takes place where the pore width ranges from 3.4nm to 2.4nm. When the temperature goes up to 100℃, there is obvious rise in specific surface area where the pore width ranges from 2.1nm to 1.6 nm. On the whole, with the rise in temperature, the obvious change of specific surface area tends to take place in smaller micro-pores rather than in larger pores.

The BET specific surface area, pore volume, average pore diameter and median pore diameter of five coal samples under different temperature are shown in Table 1.

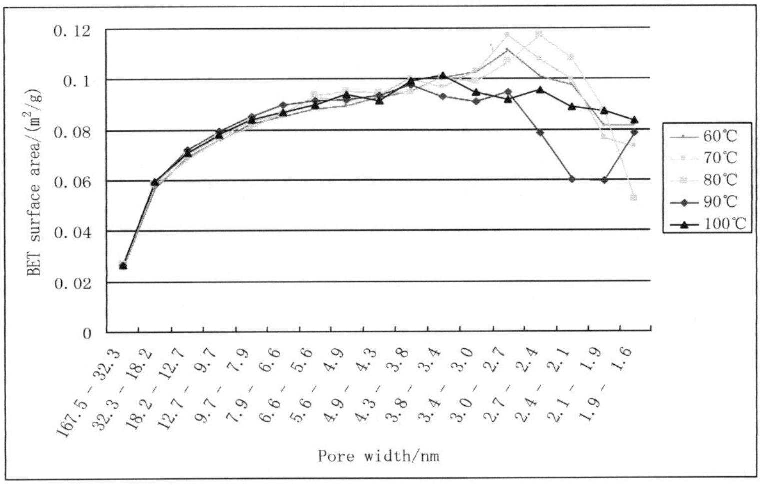

Figure 3. Specific surface area distribution in different pores under different temperature.

Table 1. Parameter values of coal matrix pore under different temperature.

Tem/℃	BET specific surface area /(m2/g)	Pore volume /(cm3/g)	Average pore diameter /nm	Median pore diameter /nm
60	1.6700	0.002900	6.94667	10.4670
70	1.5512	0.002873	7.40713	10.9979
80	1.5200	0.002910	7.65765	11.6139
90	1.4828	0.002883	7.77639	12.0830
100	1.4758	0.002877	7.79836	12.0485

According to Table 1, the BET specific surface area of coal matrix pore decrease with the rise in temperature. There is no obvious change in pore volume. There is increase in average pore width and median pore width. This is because, on conditions where the heating temperature is low and the heating time is short, the effect of coal metamorphism can be ignored and considerations should be given to the influence of thermal expansion and extraction on coal matrix pores only. When the temperature rises, intermolecular distance of coal matrix grows and the coal matrix will expand inward and outward. As a result, the pore-shrinking effect arises during the inward expansion. This results in the evolution of larger pores into smaller ones. Micro-pores with small diameter are more likely to be blocked by liquid organic matters as the pores shrink. Besides, as micro-pores contribute significantly to specific surface area, large quantities of blocked and sealed micro-pores after the heating process will impair the connectivity. Therefore, the number of effective pores decreases, the specific surface area shrinks, and the average pore diameter and median pore diameter grow accordingly.

It should be noted that the study focused on inner pores of coal matrix rather than cleat and fracture. Therefore, outward expansion of coal matrix was not involved herein. Outward expansion of coal matrix will result in compressed fracture between coal matrixes and weakened coal permeability [14].

3.2. Results and analysis of the stress-elevating experiment

The change in specific surface area of five stress-elevated coal samples with the variation of pore diameter is shown Figure 4.

According to Figure 4, the pores no more than 200nm in diameter grow rapidly in specific surface area when the stress goes up from 0 MPa to 10 MPa. Later, the specific surface area of all pores grows continually until the stress goes up to15MPa. However, it grows at a relatively slow pace. When the stress goes up to 20 MPa and 25 MPa, the growth in specific surface area of all pores is not obvious.

The BET specific surface area, pore volume, average pore diameter and median pore diameter of five coal samples under different stress are shown in Table 2.

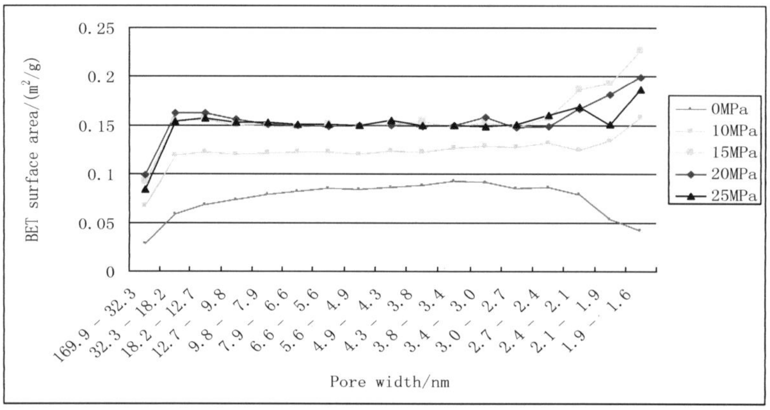

Figure 4. Specific surface area distribution in different pores under different stress.

Table 2. Characteristic parameter values of coal matrix pore under different stress.

Stress/MPa	BET specific surface area /(m2/g)	Pore volume/(cm3/g)	Average pore diameter /nm	Median pore diameter /nm
0	1.2635	0.002749	8.70238	13.9422
10	2.2177	0.005165	9.31576	20.8997
15	2.7719	0.006627	9.56263	22.9060
20	2.7929	0.006933	9.92930	24.8504
25	2.8377	0.006474	9.12514	20.5297

According to Table 2, the coal matrix pores grow rapidly in pore volume, specific surface area, average pore diameter and median pore diameter. When the stress goes up to a certain level, there is a decrease to some extent in pore volume, specific surface area, average pore diameter and median pore diameter. This is because majorities of organic matters in coal belong to brittle organic component, e.g. Vitrinite and Inertinite. Brittle organic matters are easy to fracture under destructive stress. The fracture gives rise to more effective micro-pores and fracture. Newly generated fracture will form more absorption space. With the increase in the proportion of absorption space resulting from the fracture, the average pore diameter and median pore diameter increase accordingly. When the stress goes up to a certain level, e.g. 25 MPa in this experiment, the increase in fracture will be no longer obvious. Meanwhile, the volume of the formed absorption space will be further compressed. This leads to the decrease in pore volume, specific surface area, average pore diameter and median pore diameter of coal matrix pores.

4. Conclusions

After heating and stress-elevating experiments on coal gas from No. 8 coal bed in Taoyuan coal mine, the following conclusions can be drawn:

(1) The specific surface area decrease, but the average pore diameter and the median pore size of coal matrix increase with the rise of the geo-therm.

(2) The specific surface area, pore volume, the average pore diameter and the median pore diameter of coal matrix increases firstly with the increase of ground stress, followed by the decrease to some extent after the peak.

(3) Specific surface area does a better job in making a distinction between the effect of terrestrial heat and ground stress.

Acknowledgements
The study was supported by the National Nature Science Foundation of China (41302274), Doctoral research project in Suzhou University (2011JB05) and the opening scientific research platform in Suzhou University (2014YKF02).

References

[1] Close J C 1993 Natural fracture in coal in Hydrocarbons from coal *AAPG* 38 119-32.
[2] Gamson P, Beamish B and Johnson D 1998 Effect of coal microstructure and secondary mineralization on methane recovery *Geol. Spec. Publ.* 199 165-79.
[3] Yiwen J, Bo J, Guiliang W and Quanlin H 2005 *Tectonic Coal Structure and Physical Properties of Coal Reservoirs* (Xuzhou: China Mining University Press).
[4] Ping C and Xiuyi T 2001 The research on the adsorption of nitrogen in low temperature and micro-pore properties in coal *J. China Coal Soc.* 26 552-6.
[5] Yanbin Y, Dameng L, Wenhui H, Dazhen T and Shuheng T 2006 Research on the pore-fractures system properties of coalbed methane reservoirs and recovery in Huainan and Huaibei coal-field *J. China Coal Soc.* 31 163-8.
[6] Xuehai F, Yong Q, Wanghong Z, Chong T and Rongfu Z 2005 Fractal classification and natural classification of coal pore structure based on migration of coalbed methane *Chin. Sci. Bull.* 50 51-5.
[7] Shuheng T, Chao C, Baocun Z, Lijiang D and Jiazan Z 2008 Control effect of coal metamorphic degree on physical properties of coal reservoirs *Nat. Gas Ind.* 28 30-3.
[8] Yiwen J and Xiaoshi L 2009 New progress in studying on ultra microstructure of tectonic coal *Prog. Nat. Sci.* 19 131-141.
[9] Xianghao W, Yanbin W, Shasha G, Pengfei H and Meijuan Z 2012 Differences in pore structures and absorptivity between tectonically deformed and undeformed coals *Geol. J. China Univ.* 18 528-32.
[10] Guangwu X, Hongfu L, Huifang Y and Wei L 2011 The types of tectonic coals and pore characters in Hancheng *J. China Coal Soc.* 36 1845-51.
[11] Wenping J, Xiaozhong S, Lingwen Z 2011 Research on the pore properties of different coal body structure coals and the effects on gas outburst based on the low-temperature nitrogen adsorption method *J. China Coal Soc.* 36 609-14.
[12] Zhaoping M, Shanshan L, Baoyu W, Yongdong T and Jie W 2015 Adsorption capacity and its pore structure of coals with different coal body structure *J.China Coal Soc.* 40 1865-70.
[13] No.3 exploration team of coalfield geology bureau in Anhui province 2011 *Geological report of Taoyuan coal mine* (Huaibei: Huaibei Mining Co. LTD).
[14] Jianlou L 2010 *Study of Desorption and Emission of Coal Gas under Sonic Waves* (Huainan: Anhui University of Science and Technology).

IOP Conf. Series: Materials Science and Engineering **191** (2017) 012051 doi:10.1088/1757-899X/191/1/012051

New integrated monitoring and control system for disaster gas application in coal mine

Zhao Deng-yu[1,2], Lai Xing-ping[1,2], Ji Chang-fa[1,2], Xi Hong-jun[1], Zhang Bo[1], Chen Liu[1]

[1] Energy School, Xi'an University of Science and Technology, Xi'an 710054, China
[2] Key Laboratory of Western Mines and Hazard Prevention, Ministry of Education of China, Xi'an 710054, China

E-mail: 418477634@qq.com

Abstract. This paper introduces the composition, working principle and characteristics of the new gas monitoring system, which are used to monitor the gas in the gob area and working face of the mine in real time. According to the oxygen concentration, the spontaneous combustion of gob area "three zones" was determined and divided. In order to carry out disaster early warning, the methane and carbon monoxide super gas concentration changes were studied and analyzed. And the variation law of gas concentration in gob area and coal face in coal mining was obtained. This study provides reference for the follow-up fire prevention and production.

1. Introduction

Coal contains a variety of harmful gases. During the process of coal mining, the oxidation, spontaneous combustion of coal itself, and heat, combustion of a variety of materials will produce a variety of toxic and harmful gases. Therefore, the detection and analysis of coal mine gas is the basis of coal mine safety production, and is the fundamental basis for coal mine disaster prevention, management and disaster relief decision-making. And so, the coal-seam methane-induced gas monitoring, the development and use of the new disaster-proof gas comprehensive monitoring system are particularly important [1-4]. This paper will introduce the application of new disaster-prone gas integrated monitoring system in mine disaster prevention.

2. Presentation of new integrated monitoring and control system for disaster gas

2.1. Composition of new integrated monitoring and control system for disaster gas

Based on the existing KJ352 mine monitoring system, combined with JSG6-F beam tube sampling and analysis system, the new integrated monitoring system for disaster-causing gas is developed with the functions of tube sampling, infrared analysis of underground gas sample, seamless information transmission and equipment control with other monitoring systems. Its structure is shown in Figure 1.

2.2. Principle of new integrated monitoring and control system for disaster gas

When the system works, the pump is started, the tube forms a negative pressure which is less than the underground pressure outside the beam tube, then the gas undermine was sucked into the beam tube and the analysis unit, where it is analysed real-time; at the same time undermine staff can monitor the gas concentration of sample points. The analysis unit transmits the results of the analysis to the ground via a communication line. Ground workers can simultaneously monitor the gas concentration of the measuring points in the ground monitoring room. The early warning software provided analysis to make early warning.

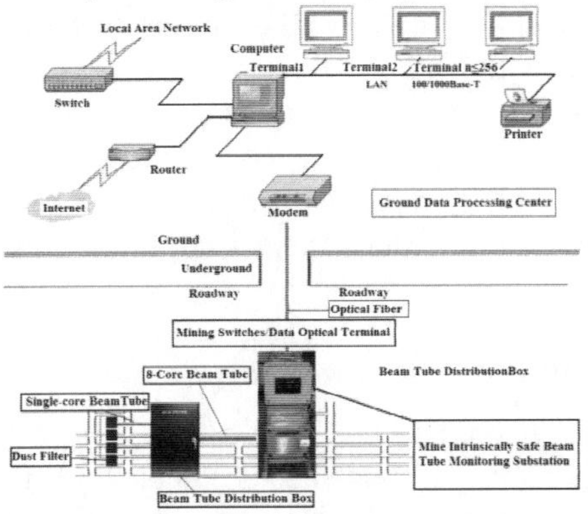

Figure 1. Schematic diagram of the new integrated monitoring and control system for disaster gas

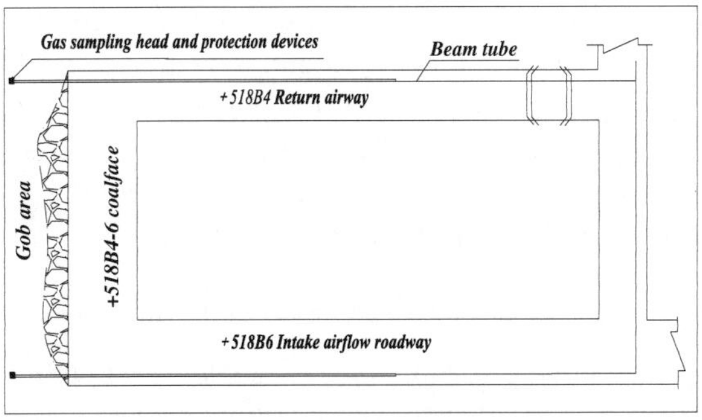

Figure 2. Schematic diagram of measuring points

2.3. Features of new integrated monitoring and control system for disaster gas

Compared with other tube monitoring systems, the system uses infrared chromatography instead of gas chromatography. Infrared chromatograph can be used underground without explosion danger, as it does not need H2 as a carrier gas. The gas chromatograph is flawed because it needs carrier gas, long beam tube pipeline, and much maintaining work. During the period of gas transfer from the underground to the ground, pipeline leakage or pipeline clogging often occurs due to the great length of the pipeline and the existence of a middle connector. It will lead to erroneous analysis results if troubleshooting cannot be detected in time [5-9]. Monitoring system puts the analysis unit in the chamber underground close to the sampling point. The analysis unit directly analyzes the gas sample collected in the beam tube underground, and then transmits the analysis result to the ground center station by the electric signal, realizing the rapid prediction of the harmful gas disaster.

Sending and analyzing gas samples will take some time so that the monitoring results cannot be punctual. Therefore, during data analysis, the gas analysis time should be advanced depending on the specific conditions used in the system. According to the gas sample transmission time T and time required for each single sample analysis T1, the time forward (T + T1) is the exact time of the gas sample analysis.

The transmission time T can be calculated by the following formula,

$$T=\sum L/v \tag{1}$$

Where, L is the length of the tube and v is the average velocity of the fluid in the pipe, the average velocity can be calculated by the following formula.

$$v = Q/A \tag{2}$$

Where, Q is the gas flow rate and A is the pipe cross-sectional area.

When tube diameter is 8mm, flow rate is 0.6L / min, v can be calculated v= 1.99m / s. When ordinary monitoring system analyzer is placed on the ground, the distance from the ground through the wellbore to the sub-station of the monitoring system generally is increased by at least about 3.5 km or more. The new integrated monitoring system is at least 6 minutes faster than the ordinary system.
According to the above, the accurate time required to move forward is calculated as at least 35min for the general monitoring system. The present monitoring system can issue the disaster warning 35min earlier than ordinary monitoring system. This time is critical for security.

3. Measurement of gas concentration

We arranged two measuring points into air inlet and outlet ends in the + 518B4-6 coal face. With the normal mining work, oxygen, carbon monoxide, methane and other gas concentrations of gob area are monitored continuously. The measuring point arrangement is shown in Figure 2.

① pipeline laying: B6 roadway is located 0.5m away from the north wall, 0.3m away from the roadway floor; B4 roadway is located 0.5m away from the south wall, 0.5m away from the roadway floor; The tube beam laying length of the tunnel is 80m,it is protected D50 by steel pipe .

② Gas collection: the system conducts a gas composition data analysis every 30 minutes, while recording the scale of the daily coalface and the distance of probe from the coalface.

③ monitoring cycle: With the normal mining work proceeding, the system stopped monitoring until the beam tube all go deep within the gob area.

4. Data analysis of new integrated monitoring and control system for disaster gas

4.1. Determination of spontaneous combustion "three zones" in gob area

According to the oxygen concentration, the area of coal oxidation and spontaneous combustion of gob area can be divided into three zones (Table 1). They are cooling zone, spontaneous combustion zone and suffocation zone.

Table 1. Oxygen concentration standard of the three zones [3]

Cooling zone	>18%
Spontaneous combustion zone	10%~18%
Suffocation zone	<10%

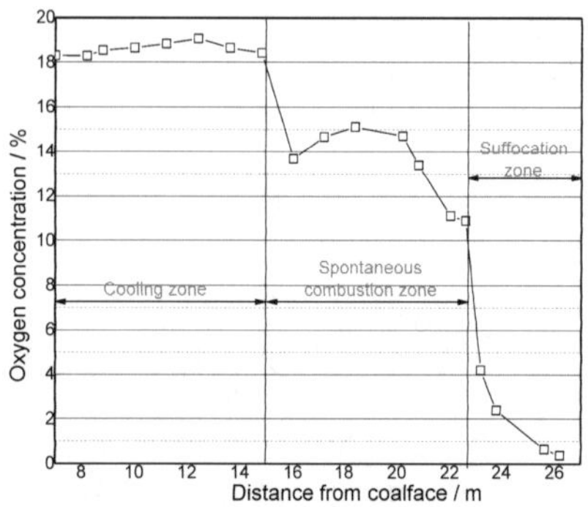

Figure 3. B4 roadway oxygen concentration curve

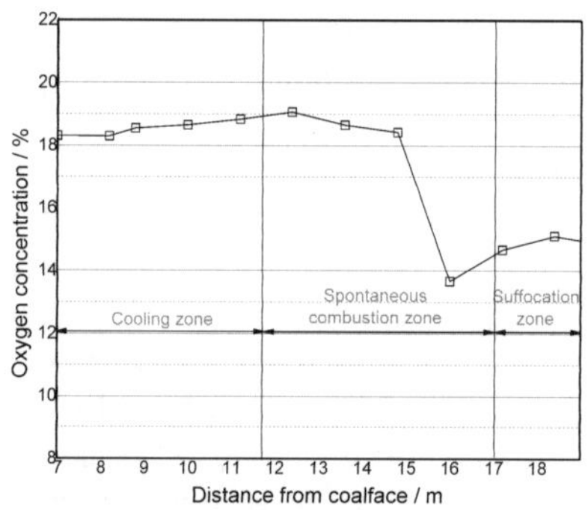

Figure 4. B6 roadway oxygen concentration curve

Figure 3 reflects the oxygen concentration changes in the coalface with the advance of B4 roadway embedded beam tube probe. The beam tube probe is 0~14.8m away from the coalface, the oxygen concentration remained above 18%. This is the cooling zone; the gob area oxygen concentration gradually decreases from 14.8m. Beam tube probe is 14.8 ~ 22.6m away from the face area, and the oxygen concentration is 10% -18%, thus the region is the spontaneous combustion zone; With the coalface continues to advance, the compaction of top-coal coal and air leakage disappears, the oxygen concentration quickly is reduced to 10% or less, which is the suffocation zone. Gob area belongs to the suffocation zone when the beam tube probe is more than 22.6m away from the coalface.

Figure 4 reflects the oxygen concentration changes in the coalface with the advance of the B6 roadway embedded beam tube probe. The beam tube probe is 0~11.2m away from the coalface, the oxygen concentration remains above 18%, and the gob air oxygen concentration is relatively high. This is the cooling zone; although the oxygen concentration at 11.2m suddenly rises to 20.84%.It is estimated to be due to air leakage caused by gob area. But the overall curve is still in a downward trend; the gob area oxygen concentration gradually decreases from 11.2m. Beam tube probe is

11.2~16.6m away from the face, the oxygen concentration is 10% -18%, thus the region is spontaneous combustion zone. When the beam probe is more than 16.6m away from the coal face, the oxygen concentration drops rapidly. That is suffocation zone. Spontaneous combustion danger zone comprehensive division shows in Figure 5 and Table 2.

Table 2. Spontaneous Combustion Hazard Zone

measuring point	Cooling zone	Spontaneous combustion zone	Suffocation zone
B4 roadway	0~14.8m	14.8~22.6m	>22.6m
B6 roadway	0~11.2m	11.2m~16.6m	>16.6m

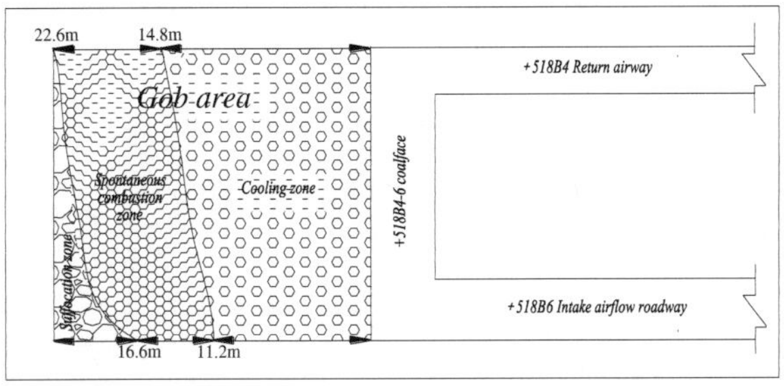

Figure 5. Spontaneous Combustion Hazard Zone

In the spontaneous combustion zone, the temperature of the floating coal increases; however, if the coalface advances rapidly to the suffocation zone before the coal temperature rises to the critical temperature, spontaneous combustion will not occur, otherwise, the spontaneous combustion will occur. Accordingly, the minimum advancing speed of the coal face can be calculated by the following formula.

$$V_{min}=L_{max}/T \tag{3}$$

Where, L_{max} is the length of the coal seam and T is the mining time.

The +518B4-6coal is a high volatile long flame coal; the fire period is generally 3 to 6 months, the shortest fire period is only 27 days.

B4 roadway return air side: V_{min}=0.8m/d

B6 roadway inlet air side: V_{min}=0.6m/d

Therefore, +518B4-6 coal face advances at a minimum speed of 0.8m/d, in fact,+518B4-6 coal face advances 2.4m/d. Comprehensive analysis shows that: spontaneous combustion will not occur.

4.2. Disaster prevention and control

Figure 6 reflects the curves of CH4 and CO concentration in B4 roadway. According to the curve, the CH4 concentration is kept below 0.04% within 15m range of the gob area. The concentration of CH4 increases gradually, and the CH4 concentration reaches 0.97% at 23.8 m in gob area, but it does not reach the explosion limit of 5% -16%, and 23.8m position has been in the suffocation zone, so CH4 in the gob area will not explode.

From the curve of CO concentration in Figure 6, it can be seen that the CO on the return air side cannot be detected, but it does not indicate that the coal in this area has not been oxidized. On the one hand, due to the high negative pressure on the return air side, the air leakage intensity of the gob area is large. On the other hand, the oxygen concentration monitored on the return air side shows that the products of coal oxidation are CO2. Based on the oxygen concentration monitored at the return air side, it is assumed that the product of coal oxidation is basically CO2, so the CO concentration on the return air side is substantially 0. Although the CO concentration rapidly increases to 73 PPM at 20.2 m position, it was due to a sudden spill of gas in the abandoned mine. Therefore, there was no evidence of ignition in roadway return air side.

Figure 7 reflects the curves of CH4 and CO concentration in B6 roadway. According to the curve, the CH4 concentration is kept 0 within the 13.6m range of the gob area. The concentration of CH4 increases gradually, and the CH4 concentration reaches 0.63% at 17.2 m in gob area. Although this concentration is in the gob spontaneous combustion zone, it does not reach the explosion limit of 5% -16%, so CH4 in the gob area cannot explode. According to the trend of the curve, although CH4 concentration is in the ascending stage, but the oxygen concentration with the CH4 concentration gradually decreases, and it greatly reduces the possibility of CH4 explosion.

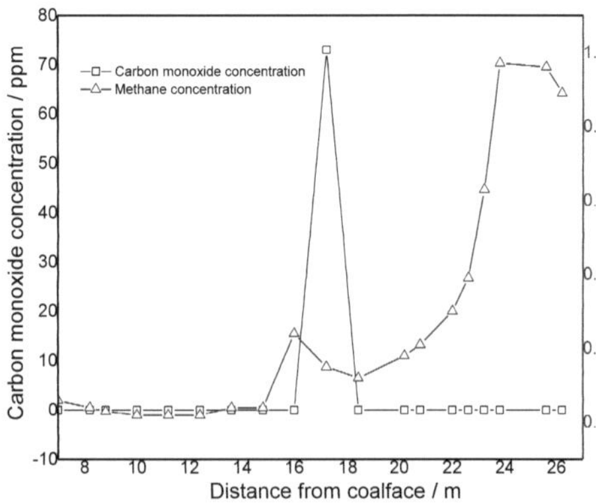

Figure 6. CH4 and CO concentration curves of gob area in B4 roadway

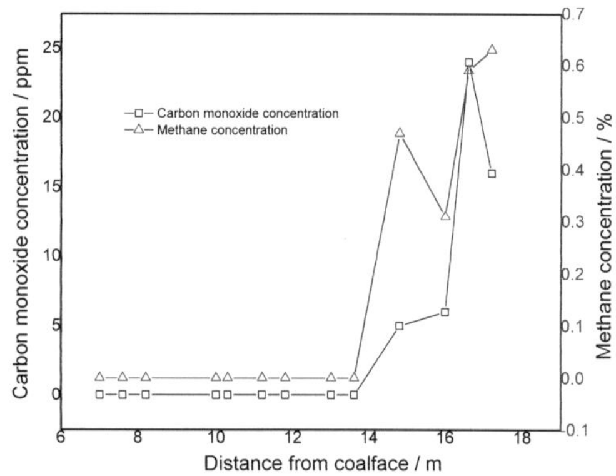

Figure 7. CH4 and CO concentration curves of gob area in B6 roadway

For CO, coal oxidation of the product is basically CO2, due to the fresh air, gob air leakage, higher O2 concentration, so the CO concentration is zero. With the coalface advancing, it gradually moves into the spontaneous combustion zone with the advance of the coalface. Gob air leakage is very small, and coal and O2 produces CO gas; and at the same time, due to top coal cave off strict and the accumulation of CO, the CO concentration reaches a maximum of 24ppm at 16.6m position. Then it moves into the suffocation zone, where due to O2 concentration decreasing, CO concentration decreases rapidly. Therefore, there was no possibility of ignition in roadway return air side.

5. Conclusion

1. The range of "three zones" in gob area was obtained by dividing the spontaneous combustion danger zones in the gob area by the concentration of O2. ① The range of cooling zone: it is less than 11.2m away from the coalface the in inlet air side and less than 14.8m away from the coalface in the return air side; ② The range of spontaneous combustion zone: it is within 11.2 ~ 16.6m away from the coal face in the air inlet side and it is within 14.8 ~ 22.6m away from the coal face in the return air side. ③ The range of suffocation zone: it is behind the edge of the spontaneous combustion zone in the air inlet-side wind and the return air side and direct to the gob area.

2. Although the CO concentration reaches 24ppm, 73ppm in the air inlet side and the return air side of gob area, respectively, as long as the daily advancing speed of coal face is greater than 0.8m, spontaneous oxidation zone will go to the suffocation zone, quickly. The gob area will not combust spontaneously.

3. The CH4 concentration in the air inlet side and the return air side of gob area is 0.63% and 0.97%, respectively, which does not reach the critical value of 5% ~ 16%. Therefore, the possibility of CH4 explosion in gob area was eliminated.

The new integrated monitoring and control system for disaster gas can detect the hidden danger of coal spontaneous combustion, and determine whether the hidden danger of spontaneous combustion occurs. In the process of fire control, it can provide a reliable basis for fire prevention and disaster prevention, ensuring the timeliness of fire prevention and control work. Aspects about correlation, maintenance and viability issues of the monitoring system will be studied in the future.

IOP Conf. Series: Materials Science and Engineering **191** (2017) 012051 doi:10.1088/1757-899X/191/1/012051

Acknowledgments
The authors would like to acknowledge the support of Preliminary Research Project of National "973" Project (Number 2014CB260404) and National Natural Science Foundation of China (Grant Number 51404191/E041001).

References
[1] Yang Qing. Research on Comprehensive Fire Prevention Technology at 1272(3) Working Face in Huainan Dingji Coal Mine [J]. Coal Technology 2016, 35(4):170-171.
[2] Yu Tao, Lu Ping, Yao Shang-wen, et al. Distribution regularity of the spontaneous combustion and carbon dioxide injection into the gob of the grand mining height panels [J]. Safety and Environment2016, 16(1):29-32.
[3] Zhou jia-ri, deng Cun-bao, et al. Research on spontaneous combustion "three zones" of fully mechanized working face's goaf [J]. Computer and Applied Chemistry, 2015, 32 (7):830-834.
[4] Li Dan. Determination and Application of Goaf Spontaneous Combustion "Three Zones" of Fully Mechanized Working Face [J]. Coal Technology, 2015, 34(8):143-145.
[5] Cai Zhou-quan, Wang Long, et al. Problems and Solutions for Application of Tube Monitoring System in Coal Mine [J]. Coal Technology,2016, 35(7):150-152.
[6] Xin Cheng-peng, Liu Yi-lei. Research on Indexity Gas of Spontaneous Combustion of Coal Prediction [J].Coal Technology. 2015, 34(8):145-147.
[7] Shao Hao, Jiang Shu-guang, Wu Zheng-yan. Numerical simulation on fire prevention by infusing carbondioxide into goaf [J]. Mining & Safety Engineering2013, 30(1):154-158.
[8] Li Zong-xiang, Gu Run-hong, et al. Simulation of gas migration in 3D goaf based on RNG κ-ε turbulence model [J]. China Coal Society, 2014, 39(5) : 880-885.
[9] TARABA B, MICHALECZ. Effect of longwall face advance rate on spontaneous heating process in the gob area -CFD modelling [J]. Fuel, 2011, 90(8) : 2790-2797.

AUTHOR INDEX

AUTHOR INDEX